PROBABILISTIC METHODS FOR STRUCTURAL DESIGN

SOLID MECHANICS AND ITS APPLICATIONS
Volume 56

Series Editor: **G.M.L. GLADWELL**
Solid Mechanics Division, Faculty of Engineering
University of Waterloo
Waterloo, Ontario, Canada N2L 3G1

Aims and Scope of the Series

The fundamental questions arising in mechanics are: *Why?*, *How?*, and *How much?* The aim of this series is to provide lucid accounts written by authoritative researchers giving vision and insight in answering these questions on the subject of mechanics as it relates to solids.

The scope of the series covers the entire spectrum of solid mechanics. Thus it includes the foundation of mechanics; variational formulations; computational mechanics; statics, kinematics and dynamics of rigid and elastic bodies; vibrations of solids and structures; dynamical systems and chaos; the theories of elasticity, plasticity and viscoelasticity; composite materials; rods, beams, shells and membranes; structural control and stability; soils, rocks and geomechanics; fracture; tribology; experimental mechanics; biomechanics and machine design.

The median level of presentation is the first year graduate student. Some texts are monographs defining the current state of the field; others are accessible to final year undergraduates; but essentially the emphasis is on readability and clarity.

For a list of related mechanics titles, see final pages.

Probabilistic Methods for Structural Design

Edited by

C. GUEDES SOARES

Instituto Superior Técnico,
Lisbon, Portugal

SPRINGER SCIENCE+BUSINESS MEDIA, B.V.

A C.I.P. Catalogue record for this book is available from the Library of Congress.

ISBN 978-94-010-6366-1   ISBN 978-94-011-5614-1 (eBook)
DOI 10.1007/978-94-011-5614-1

Printed on acid-free paper

In Memory of

Prof. Júlio Ferry Borges

and

Prof. João Tiago de Oliveira

FOREWORD

This book contains contributions from various authors on different important topics related with probabilistic methods used for the design of structures.

Initially several of the papers were prepared for advanced courses on structural reliability or on probabilistic methods for structural design. These courses have been held in different countries and have been given by different groups of lecturers. They were aimed at engineers and researchers who already had some exposure to structural reliability methods and thus they presented overviews of the work in the various topics.

The book includes a selection of those contributions, which can be of support for future courses or for engineers and researchers that want to have an update on specific topics. It is considered a complement to the existing textbooks on structural reliability, which normally ensure the coverage of the basic topics but then are not extensive enough to cover some more specialised aspects.

In addition to the contributions drawn from those lectures there are several papers that have been prepared specifically for this book, aiming at complementing the others in providing an overall account of the recent advances in the field.

It is with sadness that in the meanwhile we have seen the disappearance of two of the contributors to the book and, in fact two of the early contributors to this field.

Prof. Ferry Borges, who had his career at the National Civil Engineering Laboratory, in Lisbon and later became also Professor at the Technical University of Lisbon, passed away as a consequence of a continued illness. However, he has been active with lectures and conference participation until the later moments of his life. Ferry Borges was co-author of one of the earliest textbooks in structural reliability and he has been very active in the field of civil engineering codes in Europe, participating in many international bodies in that area.

Prof. Tiago de Oliveira, very much known by his contributions on the statistical theories of extremes has always shown an interest on how those models and methods that could be used in engineering and in particular in structural reliability analysis. One can find this association for example in the first ICOSSAR Conference organised by Freudenthal in 1972. Tiago de Oliveira, who started his academic career at the University of Lisbon, moved later to the New University of Lisbon and he was very committed to the Academy of Sciences in Lisbon, when he passed away from a heart attack.

These two gentlemen had an important impact on their generations in their specific field of activity and they have been the inspiration of several important contributors who came later. Therefore I am very pleased to dedicate this book to their memory.

The book starts with the basic aspects of structural design, which set the stage to later contributions and define the boundaries in which reliability based methods, can be used as a tool for design. With the continued improvement of the computational techniques, the emphasis in real problem solving has shifted towards the modelling and in this connection the quantification of model uncertainties is very important. Response surfaces are being used more widely for situations in which the limit state functions require relatively heavy computational schemes.

The limit state conditions are a result of different modes of collapse. An important one results from fatigue, which is dependent on many random factors and thus can be described by different probabilistic formulations. Another important mode of collapse is the buckling collapse, which can occur, in different types of components. The case of columns and plates is dealt with here. Finally, reference is also made to more complicated collapse modes that are represented by implicit formulations, which require special techniques to handle.

Having considered the reliability of components, it is necessary to analyse the case of structural systems. To deal with structures it is necessary to model also the loads, in particular their extreme values and their combined values, and these are the topics of the next two contributions.

A recent and rapidly expanding field of activity is the application of stochastic processes to model the variability of loading and material properties, as well as the assessment of reliability. Stochastic finite elements are able to cope with the variability of these properties and they are being applied to different types of problems.

Having covered various tools for reliability analysis of structures, design and maintenance are the next type of subjects of interest. Reliability methods are very useful to design and calibrate design codes, and an overview is provided of several developments in the field. Also, the specific aspects of seismic design are presented as well as reliability based maintenance.

Several topics are covered in this book, including modelling of uncertainty, prediction of the strength of components, load modelling and combination, assessment of structural systems, stochastic finite elements and design considerations. It is hoped that such a series of contributions will be found useful for practitioners as well as for researchers.

Carlos Guedes Soares

TABLE OF CONTENTS

BASIC CONCEPTS OF STRUCTURAL DESIGN

J. FERRY BORGES
Laboratório Nacional de Engenharia Civil
Lisboa, Portugal

1. Introduction

In the last 15 years several international documents have been published dealing with the basic concepts of structural design.

The Joint Committee on Structural Safety, JCSS, approved in November 1976 the *Common Unified Rules for Different Types of Construction and Materia* which were published in 1978 as volume 1 of the *International System of Unified Standard Codes of Practice for Structures* (Bulletins d' Information n° 124/125 of the Euro-International Committee for Concrete, CEB). According to a recommendation of the Economic Commission for Europe of the United Nations Economic and Social Council, the JCSS prepared the *General Principles on Reliability for Structural Design* which were used by ISO in the revision of ISO 2394. These *General Principles* were published with the *General Principles on Quality Assurance for Structures* in Volume 35 of the Reports of the International Association for Bridge and Structural Engineering, IABSE, in 1981.

The Commission of the European Communities in a first draft of the Eurocode n° 1 (EUR 8847, 1984), followed these general principles, which were further used in the drafting of Eurocodes 2 to 8.

In 1988 the JCSS, recognizing the need to update the existing documents, prepared a *Commentary on ISO 2394* published by CEB in Bulletin d'Information n° 191, July 1988. The CEB-FIP Model Code 1990 (CEB Bulletin d'Information n° 203, July 1991) adopted the new concepts without deviating from the design operational rules of the previous *Recommendations*.

In the framework of the Construction Products Council Directive 89/106/EEC, *Interpretative Documents* on the essential requirements adopted in this Directive have been prepared. The Interpretative Document ID 1 concerns *Mechanical Resistance and Stability*. Intended to be a guideline to the preparation of CEN Standards, ID 1 expresses basic concepts directly related to structural design.

The main steps of the evolution of the methods adopted in the design of buildings and other civil engineering works correspond to the introduction of the following general concepts:

- ultimate and serviceability limit states
- probabilistic formulation of structural safety
- essential requirements and performance criteria

1

C. Guedes Soares (ed.), Probabilistic Methods for Structural Design, 1–15.
© 1997 *Kluwer Academic Publishers.*

- quality assurance and quality management
- hazard scenarios and risk analysis
- risk management and technical insurance
- basic variables. Types of knowledge.

The concepts of limit state and of probabilistic safety were first presented in a thesis by Max Mayer published in 1926 entitled *Safety in Constructional Works and its Design According to Limit States Instead of Permissible Stresses* (1). Although the fundamental ideas were well expressed in this thesis, the concept of limit state was introduced in codes only in the middle forties, (in the Soviet Union). Pioneer work on probabilistic safety in structural engineering is due to Freudenthal (2) and Torroja (3) in the late forties. The design method based on partial factors, suggested by Torroja, was first implemented in the CEB recommendations of 1963 (4).

The fundamentals of the probabilistc methods for structural design are not presented in this lecture, which covers concepts of a more general character. However the concepts which are presented form a convenient introduction to the probabilistic approach.

2. Requirements and Performance Criteria

The notions of requirement, performance criteria and limit state are intimately related. Requirements are general conditions imposed on the behaviour of the construction by the owner, the user or the authorities. In order to derive methods of evaluation (analytical or experimental) to be used in design, requirements have to be transformed into performance criteria. Requirements indicate the needs in general terms; performance criteria are technical conditions which express the requirements. Limit states define the borders between acceptable and unacceptable performance.

As in many other technical activities, the guidance to building engineers is usually given by means of standards. They are usually expressed in a prescriptive way, indicating how things should be done, and not justifying the reasons for doing so, and not stating the aims to be attained. The drawbacks of this presentation were first recognized in the aviation industry which in 1943 recommended that codes should be stated in terms of objectives rather than specifications. That is, the code should spell out what is to be achieved and leave the designer to choose how this will be achieved (5).

Lists of human requirements in housing have been presented by Blachere in 1966 (6). In 1970, the CIB set up the Working Commission W60 - *Performance Concept in Building*; in 1982 they published a comprehensive state-of-the-art review of the performance approach in building practice.

One of the first wide applications of the performance concepts to housing was carried out in the *Operation Breakthrough* by the National Bureau of Standards in 1970 (7). This operation, sponsored by the Department of Housing and Urban Development, was based on two main concepts:

The adoption of performance criteria for evaluation of prototype innovative and technologically advanced housing systems, and the use of quality assurance provisions that establish general requirements and guidelines for the quality of production units.

The Committee on Housing Building and Planning of the Economic Commission for Europe of the United Nations, under the project aimed at international harmonization of standards and control rules for building and building products, has further developed the performance concept. Since 1978 this Committe has been concerned about defining human requirements. In 1981 they published (8) a list of 20 requirements related to the housing user, and 5 requirements for the limitation of harmful effects or nuisances produced by the building and affecting its surroundings. The *Compendium of Model Provisions for Building Regulations* details these requirements (9).

In the field of structural behaviour, ISO 2394 indicates three fundamental requirements: safety, serviceability and durability.

In the commentary to this document, the Joint Committee on Structural Safety enlarged the list of requirements by considering, in parallel to the safety requirement, the structural insensitivy requirement (limited damage due to expected and unexpected hazards) and further requirements on economy, adaptability, esthetics, etc.

3. Quality Assurance

According to the ISO Standards, series 9000, quality assurance is a set of planned activities which lead to the guarantee that a product or a service satisfies established requirements.

In the planning of quality assurance, four levels are usually identified:

Level 1 - Activity limited to the quality control of the final product.

Level 2 - Includes the control of the production process.

Level 3 - Extends the control to the production management, including production programming, definition of responsibilities, documentation and auditing.

Level 4 - Conducts the whole management process, including flow of information, motivation, professional upgrading, etc.

The choice of the quality assurance level should depend on the importance of the risks to be avoided.

The activities to be carried out in any quality assurance program are:

– plan the activity (written specification)
– follow the established plan (respect the planning)
– record all steps (written control)

The implementation of quality systems in an organization is influenced by many factors such as the objectives of the organization, the product or service under consideration and the specific experience of the organization. The international standards of the series 9000 aim to clarify the relationship among the principal quality concepts and to provide guidelines that can be used for internal quality management purposes (ISO 9001/3). These ISO standards are adopted as national standards by several countries. That is the case for Portugal, with the *Normas Portuguesas* NP-3000/0/1/2/3, published in 1986, and for the U.K. with the *British Standards* BS 5750, Prts 0, 1,2, and 3, published in 1987. The following comments refer to the British Standards.

BS 5750 Part 0 concerns principal concepts and applications. It is divided into Section 0.1 - *Guide to selection and use* and Section 0.2 - *Guide to quality management and quality system elements*.

A vocabulary on quality terms is given in ISO 8402. However in BS 5750 it was deemed convenient to redefine some terms considered of particular importance such as: *quality policy*: the overall quality as formally expressed by the management; *quality management*: the aspect of the whole management function that determines and implements the quality policy; *quality sytem*: the organizational structure, responsibilities, procedures, processes and resources for implementing techniques and activities that are used to fulfil requirements for quality.

As is shown by these definitions, there is an intimate relationship between quality assurance and management. Quality systems may be used in two different situations; contractual and non-contractual. In the first case, it is contractually required that certain quality system elements be part of the suppliers' quality system. In the second case, quality assurance is applied by the initiative of the producer as an adequate management policy.

Although quality assurance is extensively implemented in many different industries, its use in building and particularly in structural engineering is limited

As indicated, an early document in this field was published by the Joint Committee on Structural Safety in 1981. This document considered that the basic concepts of quality assurance are: the functional requirements, the use and hazard scenarios, the structural concept, the responsibility and the control.

4. Hazard Scenarios and Risk Analysis

Rational structural design should be based on adequate idealization of the structural system and of its behaviour. Risk analysis is a very useful tool on which to base economic and engineering decisions.

It is particularly adequate for dealing with catastrophic low probability high-consequence events, as presented in nuclear, chemical and oil industries (10). The methods of risk analysis are permeating every branch of construction.

To carry out risk analysis it is necessary to identify the assumed hazards, to estimate their probability of occurrence during a reference interval of time, and to estimate the probability of the loss, or amount of damage which corresponds to the occurrence of each hazard. The different amounts of loss are obtained by combining these probabilities. These probabilities, including information derived form other sources, mainly concerning human behaviour, should guide economic and technical decisions. In several cases, not only the central values of the risk function, but also the extreme values corresponding to small and high fractiles are paramount for guiding the decisions. This is particularly so when distinction is made between the interests of individuals and society.

In risk analysis the following five phases may be defined:

I.   identification of the hazard
II. analysis of the hazard
III. risk estimation
IV. assessment of consequences
V.   risk evaluation and control

Thus, it is necessary first to identify the hazard scenarios, to search for cause and effect relationships, what may be done within disaster planning activities. Historic information, possibly obtained by consultation of data bases, or by interviews with people involved, may be useful.

In generic risk analysis, accident logic trees are usually adopted. Giannini, Pinto and Rackwitz exemplify how to deal with them into structural problems (11).

Logic trees are formed by two parts, the first one describing alternative hypotheses concerning combination of actions, types of hazards, mechanical models, etc., and the second part describing the physical states of the sub-systems in which the structure is divided.

The tree should contain all the possible sequences through which the structure may pass, and the quantitative probabilistic assessment of each sequence. The last column identifies the different damage states (limit states) and indicates the final probabilities of these being exceeded.

The probabilistic assessment should be obtained by a combination of frequentist and subjective approaches. The human intervention in the process should be considered by including strategic decisions and procedures, as well as human errors, (due to omission and commission).

The probability distribution function which defines the risk should be obtained by combining the probability of occurrence of the hazard with the probabilities of the consequent damages.

Finally, risk evaluation and control may be based on the comparison of risk in other systems, identification of attitudes to safety and risk, impact of risk (at different social levels), and benefit studies, particularly those concerning investing in increased safety.

5. Risk Management and Technical Insurance

For a specific activity, the aim of risk management usually consists in increasing the benefit and in reducing the loss and/or in reducing the corresponding risk. To achieve these aims potential strategies may be followed (12):

I.   eliminating or avoiding the possible occurrence of the hazards at the origin.
II. avoiding the hazard acting in the system, e.g. by modifying the project concept,
III. controlling or reducing the losses, e.g. by adopting safety measures,
IV. adopting a design which corresponds to a sufficiently small risk,

V. accepting the possibility of occurrence of the loss and preparing to reduce its consequences.

These strategies may be combined.

These rules are of a general character. For structural design, their implementation may be exemplified: for a hazard consist the of the impact of a car against the column of a building possible strategies are:

    I.   prevent cars from approaching the building,

    II. install a protection for the column that would prevent the impact,

    III. create a bracing system to avoid the collapse of the building if the column fails,

    IV. design a sufficiently robust column, with a small probability of failure by impact,

    V. assume that the column may fail, but avoid installing services on it, to reduce the cost of a possible failure.

Risk evaluation should be based on overall costs. Overall costs should be obtained by adding the usual production costs and the non-quality costs including: prevention, assessment, controlling, testing, observation of the behaviour, commissioning, professional liability insurance, all risk and other insurances, and other administrative charges.

Life-cycle cost is the total cost of a system over its life time. To obtain the values over the life time, annual values should be discounted and integrated (13). Discounting involves converting cash-flows that occur over time to equivalent amounts at a common date. This common point is usually the starting point of the life-cycle cost analysis. When estimating benefits and costs in monetary terms it is necessary to consider inflation. Distinction should be made between current money (which is adjusted by taking inflation into account) and constant money (which is not adjusted by the effects of inflation or deflation).

6. Basic Variables. Types of Knowledge

As an introduction to probabilistic methods for structural design particular attention is due to the concept of probability. In probabilistic reliability, it is assumed that the basic variables which represent the actions, the mechanical properties of the materials and the geometry of the structural elements are probability distribution functions, or in more general terms, stochastic processes. If their randomness may be neglected, variables may be assumed to be deterministic. Furthermore it is generally accepted that probabilities are not only a counterpart of frequency but also a subjective measure of degree of confidance. By the adoption of the Bayesean approach, a priori information can be .... the theory of probabilities is a powerful tool to base decisions (14).

The theory of fuzzy sets is an alternative way to idealize the dispersion of variables. Criated in 1965 by Zadeh, it gives mathematical expression to the imprecision of knowledge usually expressed by language. The interest in fuzzy sets, fuzzy control and fuzzy systems is rapidly growing (15) and finding application on many multiple scientific and technical branches (16): cognitive engineering (17) and architecture (18).

From a conceptual point of view, fuzziness and randomness are completely distinct, one belonging to set theory and logic, and the other to measure theory. Fuzziness concerns the modelling of inexactnesses due to human perception processes, and randomness concerns statistical inexactness due to the occurrence of random events. However approaches for combining the two concepts have been established, leading to the theory of fuzzy random variables (15).

The theory of games, in the field of decision theory, is due to von Neumann and Morgenstern (19). The simplest and most typical problem of the theory of games is the two person game. Consider two players A and B, each one having the possibility of making a finite number of choices. The payments of player B to player A are defined by a matrix a_{ij}. The game consists in player A choosing a and player B choosing a column of the matrix. Each player tries to minimize his losses and maximize his winnings.

Both players know the matrix, but each one ignores his adversary's choice. The hypothesis that the adversary will try to minimize his losses and maximize his winnings may be adopted as a convenient base for decisions.

Let us analyse each player's choices and their consequences. When A chooses row i, he is sure to receive at least $\min(a_{ij})_j$; trying to receive as much as possible, this player will be interested in choosing that row i for which the $\min(a_{ij})_j$ is maximum. Thus will aim at $\max\left(\min(a_{ij})_j\right)_i$. Likerwise, when B chooses column j, he is sure to pay no more than $\max(a_{ij})_i$ Thus will aim at $\max\left(\min(a_{ij})_j\right)_i$.

The decision rules that consists in choosing these maxima or minima are called *max min* and *min max* respectively.

If matrix a_{ij} is such that

$$\max\left(\min(a_{ij})_j\right)_i \; = \; \min\left(a_{ij}\right)_j \; = \; v \tag{1}$$

the game is called strictly determined and has the value v. In this case it can be proved that the indicated stategies are the most convenient for both players.

In general $\max(a_{ij})_i$ is differennt from $\min(a_{ij})_i$. In this case it is no longer possible to define a deterministic choice (pure strategy) for which the game has the same value for the both players. However this aim can be secured by using a more involved strategy.

Let us consider for instance a very simple game defined by the following matrix

$$A=\begin{bmatrix} a_{11} & a_{12} \\ a_{12} & a_{22} \end{bmatrix} = \begin{matrix} a_{ij}=a_{11} & a_{12}=0 \\ a_{21}= \ ? & ? \ =0 \end{matrix} \begin{bmatrix} 1 & 0 \\ 0 & 1 \end{bmatrix} \tag{2}$$

In this case. Suppose that A decides, in a random way, to choose row 1 with probability p and row 2 with probability 1-p. If then B chooses column 1, the expected value of A's income will be

$$v_1 = p\,a_{11} + (1-p)\,a_{21} = p \tag{3}$$

If, however, B chooses column 2, the expected value of A's income will be

$$v_2 = p\,a_{12} + (1-p)\,a_{22} = 1-p \tag{4}$$

Thus, A can choose a strategy for which the expected value of his income is independent of the decisions of B. In the present case he has to take

$$v_1 = v_2 = p = 1 - p = 1/2. \tag{5}$$

B could think the same and make his payments independent of A's decisions. It is obvious that the expected value of A's income equals to the expected value of B's payments.

A strategy including a random choice is called a mixed strategy.

A strategy is optimal for A when he obtains an expected v that cannot be reduced by the choices of B. For non-strictly determined games it can be proved that optimal strategies are mixed strategies. The main object of the *Theory of Games* is to find optimal strategies.

Note that the decision rules derived from the *Theory of Games* are entirely different from those corresponding to a probabilistic idealization. If, for example, in the above instance of the game, A is convinced according to probabilities that he can estimate that B is making choices in a random way, A will no longer consider the minimax rule as convenient and may prefer to use Bayes's theorem.

The theory of games is based in strategic choices of the values of the variables and these are different from the probabilistic concepts that are the basic tools of structural safety assessmente.

Thus, four types of knowledge have been identified: deterministic, probabilistic, fuzzy and strategic. They correspond to different theoretical formulations.

7. Rational Decisions and the Concept of Utility

The design of structures on a purely economic basis corresponds to the decision rule: minimize the expected value of the overall cost. This simple economic criterion may be improved by introducing the concept of utility. The first definition of utility was presented by Daniel Bernoulli in the sentence: the utility of an additional profit is inversely proportional to the existing wealth this implies that the concept of utility is dependent on the status of the person that perceives it.

The following definition of utility is due to von Neumann (19). Consider a set of outcomes, $\theta_1 ... \theta_n$. The correspondence between these outcomes and a set of real numbers $\mu_1...\mu_n$, utilities, is established in the following way.

It is assumed that an order of preferences can be established to the outcomes θ_i . Thus the outcomes can be ordered in the following way:

$$\theta_1 < \theta_2 < ...< \theta_i < \theta_{i+1} < ...< \theta_n \qquad (6)$$

Where the sign $<$ means that if it is possible to choose between outcomes θ_i and θ_{i+1}, the second one is chosen. The concept of utility allows one to quantify these preferences. This quantification assumes a personal rational behaviour; the scale of utilities is arbitrary and only measures the relative value of the preferences.

Given the outcomes θ_1 and θ_2, the utility μ_2 expressed as a function of utility μ_1 is given by

$$\mu_2 (\theta_2) = \mu_1 (\theta_1) \frac{1-p}{p} \qquad (7)$$

Here p is a probability in a game which consists in obtaining the outcome θ_1 with probability p and outcome θ_2 with probability 1-p. The value of p is chosen by the person who is defining the scale of utilities which reflects his preferences.

The type of outcome θ_i is not specified. For economic problems outcomes are usually amounts of money. In this case the indicated definition allows one to relate a scale of monetary values to a scale of utilities. In general this relation is non linear.

The shape of the curve reflects the utility for a given person of a given amount of money, Fig. 1. This utility depends in a large measure on the fortune owned. The scale of amounts of money should be expressed in national or regional currencies, such as dollar or European units.

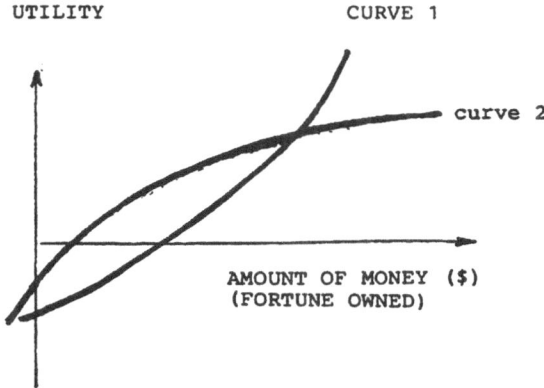

Figure 1. Relationship between utility and amount of money (fortune owned).

We may illustrate the relationship between utility and money shown in Fig. 1 by considering the behaviour of people agreeing or refusing to participate in a probabilistic game. One player may decide to gamble, even if he knows that the expected value of the winnings is smaller than the cost of participating in the game (such gambler considers the utility of large amounts of money more than proportional to the utility of small amounts (curve 1). On the other hand a player may decide not to gamble even knowing that the expected value of the winnings is larger than the cost of participating in the game (considers the utility of large amounts of money less than proportional to the utility of small amounts (curve 2).

The usefulness of the concept of utlity implies that those who have a decision to take are able to assign utilities to the various possible consequences of their choices. In many areas of human activities this may not be possible. However the concept of utility is a step toward the quantification of decision rules as compared with simple monetary terms. The theory is particularly useful in formulating trade-offs in problems which deal with engineering compromises.

Sometimes, utility and monetary gain are equivalent: expected utility equals expected gain, and optimizing the gain coincides with optimizing utility. After a number of cases, the average gain per case will approach the expected gain. The decision rule is rational. In the long run any other rule will give a smaller gain.

On the other hand, the decision maker may be in a situation where only once or a few times in a life time, has he to deal with a case of personal importance for him. A prudent decision maker will prefer the safe to the unsafe, while the daring decision maker will prefer the opposite if there is a greater gain to be made by this choice. The choice of the prudent will tend to move to a smaller gain than the expected one, but with a smaller probability of the expected value being exceeded. The choice of the daring one will do the opposite.

These types of choices may be expressed by non-linear relationships between utility and gain, such as shown in figure 1.

8. Implementation of Basic Concepts

The vast Eurocodes programme and the great amount of work involved in it, gives it a special role in the guidance of other structural codes.

A first example of this assertion is the draft: D 01 *European Greenhouse Standard* prepared by the International Society of Horticultural Science, dated June 1991 (20). This document is based on the draft of Eurocode 1, June 1990 and on the drafts of the other Eurocodes available to this date.

TABLE 1. Document D 01 introduces three classes of greenhouses according to Table 1:

Design Classification	
Class	Minimum Design Life, R (years)
G1	15
G2	10
G3	5

TABLE 2

Factors		
Adjustment Factor		Classification
Wind	Snow	Factor
0.906	0.710	0.925
0.873	0.612	0.900
0.815	0.445	0.850

For these classes Table 2 shows the adjustment factors on the 50 year recurrence of interval wind velocities and snow loads. The transformation of wind velocities to different recurrence intervals is carried out according to Eurocode 1. The characteristic actions thus obtained are multiplied by a classification factor, also indicated in the table 2.

According to D01, after applying these factors, the designer should follow the usual Eurocode rules.

It is possible to compute the theoretical probabilities of failure which correspond to each class by adopting the methods and tables presented in (21 and 22). This computation leads to the following values (Table 3):

TABLE 3

Class	Min. design life, R, Years	Yearly probability of failure	Probability of failure in R years
G1	15	5×10^{-4}	75×10^{-4}
G2	10	10×10^{-4}	100×10^{-4}
G3	5	43×10^{-4}	215×10^{-4}

Glasshouses G2 have the probability of failure in 10 years of 10^{-2} which would correspond to a theoretical mean life time of the order of magnitude of 1000 years and not 10 years as suggested by their label. The situation is similar in the other cases. The reference to life time is meaningless.

The aim of introducing safety classes should be to optimize the overall cost. This optimization should not be based on probabilistic reliability alone, and should include quality assurance and economic considerations. The reference to design life is misleading. Designing for useful life is usual for electric components and similar products (23), and involves the concept of durability. Present knowledge in civil engineering is insufficient to quantify this criterion.

A further comment on D 01 concerns the numerical values of the adjustment factors. It is excessive to define these factors by three significative figures. It is considered that the drawbacks and omissions of D 011 mainly follow from omissions of Eurocode 1. Recently the decision was taken to prepare a first section on Basic Principles to be included in EC 1. This section should deal with risk analysis, quality assurance and safety differentiation.

A second example of draft of standard which is based on the Eurocodes is the document CEN/TC 152 N30 - *Fairground and Amusement Park Machinery and Structures - Safety* dated June 1991 (24). This draft covers a broad variety of structures and machinery. In contrast to the standard on greenhouses, this standard is concerned mainly with the safety of persons, public, passengers and personal, and not with an economic optimization at the expense of reduced safety. The document produced by TC 152 is based on past experience and risk analysis. The list of references includes the set of Eurocodes, a large number of European norms and ISO Standards. However it excludes the ISO basic standards on Quality Assurance, series 9000. As the principles of quality assurance are in fact followed, the reason for this omission is not understood.

The draft deals with hazard identification, risk assessment and risk reduction in general terms. Risk reduction is detailed for the different types of machinery and structures. Static structural design follows the general principles of the Eurocodes. Design actions are subject to adaptations due to the special nature of the devices. Aspects such as location, duration and period of installation, supervision by an operator and possibilities of protecting and strengthing are duly considered. The document covers design, manufacture and supply, operation and installation, and approval, examination and test procedures. The style concerning design is prescriptive.

The publication of these drafts was the main reason for the approval by CEN TC 250, in the London September 1991 meeting, of Resolution n° 32 requiring Structural Eurocodes design rules to be dealt with in TC 250 only.

They two examples mentioned call the attention for the need for defining a policy concerning the coordination of the Eurocodes, not only between them, but also with other documents to which they are related.

9. Conclusions

Probabilistic methods of structural design have a firm theoretical supprot, and may be implemented at different levels compatible with the importance of the problems to be solved.   However they should be considered as one of several means of promoting safety.   Risk management and quality assurance, associated with probabilistic reliability, lead to improved solutions and permit the adequate treatment of problems, such as design optimization and safety differentiation.   The strict formulation within probabilistic reliability of these two problems is in general unsatisfactory, from both technical and economic points of view.

Eurocode 1 should include the presentation of the most important basic concepts according to a current formulation. In the future it would then be easy to improve the specific drafts of the various standards in this field. There is no practical difficulty in introducing these concepts, as the present method of partial factors of safety is compatible with the broad formulation indicated.

Cranston, in his paper *Reflections on Limit State Design* (25) summarizes the controversy concerning the use of states design and probabilistic methods which has been taking place in the UK during the last 25 years. This paper shows how difficult it is to introduce new concepts in engineering practice, even if these are simple, logic al and clear. It is recognized that most of critical comments were pertinent when first presented. However they are presently superseded, not by the return to the allowable stress design method, but by the new basic concepts which have been described here.

At the TC 250 London meeting, a resolution was adopted expressing the need for all CEN/TC's to be made aware of the Agreement clauses requiring Structural Eurocodes design rules to be dealt with in- TC 250 only. The aim of this resolution is to stop work on structural design rules outside TC 250, in particular the work concerning greenhouses.

However, from our point of view, the main reason for the unsatisfactory orientation of the draft concerning greenhouses is due to the omission in the Eurocodes of guidance on how to deal with safety differentiation problems which are basically economic.

At the London meeting. Breitschaft presented a list of arguments expressing the importance of the Structural Eurocodes for the European Community and for CEN. In this list it is indicated that approximately 35 TC's are concerned with questions of structural design. To give guidance to these various problems, and to deal with the numerous mandates received from the Communities, it is imperative to draft the new *Basic Principles* less strictly than in the past, making use of the most important concepts dealt with here. Franco Levi (26), expresses the point of view that civil engineering codes should not serve merely as a basis for day-to-day design and execution" but they should also work as an objective record of the significant progress made by research everywhere, and as a carefully compiled compendium of what has been done to meet social needs in the field of building".

However we should not forget the convenience of preparing, in parallel to design standards, guidelines on design, execution and use which may include information from different sources, outside the strict field of standardization. Design standards should be concise, open and covering the fundamentals, to the guidelines: design aids, details and general information. Recent progress in information technology allows an efficient storage and retrieval of large volumes of information: multilingual texts, drawings, photographs, and videos; this material can be supported by advanced expert systems. The hardware and the software needed to progress along these lines is available. The problem is to share this task in the most efficient way.

10. Acknowledgment

The collaboration in the revision of this paper by. Castanheta and E. Cansado de Carvalho, respectively research coordinator and principal researcher of LNEC, is kindly acknowledged.

11. References

1. Mayer, Max: *Die Sicherheit der Bauwerke und ihre Berechnung nach Grenzkraften instate nach Zulassigen Spanungen*, Vorlag von Julius Springer, Berlin 1926 (English and Spanish translations published by ANATOMIC, Madrid, 1975).
2. Freudenthal, A. M.: The safety of structures, *Proceedings of the American Society of Civil Engineers*, Vol. 71, no 8, October, 1945.
3. Torroja, E. et Paez, A.: *La Determinación del Coeficiente de Securidad en las Distintas Obras*, Instituto Técnico de la Construccion e del Cemento, 1949.
4. Recommandations *Pratiques a l'Usage des Constructeurs*. Bulletin d'Information nQ 39. Comité Europeen du Beton, Paris 1963.
5. Sande, T: Risk in Industry, Chapter 12 of *Risk and Decisions*, edited by Singleton, W.T. and Hovden, J., John Wiley and Sons, Chichester, 1987.
6. Blachere, G.: *Savoir Batir*, Eyrolles, Paris, 1966.
7. Pfrang, E. O.: *Operation Breakthrough*, National Bureau of Standards, Washington, December 1970.
8. *International Harmonization of the Technical Content of olden Regulations*, Committee on Housing Building and Planning, Economic Commission for Europe, HBP/WP.2/R.144, Geneva, March 1981.
9. *Compendium of Model Provisions for Building Regulations*, Economic Commission for Europe, ECE/HBP/55, Add. 1 and 2, New York, 1984, 1987.
10. Waller, R.A. and Covello, V.T. (Editors): *Low-Probability High-Consequence Risk Analysis, Advances in Risk Analysis*, Plenum Press, New York 1984.
11. Giannini, R., Rackwitz, R., Pinto, P.E.: *Action Scenarios and Logic Trees*, Joint Committee on Structural Safety, JCSS, Working Document, IABSE, Zurich, March, 1991.
12. Borges, J. F.: *The Concept of Risk in Building Pathology*, CIB Was, Building Pathology, Llsbon, October, 1991.
13. Borges, J. F. and Castanheta, M.: *Structural Safety*, Course 101, 3rd edition, Laboratorio Nacional de Engenharia Civil, Llsbon, 1985.
14. Ditlevsen, O. *Bayesian Decision Analysis as a Tool for Structural Engineering Decisions*, Joint Committee on Structural Safety, JCSS, Working Document published by IA8SE, Zurich, February 1991.

15. Pedrycz, W.: *Fuzzy Control and Fuzzy Systems*, Control Theory and Applicatlon Series, John Wiley & Sons Inc. New York 1989.
16. Gupta, M. M. and Sanchez, E. (Editors): *Fuzzy Information and Decision Processes*, North-Holland Publishing Company, Amsterdam 1982.
17. Blockley, D.I.: Analysis of structural sailures, *Proceedings of the Institution of Civil Engineers*, Part I, Vol. 62, February 1977.
18. Gero, J.S. and Oguntade, 0.0.: *Fuzzy Set Evaluators in Architecture*, Computer Report CR 29, University of Sydney, 1978.
19. Von Neumann, V. and Morgenstern, O.: *The Theory of Games and Economic Behavior*, Princeton University Press, Princeton, 1953.
20. *Greenhouses: Design, Construction and Loading*, Draft D 01 European Greenhause Standard, International Society for Horticultural Science, Scientific Working Group, June 1991 .
21. Borges, J. Ferry and Castanheta: M. - *Structural Safety*, Course 101, 2nd Edition, Laboratorio Nacional de Engenharia, Lisbon, 1971.
22. Borges, J. Ferry and Castanheta, M.:. *Reliability and Structural Risks in Codifying Wind Actions*, Laboratory Nacional de Engenharia Civil, Lisbon, September 1991.
23. Schwob, M. et Peyrache, G: - *Traite de Fiabilite*, Masson & C., Paris 1969.
24. *Fairground and Amusement Park Machinery and Structures*, Draft European Standard CEN/TC 152 N3o, June 1991.
25. Cranston, B.: *Reflections on Limit State Design*, Structwirl Safety Session, ACI Fall Meeting, November 1990, Revised March 1991, Paisley College, Scotland.
26. Levi, F.: *The World of Civil Engineering and the Eurocodes*, L'lndustria Italiana del Cemento. no 658, Settembre 1991.

QUANTIFICATION OF MODEL UNCERTAINTY IN

STRUCTURAL RELIABILITY

C. GUEDES SOARES
Instituto Superior Tecnico
Universidade Técnica de Lisboa
1096 Lisboa, Portugal

ABSTRACT The different types of uncertainties are considered and their differences are identified. Various methods of statistical analysis of data are reviewed and their usefulness and domain of applicability are identified. The common methods of representing model uncertainties are indicated and several examples of assessment of model uncertainties indicate how the principles described can be applied.

Introduction

The formulation of reliability based structural design implies the recognition that the physical variables considered in engineering are subjected to some variability and thus should be treated as random variables. The earlier treatments of reliability dealt with the uncertainty due to the randomness of the physical variables i.e., with their fundamental or inherent variability.

Soon after it was realised that the description of the random variables was made through parameters that had to be estimated from the analysis of samples of data. However the statistical methods for estimating parameters from samples also yield confidence intervals to the estimated values i.e., the parameters can also be considered as uncertain quantities and the uncertainty of the estimates depends on the size of the sample that is used. The scope of reliability analysis was expanded then, including both the fundamental uncertainty of the variables as well as the statistical uncertainty of the parameters that describe those variables. In this context the use of Bayesian formulations was particularly useful (Cornell, 1972).

A source of differences in the calculated values of the reliability index of a given problem are the different methods of making the relevant engineering predictions. For any engineering problem there are often several methods available of different degree of sophistication and accuracy, ranging from the simple ones based on analytical formulations to the complicated ones based on numerical methods. Although the

17

C. Guedes Soares (ed.), Probabilistic Methods for Structural Design, 17–37.
© 1997 *Kluwer Academic Publishers.*

physical problem is the same, the predictions made with the different engineering theories will be different. This has been recognised as an additional source of uncertainty, which has been called model uncertainty and has been incorporated in reliability formulations.

While the inherent or fundamental variability of a physical phenomenon cannot be changed, the model uncertainty of a mathematical model can be reduced by improving the model i.e. by making more realistic assumptions and by including more physical effects in the model.

While in earlier formulations (Ang and Cornell, 1975), this uncertainty was called subjective and was generally estimated by engineering judgement, more recently quantitative methods have been adopted and it has became clear that this source of uncertainty needs to be represented systematically when developing reliability formulations, together with the models of fundamental uncertainty.

This work will address the topic of model uncertainty dealing with methods to quantify it and to incorporate it in a structural reliability analysis. The assessment of model uncertainty is very much dependent on the problem at hand so that various examples will be included from load modelling, response analysis and strength modelling, to illustrate possible ways of handling its assessments.

Uncertainty Classification

While physical laws describe the regularity of phenomena, relating the expected values of the different quantities of interest, the uncertainty analysis describes the fluctuations that can be superposed on that regularity. In developing and applying physical laws it is always important to assess their limits of applicability and to compare them with experimental evidence. In doing so it is vital to be able to distinguish between the systematic mismatches that result from inapplicability of the theories and those that are simply a result of natural variability.

A better knowledge of the effect of some variables in the physical processes and a detailed analysis of the measuring methods led to a more widespread treatment of different types of uncertainties. Three basic types have been considered, namely the fundamental or intrinsic variability of physical phenomena, the uncertainty associated with the models used in the analysis and the statistical uncertainty associated with the estimation of the values of the parameters of probabilistic distributions as described by several authors in relation with structural reliability (Benjamin and Cornell, 1970; Ferry Borges and Castanheta, 1971; Ang and Tang, 1975, 1984; Thoft-Christensen and Baker, 1982; Augusti, Baratta and Casciati, 1984; Madsen, Krenk and Lind, 1986 ; Melchers, 1987).

The fundamental uncertainty concerns the random nature of some physical phenomena, which is described by representing the physical quantities as random variables or as random processes, depending on whether it is important or not to account for time variation.

Model uncertainty describes the limitations of the theoretical models used in the analysis. They are the mathematical models of the physics of load, load effect or of structural capacity assessment, whose uncertainty is often related with the level of detail used in describing the phenomenon studied.

Finally, statistical uncertainty results from the estimation of the parameters of the probabilistic models from limited samples of data. It can be quantified with the methods of classical statistics, although the inclusion of statistical uncertainty in parameter estimation is the subject of Bayesian statistics (Cornell, 1972).

Often the values of the parameters that govern probability distributions are estimated from samples of data. Different assertions can be made about a population assuming that it is described by a given probabilistic model with the estimated values of the parameters. Bayesian statistics recognise that the parameter values are themselves random and include the effect of that uncertainty on the uncertainty of the predictions. Bayesian analysis also provide the framework to incorporate model uncertainty in the analysis.

Statistical uncertainty and Bayesian formulations can also cover the uncertainty of the probabilistic model adopted instead of only its parameters. Increased amount of data will generally decrease this type of uncertainty and clarify which probabilistic model better describes a set of data.

The method of assessing uncertainty depends on the type under consideration and in the case of model uncertainty it depends very much on the type of problem under study. Generally speaking fundamental and statistical uncertainty can be assessed by applying classical statistical methods to the analysis of data. However model uncertainties, although using also classical statistical methods, have to apply them to different types of situations. Of particular interest are the cases in which model uncertainties cannot be quantified objectively but are assessed on the basis of expert opinion (Cooke, 1991).

Formulations of Model Uncertainties

An important part of the work on model uncertainty is concerned with the mathematical models adopted to describe loads, load effects or strength capacity. These problems are more important in the design situation than in the analysis case because the former methods are of a more simplified nature. Furthermore, since at the design stage often many parameters that are the result of fabrication procedures are not known, the design formulations represent explicitly only a limited number of variables, becoming less accurate when the unrepresented variables exhibit values out of their normal range. The differences between the formulations in these situations have been explored in a specific example dealing with analysis and design of a plate subjected to collapse by compressive loading in Guedes Soares, (1988a).

To assess the model uncertainty of the mathematical models of physical phenomena it is necessary to compare the results of the method under consideration with the predictions of a more sophisticated one or with experimental results. In the later case it must be ensured that the experimental variability is quantified and deducted from the total uncertainty.

The first formal treatment of this type of uncertainties is due to Ang and Cornell (1975), who represented it by a random variable B which operates on the model prediction $\hat{X}$ to yield an improved estimate of the variable X :

$$X = B.\hat{X}$$

The variable B represents the model error, so that its mean value and standard deviation quantifies the bias and the uncertainty of the model. Ang and Cornell used to call these the subjective uncertainties because most often they were based on engineering judgement.

More recently Ditlevsen (1982) dealt with the incorporation of model uncertainty in advanced first order second moment calculations and showed that a representation invariant to mathematical transformations of the limit state function is of the form:

$$X = a\hat{X} + b$$

where a and b are normally distributed random quantities. Comparison with the previous equation shows that it is basically a generalisation that adds a constant term.

Lind, (1976) dealt with model uncertainty in strength calculations emphasising that the choice between two calculation methods of different degree of sophistication should be made on the basis of economic considerations. This means that the model uncertainty of an approximate calculation method should be weighted against the extra benefits and costs of a more refined model. Lind determined the model uncertainty by comparing the predictions of two calculation methods of different degree of sophistication.

However, in most cases the model uncertainty has been derived from comparisons between model predictions and experimental results as for example with beam columns (Bjordhovde, Galambos and Ravindra, 1978); with the collapse of stiffened cylinders (Das, Frieze and Faulkner, 1982); with the fatigue capacity of welded joints (Engesvik and Moan, 1983); with the compressive strength of stiffened plates (Guedes Soares and Soreide, 1983); with the compressive strength of plate elements (Guedes Soares, 1988b); or with the collapse strength of different types of structural components (Faulkner, Guedes Soares and Warwick, 1987 and Smith, Csenki and Ellinas, 1987).

Most of the initial treatments of model uncertainty dealt with strength formulations. However applications to load effect predictions have also been presented. In Guedes Soares and Moan, (1983) the model uncertainty in the theories of wave kinematics were derived from comparisons with measurements. The uncertainty in wave spectra was examined by Haver and Moan, (1983) and its effect on the responses was studied by Guedes Soares, (1991). The uncertainty in long term formulations of wave heights and wave induced responses was considered by Guedes Soares, (1986) and Guedes Soares and Moan, (1991) respectively. The uncertainty in the calculation of transfer functions for motions and loads was studied in the case of ships by Guedes Soares, (1991) of TLP platforms by Eatock Taylor and Jefferys, (1986) and of semi submersible platforms by Incecik, Wu and Soylemez, (1987).

Very often the design methods involve several parameters so that the model uncertainty can only be adequately represented in its whole range of variation by multiple regressions. Bjordhovde, Galambos and Ravindra, (1978) described the bias of a design equation for axially compressed tubular columns as a function of column slenderness. Guedes Soares, (1991) modelled the bias of the transfer functions of wave induced loads in ships as a regression of ship heading, block coefficient and speed. Therefore the basic formulations of the model uncertainty indicated in the two previous equations have very often to be extended to the multivariable case using a problem dependent formulations.

In these cases the observed variability is a result of both the model uncertainty and of the measurement uncertainty of the experimental method. The model uncertainty can only be isolated if the measurement uncertainty is identified and quantified.

Measurement Uncertainty

The problem of experimental errors has been considered for a long time and methods are available to quantify it (Mandel, 1964). Generally one wants to measure a material property that has an inherent variability with a measurement equipment which is in fact an engineering system that responds to some external effect. As such, even if the external effect would be absolutely identical, in repetitive trials one would expect that some variability of measurements would be apparent. This is the measurement uncertainty which cannot be eliminated but which can be reduced by repetitive measurements of the same physical quantity which is known to be constant.

Using this procedure, one is able to quantify the uncertainty of a given measurement method and when applying it to a series of measurements in a sample of specimens with physical variability, one can often separate the uncertainty of the measurement method from the variability of the measured physical property.

Many of the variables used in engineering models are evaluated trough measurements. However, measurements may be subjected to random and to systematic errors which need to be quantified. Measurements can be described by their accuracy and precision which can only be assessed in repeatable and reproducible measurements.

The repeatability of measurements is determined from the closeness of the agreement between the results of successive measurements of the same quantity subjected to the same conditions as regards the method of measurement, the observer, the measuring instrument, the location, the conditions of use, and in addition the measurement must be made over a short period of time. The reproducibility of measurements is defined as the closeness of the agreement between the measurements of the same parameter, when they are carried out under changing conditions, as regards the factors mentioned previously. Both concepts can be quantified by the dispersion of the results.

The precision of a measurement is the closeness of the agreement between the results of applying the experimental procedure several times under the same prescribed conditions. The accuracy is the closeness of the agreement between the results and the true value of the measured parameter. Thus, while imprecision is quantified by the dispersion of the results about the sample mean, inaccuracy is reflected by the shift of the sample mean from the true mean. The quantification of the bias and uncertainty of a measuring process is a prerequisite to the correct derivation of the natural variability of the quantities that are being measured.

Measurement uncertainties can be represented by an uncertain factor B that multiples the correct value of the variable X to yield the measured value $\hat{X}$:

$$\hat{X} = B \ X$$

If the measurement error and the measured variable are statistically independent, the mean and the uncertainty of the variables are given by:

$$\overline{X} = \overline{B} m_x$$

$$\hat{V}_x^2 = V_B^2 + V_x^2$$

where m_x is the true mean of X, $\overline{X}$ is the mean of the measurements, $\hat{V}_x^2$ is the coefficient of variation of the measured values, V_B and V_x are the coefficient of variation of the measurement error and of the variable itself. If the mean of the variable is known, we can define the bias of the measurements from the expected value of the variable i.e.:

$$\overline{B} = \overline{X} / m_x$$

In this case the variance of B is given by:

$$V[B] = \frac{V[\overline{X}]}{m_x^2} = \frac{V[X]}{Nm_x^2}$$

where the variance of the mean is the variance of the variable divided by the number of observations N. This expression shows clearly that the measurement uncertainty will decrease with increasing number of observations.

An interesting situation that happens often is the indirect observation in which case the value of $\hat{X}$ is estimated from measuring $\hat{Y}$:

$$\hat{X} = a + b\hat{Y} + \varepsilon$$

where a and b are regression coefficients and ε is a normally distributed random variable with zero mean and standard deviation σ_ε, which quantifies the dispersion relative to the regression line. Considering a and b and σ_ε as constants, the statistical moments of X are given by:

$$\overline{X} = a + b\overline{Y}$$

$$V[X] = b^2\sigma_y^2 + \sigma_\varepsilon^2$$

Guedes Soares (1990b) discusses of this problem, while Guedes Soares, (1988a) provides an example of an analysis of experimental results in which the measurement uncertainty is separated from the randomness of the parameters.

Statistical Analysis of Data

The various examples referred here indicate clearly that in most cases it is not necessary to assess model uncertainty based only on engineering judgement and that in most cases a quantitative approach can be adopted. Therefore, model uncertainties are objective uncertainties that are associated with physical models and which can be quantified based on the results of experimental or numerical studies. Model uncertainty can in fact be quantified by traditional methods of statistical analysis of data, as briefly highlighted hereafter.

Statistical analysis of data is the basic approach of characterising the fundamental uncertainty of physical variables. Often model uncertainty is assessed from comparisons between theoretical predictions and measurements. These comparisons produce a set of data which can be analysed by various methods of statistical analysis that are covered in textbooks on statistics.

A brief overview is provided here of the main techniques so that can be used to quantify model uncertainty, referring where appropriate examples of aplication in the assessment of model uncertainty.

Sets of data can be appropriately characterised by descriptive statistics, which are ways of summarising data. Common ones are the mean, median, variance, range and the histogram. However, if the sets are random samples from a certain population, the descriptive statistics can also be used to draw inferences about the population. The two main problems in statistical analysis are the estimation of the parameters and the testing of hypothesis. An important inference is the specification of the probabilistic model of a population, which is a function of parameters to be estimated from samples of observed data.

The inferences that can be extracted from data are point estimates, interval estimates or tests of hypothesis. An estimate may yield different values for different samples, according to a sampling distribution. One is generally interested in estimators that are consistent, unbiased and have minimum variance. They should also extract the maximum information from the data available as do maximum likelihood estimators.

The classical statistical inference is based on assumptions concerning the form of the population distribution, providing methods to quantify the population parameters. Generally the assumption of normal distribution is underlying the statistical tests but in some cases the previous experience may be too limited or the sample size may be too small to justify the choice of the distribution type.

Another useful field of statistics is concerned with distribution free inferences both for testing and for estimating (Siegel, 1956; Gibbons, 1971; Conover, 1978). These methods are based on functions of the sample observations which do not depend on the specific distribution function of the population from which the sample was drawn. Inferences that do not depend on the value of parameters are called non-parametric, although this name is also used to denote all distribution-free inferences to distinguish them from the parametric inferences of classical statistics.

The main requirements for non-parametric tests are only that the population be continuous and that the sample is random. The choice between the different methods must be based on the their power, that is on their sensitivity to changes in the factors tested and on their robustness or sensitivity to reasonable changes of magnitude of extraneous factors. In general parametric tests are more precise and non-parametric ones are more robust, which is particulary important for small samples.

A fundamental assumption in statistical analysis is that the data analysed results from random sampling the population. Thus in analysing a new type of data it may be worthwhile to conduct a preliminary analysis to check the randomness and the independence of the samples of data. For this purpose there are different non-parametric tests based on runs. The data must be transformed in a succession of dichotomous symbols which can even be + and -, depending on whether the observation is larger or smaller than the preceding one. A run is defined as a succession of one or more identical symbols which are followed and proceeded by a different one. There are tests based on the total number of runs or on the length of the longest run. Whenever these are too small or too large the hypothesis of randomness must be rejected although due to different types of lack of randomness.

Another hypothesis in which statistical analysis is often based is that the data results from independent observations. This assumption can also be tested with non-parametric tests based on rank order statistics. These are defined as sets of numbers which result when each original observation is replaced by the value of an order preserving function. Still an important problem is to ensure that all samples that are used in a study are drawn from the same population. In classical statistics use can be made of the F test to check that all samples are from a normal distribution having the same variance but different means. An application of several non-parametric tests to analyse load effect data can be found for example in Guedes Soares and Moan, (1982).

To check the adequacy of a probabilistic model it may be necessary to estimate the location, scale and shape parameters. The estimation of parameters can be made by the method of moments of regression or of maximum likelihood. The latter is generally more efficient but also more complicated to apply. In Guedes Soares and Henriques (1994) a case was presented in which the estimation of the parameters of the Weibull distribution describing wave data by the method of moments was providing clear differences from the results of the other methods. Non-parametric tests can also be used to check for the symmetry of a distribution, as well as for the goodness of fit.

The Chi-square and the Kolmogorov-Smirnov tests, which are commonly applied, are also non-parametric tests. While the first one deals with histograms the latter is based on cumulative distribution functions. Although the Chi-square test is used more often, the Kolmogorov-Smirnov test is often preferable, specially for small samples. The Chi-square test groups the data, loosing thus some information. In fact the choice of the number of intervals can change the result of the test. Furthermore it is limited to sample sizes that give an adequate number of observations in each interval.

Interest is often not limited to the analysis of only one variable but is directed to several variables or to the effect that they can have on the outcome of a given process. In these cases regression analysis provides a framework to determine the relationship between random variables (Draper and Smith, 1966; Morrison, 1969). The regression equation indicates the expected value of the dependent variable conditional on the value of the regressed or independent variable. A measure of the uncertainty is provided by the standard deviation of the residuals or the standard error.

Most common are the linear regressions which can be simple or multiple if applied to one or to several variables respectively. However, non-linear regressions can also be used whenever applicable. Examples of applications of simple regressions are found in Guedes Soares, (1986) and in Jastrzebski and Kmiecik, (1986) while the results of multiple regressions can be found in Antoniou, Lavidas and Karvounis, (1984) and in Guedes Soares and Moan, (1988).

Regression methods are frequently used to analyse data from unplanned experiments such as might arise from observation of uncontrolled phenomena or from historical records, examples of which can be found in Guedes Soares, (1986); Jastrzebski and Kmiecik, (1986); Antoniou, Lavidas and Karvounis, (1984) and on Guedes Soares and Moan, (1988).

Statistical design of experiments refers to the process of planning the most efficient way of data collection for a given problem so that the statistical analysis may lead to the maximum amount of relevant information (Winer, 1970; Montgomery, 1984). It involves the three basic principles of replication, randomisation and blocking.

Replication means the repetition of one basic experiment or process outcome, which allows one to identify the error due to that effect. This is important to allow conclusions about whether the observed differences in the data are really statistically different. In addition it allows an improved estimate of the effect under study because the sample mean has a smaller variance.

Randomisation is the cornerstone underlying the use of statistical methods, requiring that the allocation of the collected data and the sequence in which it is done be random.

Blocking intends to increase the precision of the experiment by selecting the portions of the data that are more homogeneous than the whole set. Blocking involves making comparisons among the conditions of interest of the effects within each block.

Statistical design of experiments has been used recently with the planning of fatigue tests (Sorensen et al, 1992, Engelund et al, 1993).

Examples of Applications

The assessment of model uncertainty is made with procedures that are developed on a case by case basis. Depending on the problem at hand, on the information available and in the uncertainties to deal with, different formulation are advisable. Therefore, in this section reference is made to some applications in order to illustrate a spread of possibilities.

MODEL UNCERTAINTY IN SPECTRAL FORMULATIONS

There are different types of structures that are subjected to dynamic excitation and response. Since the environmental excitation is normally of random nature, such as in the case of wind load, earthquakes or ocean waves, a common approach to the solution of this kind of problem has been to adopt a spectral formulation. The excitation is modelled as a Gaussian process and as a such is liable of being represented by a spectrum. The linear response to this excitation is a response spectrum that is obtained as the product of the input spectrum and the transfer function which represents the systems characteristics.

These are different models of theoretical spectra adopted to describe the wind velocity (Forristal, 1988). The case of ocean waves, will be considered in detail having however in mind that this illustrates a methodology that could also be applied to other excitations and other structures.

Short term sea states are usually modelled as ergodig random processes which become fully described by a variance spectra. These spectra, which are estimated from records of wave surface elevation, have a shape that depend on the characteristics of the sea state.

Developing sea states are described by a Jonswap spectrum (Hasselman, et al., 1973) and fully developed sea states by a Pierson-Moskowitz model (Pierson and Moskowitz, 1964). Whenever in an ocean area there coexists two wave systems the spectrum often exhibits two peaks in which case one can adopt the 6 parameter formulation of Ochi and Hubble, (1976), or the 4 parameter proposal of Guedes Soares, (1984).

When performing predictions for design, which is a main interest in engineering, one does not know which type of sea state will occur and thus, which type of wave spectrum is applicable. By choosing one of them, the prediction may be affected by a model error.

In addition to the type spectrum one can also account for the variability of the spectral ordinates around the mean spectrum which is described by the theoretical models. Haver and Moan (1983) have studied the uncertainty of the spectral formulations of single peaked sea states and analysed the variability of the spectral ordinates showing that they were generally independent of the frequency.

This variability can be represented by a unit mean random variable i that multiplies the mean spectral ordinate at the frequency $S(\omega_i)$:

$$S(\omega_i) = \bar{S}(\omega_i)(1 + \varepsilon_i)$$

This formulation, which was adopted by Guedes Soares, (1991) for the case of ocean spectra, can be generalised for any other spectral formulations of structural response, and thus it is described hereafter for illustration purposes.

Wave spectra are of interest to calculate the response of a marine structure to a given sea state. Thus, often the quantity of interest is the variance of the response R which is given by the area under the response spectrum $S_R(\omega)$:

$$R = \int_0^\infty S_R(\omega)d\omega = \int_0^\infty S(\omega)H^2(\omega)d\omega$$

where $H(\omega)$ is the response amplitude operator.

That value of variance is obtained for a specific type of spectrum that is denoted by $\phi \theta$. It can be considered as an uncertain quantity because the exact shape of the spectrum depends on the random variable R and in addition it depends on the model uncertainty about which type of spectrum is the correct one. The model uncertainty can be separated from the fundamental uncertainty of R by conditioning. Thus the marginal distribution of response variance $f_R(r)$ is given by:

$$f_R(r) = \int f_R(r|\theta)f_\theta(\theta)d\theta$$

where the conditional distribution is assumed to be Gaussian with a mean value of $\mu_{R|\theta}$ and a variance of:

$$\sigma_{R|\theta}^2 = \sigma_\varepsilon^2 \int_0^\infty \left[S(\omega) H^2(\omega) \right]^2 d\omega$$

where the estimates of the spectral ordinates are assumed to be independent and identically distributed with a variance σ_ε.

The variance of the marginal distribution of R, is given by:

$$\sigma_R^2 = \iint \left[\sigma_{R|\theta}^2 + \left(\mu_\sigma - \tilde{\mu} \right)^2 \right] f_\theta(\theta) \, d\theta$$

where the first term is the variance conditional on the type of spectrum and the second term represents the contribution of model uncertainty. The mean of the conditional distribution is μ_θ and the mean of the marginal one is:

$$\tilde{\mu} = \int r \, f_R(r) \, dr = \int \mu_\theta \, f_\theta(\theta) \, d\theta$$

If one decides to use one model of spectrum for the whole analysis, this implies adopting the expected value of μ_θ instead of $\tilde{\mu}$. Thus, the model error in that formulation can be described by an uncertain quantity:

$$\phi_\theta = \frac{\tilde{\mu}}{\mu_\theta}$$

which has a model uncertainty of σ_R^2.

In some situations of assessing the wave loading in offshore structures, it is not enough to have a description of the sea surface elevation process as described by a wave spectrum. In fact in the estimation of the hydrodynamic loads on offshore structures it is necessary to use a wave theory to predict the distribution of velocity and acceleration of the wave particles at different water depth. The wave particle movement is then transformed into the forces that are being induced in infinitesimal elements of the structure, through the Morison equation.

There are various wave theories available, from the Airy theory, which is a linear one valid for small amplitudes to the Stokes fifth order or the Stream function, which are applicable to large wave amplitudes. Outside the two extreme situations of small and large amplitudes there is a range of wave heights and water depths in which various theories could be applicable with different degrees of error. The model uncertainty of a specific theory could be established from comparison with measurements when they are available, as was done for example in (Guedes Soares and Moan, 1983).

Although this problem was originally formulated for the response of marine structures to wave spectra, the general framework is applicable to other cases such as for example the response of tall buildings or bridges to wind gust loading the response of structures to earthquake excitation.

MODEL UNCERTAINTY IN TRANSFER FUNCTIONS

The prediction of the properties of the response parameters require the knowledge of the transfer function $H(\omega)$ as indicated in the previous section. The engineering models adopted to calculate $H(\omega)$ will have different degrees of accuracy and uncertainty.

In general the model uncertainty of a theory can be assessed by comparing its predictions with experimental results. Defining the model error ϕ, as a function of frequency ω, one has:

$$\hat{H}(\omega) = \phi(\omega)H(\omega) + \varepsilon(\omega)$$

where $\hat{H}(\omega)$ is the measured value, $H(\omega)$ is the theoretical prediction and ε represents an experimental error of zero mean value.

Often the model error can have a general form of:

$$\phi(\omega) = \sum_{i=0}^{n} a_i \, \omega^n$$

where a_i, are regression coefficients to be determined from the analysis of data. The regression equation will indicate the mean of the model error. The standard deviation of the regression residuals will indicate the model uncertainty.

In the application of this formulation to the responses of ships to wave excitation, it was found that the constant model error $\phi = a_0$ would be adequate for practical purposes (Guedes Soares, 1991). However, the transfer functions depend on the relative wave direction and the adequacy of the transfer function theories depend on ship speed and geometry. Thus a global description of the dependence of ϕ on those variables was obtained by a regression analysis.

The uncertainty in the calculation of transfer functions for motions and loads of TLP platforms was studied by Eatock Taylor and Jefferys, (1986) and of semi submersible platforms by Incecik, Wu and Soylemez, (1987), although the uncertainty of the predicted transfer functions was not modelled explicitly.

Winterstein et al, (1993) and Sorensen et al (1993) have adopted models similar to those to model the uncertainty of transfer functions of offshore structure.

MODEL UNCERTAINTY IN LONG-TERM DISTRIBUTIONS

In the design of marine structures, either of ships or of ocean or offshore platforms one often requires the distribution of load effects for time spans of the order of the structure's lifetime, i.e. the long-term distribution.

To obtain them one starts from shorter periods of stationary that correspond to sea states. The probability of exceeding the amplitude x in any of them is given by the Rayleigh distribution (Longuet-Higgings, 1952, 1983):

$$Q_S\left(x|R\right) = \exp\left(-\frac{x^2}{2R}\right)$$

which is conditional on the value of the response variance R.

The probability of exceedance in the long-term is obtained as a marginal distribution:

$$Q_L(x) = \int_0^\infty Q_s(x|R)\, f_R(r)\, w(r)\, dr$$

where $w(r)$ is a weighting factor that depends on the mean period of the response in each sea state, and $f_R(r)$ is the probability density function of the response variance.

This long-term probability distribution is constructed from the evaluation of the integral at several levels x. However very often a theoretical distribution is fitted to the calculated points to make easier the future analysis. Common distributions are the Weibull and the log-normal which yield different extreme value predictions. Thus the choice of one of them will involve a model uncertainty.

The problem of the statistical uncertainty in the fitting of a Weibull distribution to wave data has been discussed by Guedes Soares and Henriques, (1994).

One can use a Bayesian approach to deal with both model and statistical uncertainty as adopted for example by Edwards, (1984) who studied the structural reliability of a simple system subjected to a loading that could be described by either a normal, a log-normal or a Weibull distribution. The probability of each of the probabilistic models was assessed from the data in the classical Bayesian way, and the posterior distribution was obtained including and nor including the effect of statistical uncertainties.

A similar problem was tackled by Guedes Soares (1989) dealing now with predictions of extreme values of significant wave height as predicted from long-term distributions or by other methods. Since in extreme predictions there is often not enough data to allow the reliable use of the classical Bayesian methods to assess the conditional probabilities of the models, it was proposed there that expert opinions could be used to assess those probabilities and to predict an estimate that accounted for the model uncertainty as assessed by experts.

MODEL UNCERTAINTY IN COMPRESSIVE PLATE STRENGTH

In addition to various sophisticated numerical methods available for the analysis of the compressive strength of plates, there are several design methods available also as reviewed in (Guedes Soares, 1988b). In the formulation of a design method one must ensure that a reasonable degree of accuracy is maintained without unnecessarily complicating the calculation procedures. This objective can be achieved if the design equations only include the most important physical variables.

The number of variables included in the design equation must be such that the strength predictions are always within a narrow scatter band independently of the value of the variables not represented in the equation. The single most important parameter that governs plate strength is the reduced slenderness:

$$\beta = \frac{b}{t} \sqrt{\left(\frac{\sigma_0}{E} \right)}$$

where b and t are the plate breadth and thickness σ_0 and E are the material yield stress and Young modulus.

After having analysed the sensitivity of plate strength to each of the variables, it was concluded that the simplest design method should include only the plate slenderness β, because the plate strength can change by as much as 60% over the useful range of slenderness. If any improvement of accuracy is desired, explicit account must be given to the variables that can produce changes of 20% in the plate strength. Thus consideration should be given to residual stresses, initial distortions and boundary conditions (Guedes Soares, 1988b).

The model uncertainty B_b of one specific method ϕ_b that depends explicitly only of plate slenderness has been assessed from comparisons with experimental results. This approach was also adopted to asses the effect of residual stresses and initial imperfections. The strength of a plate without defects is given by $\phi_b B_b$ and the degrading effect of the residual stresses and of the initial distortions are given by R_r and R_δ respectively.

The formulation adopted to predict the compressive plate strength is given by:

$$\phi = \frac{\sigma_u}{\sigma_0} = \left(\phi_b B_b \right)\left(R_r B_r \right)\left(R_\delta B_{r\delta} \right)$$

where σ_u is the ultimate stress, B_r and $B_{r\delta}$ are modelling errors which affect the reduction factors for the effect of weld induced residual stresses and initial distortions respectively. The exact form of each expression is given in (Guedes Soares, 1988b) but for the present purposes it is enough to indicate how the modelling factors were determined. While B_b and B_r turned out to be constants, $B_{r\delta}$ resulted in a linear multiple regression on plate slenderness, intensity of residual stresses and amplitude of initial distortions.

For code purposes simpler design equations are wanted and in that case the three model errors can be combined in one only as indicated in (Guedes Soares, 1988b).

To obtain a design equation that depends only on plate slenderness, it is necessary that the effect of the other parameters is taken at their expected values. Expressing the design equation:

$$\phi_G(\beta,\eta,\delta_o) = \phi_b(\beta).B(\beta,\eta,\delta_o)$$

one aims at determining the bias factor $\overline{B}$ which will depend on the distributions of the governing parameters. Its mean value and variance is given by weighting the model error B by its probability of occurrence:

$$\overline{B} = \int\int\int B(\beta,\eta,\delta_o).f(\beta,\eta,\delta_o)d\beta \ d\eta \ d\delta_o$$

$$\sigma_B^2 = \int\int\int \{B(\beta,\eta,\delta_o) - \overline{B}\}^2 f(\beta,\eta,\delta_o)d\beta \ d\eta \ d\delta_o$$

where $f(\beta,\eta,\delta_o)$ is the joint probability density function of the variables β,η,δ_o. This function represents the probability that for a given ship the typical plates of the midship section have a specified value of β,η and δ_o. It is assumed that these effects are independent so that:

$$f(\beta,\eta,\delta_o) = f_1(\beta).f_2(\eta).f_3(\delta_o)$$

Furthermore, when applying these equations to plates of aspect ratio slightly smaller than unity a further model uncertainty is introduced (Guedes Soares, and Faulkner, 1987).

MODEL UNCERTAINTY IN THE COMPRESSIVE STRENGTH OF STIFFENED CYLINDERS

As occurs with plates, there are also several simple expressions for the design of stiffened cylinders under different load combinations. The model error of any method can be expressed as the ratio of the experimental result to the predicted value.

In (Faulkner, Guedes Soares, and Warwick, 1987) there is a summary of the model uncertainty of various code design methods. It was found that the modelling error in that case can have a mean value ranging from 0.80 to 1.70 and the coefficient of variation would range from 0.13 to 0.43. However, a good design method should have an uncertainty around 10 to 15% at most.

Das, Faulkner and Zimmer, (1992) have reviewed various codified ultimate strength formulations for orthogonally stiffened cylindrical strength shell components by comparison with available experimental results in order to establish model uncertainty factors associated with these formulations.

They used experimental results produced in the mid 1960s in an aerospace research programme and obtained in the 1970s and early 1980s in the scope of offshore structures research. The aerospace tests were mainly conducted in the elastic range and the majority of them used high strength aluminium allow models. The stringers in the models were large in number and closely spaced. They were machine finished leading to minimal residual stresses and initial imperfections. The offshore models were of steel, and had considerable initial imperfections due to the fabrication procedure.

The model uncertainty of four design methods were determined from comparisons with this data. They were the methods in the ABS Model Code for Tension Leg Platforms, the API Bulletin ZU, the DnV Classification Note 30.1 and the European Convention of Constructional Steelwork recommendation. Stiffened shells have several collapse modes which makes a detailed account of their model uncertainty too long to be included here. However it was shown that the bias and uncertainty of the different methods was significantly different, and an interesting conclusion of the study was that for some methods there was a clear dependency between the modelling error and the collapse load, which resulted from an incorrect modelling of the slenderness effect. This is a feature that should be avoided in developing design methods that are calibrated probabilistically.

Concluding Remarks

There are presently available several good algorithms for calculating the reliability index for different kinds of problems in an efficient manner. The widespread application of the existing technology to practical problems requires that work be developed in probabilistic modelling of the different problems.

Of particular importance is the consideration of the model uncertainty of the different engineering theories that are used in analysis and in design.

Although the assessment of model uncertainty is very much problem dependent, this work has discussed the main problems encountered, the basic tools that are used and several examples have been described from different engineering disciplines.

References

Ang, A. H-S. and Cornell, C.A. (1975): "Reliability Bases of Structural Safety and Design", *J. Structural Division*, ASCE, Vol. 100, pp. 1755-1769.

Ang, A.H-S. and Tang, W.H. (1975, 1984): *Probability Concepts in Engineering Planning and Design*, John Wiley & Sons, New York, Vol. I and Vol. II.

Antoniou, A.C., Lavidas, M. and Karvounis, G. (1984): "On the Shape of Post-Welding Deformations of Plate Panels in Newly Built Ships", *Journal of Ship Research*, Vol. 28, pp. 1-10.

Augusti, G., Baratta, A. and Casciati, F. (1984): *Probabilistic Methods in Structural Engineering*, Chapman & Hall, London.

Benjamin, J.B. and Cornell C.A. (1970): *Probability, Statistics and Decision for Civil Engineers*, McGraw-Hill Book Co., New York.

Bjordhovde, R., Galambos, T.V. and Ravindra, M.K. (1978): "LFRD Criteria for Steel Beam Columns", *J. Structural Division*, ASCE, Vol. 104, No. ST9, pp. 1371-1388.

Conover, W.J. (1978): *Practical Nonparametric Statistics*, J. Wiley & Sons, New York, 2nd Ed.

Cooke, R.M. (1991): *Experts in Uncertainty*, Oxford University Press, Niew York

Cornell, C. (1972): "A.Bayesian Statistical Decision Theory and Reliability-Based Design", *Proc. International Conf. on Structural Safety and Reliability*, A.M. Freudenthal (Ed.), Pergamon Press, pp. 47-68

Das P.K., Faulkner, D. and Zimmer, R.A. (1992), "Selection of Robust Strength Models for Efficient Design of Ring and Stringer Stiffened Cylinder under Combined loads", *Proceedings of 10th Offshore Mechanics and Artic Engineering Conference*, ASME, Vol II, pp 417-428.

Das, P.K., Frieze, P.A. and Faulkner, D. (1982): "Reliability of Stiffened Steel Cylinders to Resist Extreme Loads", *Proc. 3rd Int. Conf. on Behaviour of Offshore Structures*, (BOSS'82), MIT, pp. 769-783.

Ditlevsen, O. (1982): "Model Uncertainty in Structural Reliability", *Structural Safety*, Vol. 1, pp. 73-86.

Draper, N.R. and Smith, H. (1966): *Applied Regression Analysis*, John Wiley & Sons, New York.

Eatock Taylor, R. and Jefferys, E.R. (1986): "Variability of Hydrodynamic Load Predictions for a Tension Leg Platform", *Ocean Engineering*, Vol. 13, No. 5, pp. 449-490.

Edwards, G. (1984): "A Bayesian Procedure for Drawing Inferences from Random Data", *Reliability Engineering*, Vol. 9, pp. 1-17.

Engelund, S., Bouyssy, V. and Rackwitz, R. (1993): Optimal Bayesian Designs for Fatigue Tests, *Reliability and Optimization of Structural Systems, V*, Thoft-Christensen, P. and Ishikawa, H. Ed.), North-Holland, pp. 55-63

Engesvik, K. and Moan, T. (1983): "Probabilistic Analysis of the Uncertainty in the Fatigue Capacity of Welded Joints", *Engineering Fracture Mechanics*, Vol. 18, pp. 743-762.

Faulkner, D., Guedes Soares, C. and Warwick, D.M. (1987): "Modelling Requirements for Structural Design and Assessment", *Integrity of Offshore Structures-3*, D. Faulkner, M.J. Cowling and A. Incecik, (Eds.), Elsevier Applied Science Publishers, pp. 25-54.

Ferry Borges, J. and Castanheta, M. (1971): *Structural Safety*, Laboratorio Nacional de Engenharia Civil, Lisboa.

Forristal, G.Z., (1988): "Wind Spectra and Gust Factors over Water", *Proceedings Offshore Technology Conference*, Vol. 2, OTC 5735, pp. 449-460.

Gibbons, J.D. (1971): *Non Parametric Statistical Inference*, McGraw-Hill Book Co., New York.

Guedes Soares, C. (1984): "Representation of Double-Peaked Sea Wave Spectra", *Ocean Engineering*, Vol. 11, pp. 185-207.

Guedes Soares, C. (1986): "Assessment of the Uncertainty of Visual Observations of Wave Height", *Ocean Engineering*, Vol. 13, pp. 37-56.

Guedes Soares, C. (1988a): "Uncertainty Modelling in Plate Buckling", *Structural Safety*, Vol. 5, pp. 17-34.

Guedes Soares, C. (1988b): "Design Equation for the Compressive Strength of Unstiffened Plate Elements with Initial Imperfections", *J. Constructional Steel Research*, Vol. 9, pp. 287-310.

Guedes Soares, C. (1988c): "A Code Requirement for the Strength of Plate Elements", *Marine Structures*, Vol. 1, No. 1, pp. 71-80.

Guedes Soares, C. (1989), "Bayesian Prediction of Design Wave Heights", *Reliability and Optimization of Structural Systems '88*, P. Thoft-Christensen (Ed.), Springer-Verlag, , pp. 311-323.

Guedes Soares, C. (1990a): "Effect of Spectral Shape Uncertainty in the Short Term Wave-Induced Ship Responses", *Applied Ocean Research*, Vol 12, N ° 2, pp. 54-69.

Guedes Soares, C. (1990b): "Uncertainty Modelling in Systems Reliability Analysis", *Systems Reliability Assessment*, A.G.Colombo and A. Saiz de Bustamente (Eds.), Kluwer Acad. Pub., Dordrech, pp. 285 - 303.

Guedes Soares, C. (1991): "Effect of Transfer Function Uncertainty on Short Term Ship Responses", *Ocean Engineering*, Vol. 18, N° 4, pp. 329-362.

Guedes Soares, C. and Faulkner, D. (1987): "Probabilistic Modelling of the Effect of Initial Imperfections on the Compressive Strength of Rectangular Plates", *Proc. Third International Symposium on Practical Design of Ships and Mobile Units* (PRADS), Trondheim, Vol. 2, pp. 783-795.

Guedes Soares, C. and Henriques, A.C. (1994), "On The Statistical Uncertainty in Long Term Predictions of Significant Wave Height" *Proceedings of the 12th Offshore Mechanics and Artic Engineering Conference (OMAE)*, New York, Vol. II, pp. 65-75.

Guedes Soares, C. and Moan, T. (1982): "Statistical Analysis of Still-Water Bending Moments and Shear Forces on Tankers, Ore and Bulk Carriers", *Norwegian Maritime Research*, Vol. 10, pp. 33-47.

Guedes Soares, C. and Moan, T. (1983): "On the Uncertainties Related to the Hydrodynamic Loading of a Cylindrical Pile", *Reliability Theory and its Applications to Structural and Soil Mechanics*, P. Thoft-Christensen (Ed.), Martinus Nijhoff Pub., The Hague, pp. 351-364.

Guedes Soares, C. and Moan, T. (1988): "Statistical Analysis of Still-Water Load Effects in Ship Structures", *Transactions Society of Naval Architects and Marine Engineers*, New York, Vol. 96, pp.129-156.

Guedes Soares, C., Moan, T. (1991): "Model Uncertainty in the Long Term Distribution of Wave Induced Bending Moments for Fatigue Design of Ship Structures", *Marine Structures*, Vol. 4, pp. 295-315.

Guedes Soares, C. and Soreide, T.H. (1983): "Behaviour and Design of Stiffened Plates under Predominantly Compressive Loads", *International Shipbuilding Progress*, Vol. 30, No. 341, pp. 13-27.

Guedes Soares, C. and Viana, P.C. (1988): "Sensitivity of the Response of Marine Structures to Wave Climatology", *Computer Modelling in Ocean Engineering*, B.A. Schrefler and O.C. Zienkiewicz (Eds.), A.A. Balkema Pub., pp. 487-492.

Hasselman, K., et al. (1973), "Measurements of Wind-Wave Growth and Swell Decay During the Joint North Sea Wave Project (JONSWAP)", *Deutschen Hydrographischen Zeitschright*, Reihe A(8), No. 12.

Haver, S. and Moan, T. (1983): "On Some Uncertainties Related to the Short Term Stochastic Modelling of Ocean Waves", *Applied Ocean Research*, Vol. 5, pp. 93-108.

Incecik, A., Wu, S-K. and Soylemez, M. (1987): "Effect of Different Mathematical Models in Calculating Motion and Structural Response of Offshore Platforms", *Integrity of Offshore Structures-3*, D. Faulkner, M.J. Cowling and A. Incecik, (Eds.), Elsevier Applied Science, pp. 115-144.

Jastrzebski, T. and Kmiecik, M. (1986): "Statistical Investigations of the Deformations of Ship Plates", (in French), *Bulletin Association Technique Maritime et Aeronautique*, Vol. 86, pp. 325-346.

Lind, N.C. (1976): "Approximate Analysis and Economics of Structures", *J. Structural Division*, ASCE, Vol. 102, pp. 1177-1196.

Madsen, H.O., Krenk, S. and Lind, N.C. (1986): *Methods of Structural Safety*, Prentice-Hall, New Jersey.

Mandel, J. (1964): *The Statistical Analysis of Experimental Data*, Dover Publications, New York.

Melchers, R.E., (1987): *Structural Reliability: Analysis and Prediction*, Ellis Holwood, Chichester

Montgomery, D.C. (1984): *Design and Analysis of Experiments*, J. Wiley & Sons, New York.

Morrison, D.F. (1969): *Multivariate Statistical Methods*, McGraw-Hill Book Co., New York.

Ochi, M.K. and Hubble, E.N. (1976): "On Six-Parameter Wave Spectra", *Proc. 15th Coastal Engineers Conf.*, Amer. Soc. Civil Engineers (ASCE), pp. 321-328.

Pierson, W.J. and Moskowitz, L. (1964): "A Proposed Spectral Form for Fully Developed Wind Seas Based on the Similarity Theory of S.A. Kitaigorodskii", *J. Geophysical Research*, Vol. 69, pp. 5181-5190.

Siegel, S. (1956): *Non Parametric Statistics*, McGraw-Hill Book Co., New York.

Smith, D., Csenki, A. and Ellinas, C.P. (1987): "Ultimate Limit State Analysis of Unstiffened and Stiffened Structural Components", *Integrity of Offshore Structures-3*, D. Faulkner, M.J. Cowling and A. Incecik, (Eds.), Elsevier Applied Science Publishers, pp. 145-168.

Sørensen, J.D., Faber, M.H. and Kroon, I.B. (1992): "Risk Based Optimal Fatigue Testing,". *Probabilistic Mechanics and Structural and Geotechnical Reliability*, ASCE.

Thoft-Christensen, P. and Baker, M.J. (1982): *Structural Reliability Theory and its Applications*, Springer-Verlag, Berlin.

Winer, B.J. (1970): *Statistical Principles in Experimental Design*, McGraw-Hill Book Co., New York.

Winterstein, S.R. , Kroon, I.B., and Ude, T.C. (1993): Fatigue of Floating Offshore Structures: Modelling Uncertainty in Hydrodynamics and Fatigue Properties. *Proceedings of ASCE Structures Congress*, Irvine, California.

RESPONSE SURFACE METHODOLOGY IN STRUCTURAL RELIABILITY

JACQUES LABEYRIE
Ifremer
BP 70 29280 Plouzane - France

1. Brief Review

We first propose to trace some streams of thought which have contributed directly to what we now call Response Surface Methodology (abbreviated to RSM). A mathematical description of RSM is given in section 2.

During the period 1930-1950 and in various practical studies such as growth rates in nutrition of pigs, probit analysis or crop yield to fertilizer levels, some revealing requirements for RSM were mentioned, [1] - [3]. The main purpose was to gain an insight into the observed behavior of the process under investigation and to obtain the determining setting of the variables involved.

The pioneering works of Box [4] - [7], produced a set of statistical procedures in two main areas dealing with the design of experiments and regression analysis. The most successful applications were obtained in the fields of chemistry and chemical engineering. The dominant assumption was clearly that the response can be approximated by polynomial functions. One reason for the popularity of polynomial models lay in their conceptual and computational tractability.

During the period 1950 - 1970 the major topics of research on RSM were marked by significant probabilistic guidelines ;

- Robbins and Monro [8] introduced the stochastic approximation for finding an optimum in the presence of outliers ; they also included multi-dimensional aspects.
- The comparative analyses of growth curves in Biometrics performed by Rao, [9] adopted a multivariate approach. The response function arose from the projection onto a family of orthogonal polynomials. The coefficients of these expansions served for subsequent analyses and prediction purposes.
- Keefer and Wolfowitz [10] laid the theoretical foundation for a concept of optimal design based almost exclusively on linear models and with an optimality criterion which used the generalized variance among parameters.

During the last decade, Response Surface Methodology has gained robustness for modelling in various technical applications such as material design, electronic error

C. Guedes Soares (ed.), Probabilistic Methods for Structural Design, 39–58.

detection and structural integrity. In particular, the introduction of non-linear models, and the increasing numerical potential of electronic computers have affected the evolution of techniques. The concept of Response Surface arose from a system description of the physical process (INPUT ---> TRANSFER FUNCTION ---> OUTPUT).The main objective was to provide flexible analytical formulations as surrogates for the original models in order to perform subsequent uncertainty and sensitivity studies. The user friendly nature of this approach was partly explained by its reliance on geometrical concepts. Moreover, for safety assessment of mechanical systems, the suggested formal modelling turned out to be effective when implementing reliability indices from FORM/SORM methods.

At the present time there are probabilistic mechanics packages such as ; RPEJ (Evaluation of Jackets from a Probabilistic Redundant Analysis), and RASOS (Reliability Analysis Systems for Offshore Structures), which contain extensive implementations of RSM (environmental loadings, limit state functions, ...).

2. Definitions and Basic Concepts

The Response Surface Methodology (RSM) is a formal representation based on geometrical ideas. It leads to the investigation of the properties of a physical process. That means the required response Y (random variable, vector or process) is considered as the output of a system, which varies in response to the changing levels of several input variables.

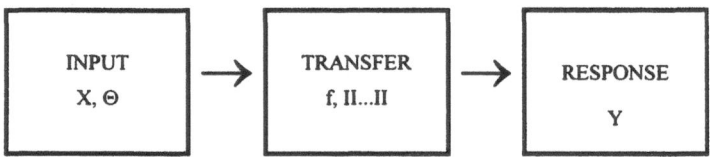

A response (hyper-) surface is selected :
- $X = \{X_1, X_2, ... X_n\}$ is a finite representative set of **stimuli** random variables or processes.
- $\Theta = \{\Theta_1, \Theta_2,..., \Theta_p\}$ is the available **statistical** information on X (free or parametric distribution functions, fourier series, normalized moments, ...),
- f is an explicit analytical function of X given Θ ; note $f(X/\Theta)$, serves as an **approximation** for Y.
- II...II is a metric in a functional space containing Y and $f(X/\Theta)$ which gives some measure of the **goodness of fit** of the approximation.

Note that it is mathematically more correct to use the terminology of response **hyper- surface** when n>2. As long as there is no ambiguity, we will use the usual term **response surface** whatever the dimension of the space of the basic variables.

2.1 CRITERIA FOR BUILDING RESPONSE SURFACES.

2.1.1 Physical Meanings

The selection of the set of stimuli must be based as far as possible on some understanding of the underlying mechanism.

As an example, the usual Morison equation for computing loads for marine structures can be viewed as a response surface,

$$f_t = \frac{1}{2}\rho D C_d\, u_t |u_t| + \frac{1}{4}\pi\rho D^2 C_m u'_t \tag{1}$$

Here f_t is the force per unit span separated into drag and inertial components, ρ is the water density, D is the cylinder diameter, and u_t is the instantaneous flow velocity.

This equation generally predicts the main trends in measured data quite well, once an appropriate joint distribution function of the drag and inertia coefficients can be provided depending of the sea-state parameters. Nevertheless, some interesting characteristics of the flow are not represented with enough accuracy (e.g. high frequency content, gross vortex shedding effects, ...). So when applying the response surface f_t , unfortunately we miss some sources of response problems for an offshore platform. That confirms the need to validate extensions by using the NARMAX modelling techniques (Non linear Auto-Regressive Moving Average with eXogenous inputs) ; this is especially important for non linear effects [13].

The concept of limit state for structural systems also needs to be formally introduced with its physical interpretation. It is a mathematical way of separating the relevant determining variables for the system into desirable and undesirable domains of the variations ; that is a way of constructing a boundary of a failure domain. Here a failure event is defined to be structural damage which has socio-economic consequences. The problem is to estimate the failure costs, its impact on the target risk, and the risk levels which society is willing to tolerate. Discussions of these issues must often take place among professionals with various backgrounds and feelings.

The detailed discussion reported in [14], illustrates the respective codified reliability meanings for the design of reinforced concrete structures based on "elastic" or alternatively "plastic" ultimate limit states.

2.1.2 Distribution Effects

Another essential criterion is given under the term **distribution effects** [15]. The output distributions change, depending on the different assumptions selected to describe the statistical properties of X, and to fit the function f. The selection should reflect how the statistics are well transferred through the response surface.

When predicting the stochastic response of offshore platforms under Morison type non linear random wave loading, many researchers suggest approximating the drag component of the Morison equation f_t by polynomial functions of the instantaneous flow velocity.

Let X be a random variable with probability density f_X. Let us consider $Y = g(X)$, where g is differentiable and bijective. Then the density f_Y exists and has the form,

$$f_Y(y) = f_X(g^{-1}(y)) \left| (\frac{1}{g}) (g^{-1}(y)) \right| \tag{2}$$

A consistent approximation of f_t with respect of the density output must also be a good approximation to its derivative with respect to u_t. If the approximations are linear, quadratic or cubic, then their derivatives will be constant, linear or quadratic, respectively. The discrepancy between these three last functions is generally high on the edges of their definition interval. Consequently, their effects in changing the upper and lower tails of the velocity density will be quite different. In particular the skewness and kurtosis coefficients may not be well estimated. The choice of a polynomial of low order, so tractable for the computational procedure, although apparently appropriate for an approximation of the response surface itself, can lead to an erroneous probability density output.

In support of this warning we summarize the conclusions of an extensive numerical analysis [16] :
- a linear approximation of the drag loading failed to predict response moments in quasi-static cases.
- a cubic expansion yielded good estimates of the response variance for any kind of excitation, but could not accurately predict fourth order moments of the response, in drag-dominated quasi-static cases.
- it was claimed that a fifth order approximation is needed in order to assess accurately the first four response moments for any kind of excitation.

This example shows that the distribution effects significantly influence the construction of response surfaces as input criteria.

For problems with several variables, one must ensure that the Jacobian,

$\left(\dfrac{D(X)}{D(Y)} \right)$ is properly modelled, in order to prevent the main distribution effects. As

is well known, the input and output p.d.f are related according to,

$$P_Y(y) = P_X\left(f^{-1}(y)\right) \bullet \left| \frac{D(X)}{D(Y)} \right|$$

2.1.3 Goodness of Fit Measure

The first concern is the selection of the basic variables and especially their ranking. The usual metrics called **sensitivity** and **uncertainty importance** measures are based on second order statistics. They are useful for obtaining the main contributors to the response by evaluating their corresponding contributions to the output variance [17], [18]. Their sensitivity to the presence of outliers of input distributions make them questionable as absolute measures of uncertainty importance. To overcome this problem we may use measures based on the shifts in the quantiles of the output distribution [19], or on information theoretical entropy [20].

Uncertainty modelling leads to the **regression techniques**. Let us consider the regression model,

$$Y = f(X/\Theta) + \mathcal{E}$$

where $\mathcal{E}$ is a random vector or process error. The function f is a prescribed function of which the parameters are to be estimated. Usually it is assumed that the random vector $\mathcal{E}$ is distributed according to the Gaussian distribution with zero mean and diagonal covariance matrix. Under this assumption, the maximum likelihood and the least squares estimate are identical. As a consequence, the metric L_2 is shown to be the most efficient. However the L_1 metric, e.g. the integral of the absolute values of the residuals, provides most likely estimates when errors are doubly exponentially distributed [21].

In the previous subsection some requirements were given in order to prevent the main distribution effects from the additional fitting of the partial derivatives of f with respect to X_i. The metric of the Sobolev space [22],

$$H^1 = \{u \mid \frac{\delta^\alpha u}{\delta X_i^\alpha} \in L^2 \text{ (square integrable) whatever } \alpha = 0,1 \text{ and } i=1,...n\}$$

in the form,

$$\|u\| = \sqrt{\|u\|_{L_2}^2 + \sum_{i=1,...n} \left\|\frac{\delta u}{\delta x_i}\right\|_{L_2}^2}$$

is appropriate as it allows the least squares error on the partial derivatives to be introduced too.

2.1.4 Complexity Reduction and Computational Tractability

Engineers prefer models that are simple and easy to compute ; for such one may use the new techniques of probabilistic mechanics, and the new and powerful computers and algorithms. The RSM integrates these objectives as criteria.

An increase of the complexity level in stochastic modelling has to be viewed with care. It is always a time consuming option when computing, and one which does not necessarily ensure more realistic results. As an example, in North Sea conditions and among the Stokes model family, the simple Airy wave model is shown to contain the essential random structure of the inputs to be considered for structural reliability of jacket platforms under quasi-static loads [23].

Another way to simplify the problem is to consider a subset D which envelops the hyper-surface defined by f and which is easier to compute. For example convex polytopes, ellipsoids, cylinders, ..., are possible candidates for describing the safety region. As a consequence this leads to a conservative approach in structural safety, provided (see 2.1.1) the physical interpretation of the new domain D remains consistent with what is called a failure domain.

2.2 SELECTION OF THE BASIC VARIABLES

For a time invariant system which survives only for realizations of a random vector X, interior to a subset D_X of a probability space, its probability of failure has the form,

$$P_f = \int_{D_x^c} dF_x \tag{3}$$

where F_X is the probability distribution function of X. The topology of the complementary failure domain D_x^c can arise from the use of RSM (e.g. limit state functions, ...).

Practically there is a choice to be made on the representative stimuli. Usually we know only a part of the underlying distribution F_X ; we call this the Θ- set throughout this presentation. So the main objective for providing a safety measure is to evaluate,

$$P_{upper} = Sup \left\{ P \left(D_x^c \right) , P \text{ verifies the } \Theta - \text{ set} \right\} \tag{4}$$

This question is closely related to the so-called General Moment Problem in its mathematics (geometrical approach, convex analysis) [24], [25]. A basic technique for finding various properties of P_{upper} is to use Chebychev's inequalities, and the possible extensions of them.

In particular the proposed bounds are attained by distributions consisting of discrete point - masses. These forms can be translated as the strongest assertion possible in the absence of any further information on the initial distribution. For example the conventional Chebychev inequality leads to the statement :

Let $D_X = [-\beta, \beta]$ and the Θ - set have $E(X) = 0$, $E(X^2) = 1$, then we have $P_{upper} = \dfrac{1}{\beta^2}$.

The selection of the basic variable X is now presented to illustrate the key factors which act upon the safety measure. They are introduced by means of examples which are intentionally academic and simple. These exercises are significant because they give concrete expression to different concepts at a preliminary stage.

2.2.1 Dimension Space

An increase in dimension makes the system dependent on supplementary variables and consequently more unsafe.

This can be illustrated as follows. Let us consider a safety domain in the form of an open hypercube,

$$\Xi_{n,\beta} = \left\{ x \in R^n \ / \ |x_i| < \beta, i = 1, \ldots, n \right\}, \ \beta \geq 1$$

given the set Θ of statistical properties $E(X_i) = 0$ and $E(X_i^2) = 1$ whatever i. The complementary $\Xi_{n,\beta}^c$ is called the failure domain. We are concerned with evaluating,

$$P_{upper} = Sup \left\{ P\ (\Xi_{n,\beta}^c) \ , \ P \ given \Theta \right\}$$

From the Chebyshev and Boole inequalities this becomes,

$$P(\Xi_{n,\beta}^c) \leq \sum_{i=1}^{n} P(|X_i| \geq \beta) \leq \min(1, \frac{n}{\beta^2})$$

Moreover there exists a probability measure of finite support, satisfying Θ, and which attains the upper bound ;

– If $\dfrac{n}{\beta^2} \leq 1$, we consider P concentrated on the origin and the 2n centers of the hypercube faces. Each center point has an identical mass equal to $\dfrac{1}{2\beta^2}$. Then P satisfies the Θ-set. As only the origin is in the safety domain, we have

$$P(\Xi_{n,\beta}^c) = 2n \ \frac{1}{2\beta^2} = \frac{n}{\beta^2}.$$

– If $\dfrac{n}{\beta^2} > 1$, it is sufficient to take P with equal distribution of the total mass concentrated on points which are the images of the 2n centers of the hypercube faces by the similarity transformation with center O and ratio $\frac{\sqrt{n}}{\beta}$. The Θ_- set properties are satisfied. All the mass points are in the failure area and then, $P(\Xi_{n,\beta}^c) = 1$.

Finally the safety measure is given by,

$$P_{upper} = \min(1, \frac{n}{\beta^2}) \tag{5}$$

This shows that the probability of failure increases quickly with the dimension n (e.g. with the number of selected basic variables).

It is thus important to find the minimum set or parameters which allow the system response to be controlled. We recommend a preliminary ranking of the importance of the various input parameters by measuring their effect on the safety measure.

2.2.2 Correlation Effects

The statistical properties of X are to be addressed with care. We emphasize correlation effects. Let us now consider the hypercube $\Xi_{n,\beta}$ and the set Θ (see 2.2.1), given the additional statistical information on the covariance matrix $V = (E(X_i X_j))_{i,j}$.

Let M^+ be the set of definite positive matrices with diagonal terms equal to 1. The trace function $Tr\ M^{-1}V$ defined on M^+ is convex and thus reaches its minimum on a unique matrix $\overline{M}$ interior to M^+, such that the product $\overline{M}^{-1}\ V\overline{M}^{-1}$ is a diagonal matrix.

A result due to Whittle [26] ensures that,

$$P_{upper} = \min(1, \frac{1}{\beta^2} Tr\ \overline{M}^{-1}\ V) \tag{6}$$

Without loss of generality an equicorrelation ($0 \leq \rho \leq 1$) is assumed for illustration. The covariance is a combinatorial matrix of the form,

$$V = (1-\rho)Id + \rho A$$

where $a_{ij} = 1$ whatever i,j. As $A^2 = nA$, the matrix $V^{1/2}$ is a linear combination of the two matrices Id and A. We obtain,

$$V^{1/2} = \sqrt{1-\rho}\ Id + \frac{\sqrt{1+(n-1)\rho} - \sqrt{1-\rho}}{n} A$$

Considering the diagonal term d of $V^{1/2}$, the matrix $\overline{M} = \dfrac{V^{1/2}}{d}$, is definite positive with diagonal terms equal to 1 ; it verifies $\overline{M}^{-1} \, VM^{-1} = d^2 \, Id$.

Thus Tr $\overline{M}^{-1} \, V = nd^2$ and from eq.(6) we have,

$$P_{upper} = \min(1, \frac{n \, d^2}{\beta^2}) = \min(1, \frac{\left[(n-1)\sqrt{1-\rho} + \sqrt{1+(n-1)\rho}\right]^2}{n\beta^2}) \qquad (7)$$

The probability of failure is a decreasing function of $\rho (o \le \rho \le 1)$, and varies from $\min(1, \dfrac{n}{\beta^2})$ to $\min(1, \dfrac{1}{\beta^2})$.

It follows that the more the control variables are correlated, the safer is the system. Compared with the previous one, this new case contains additional statistical information which reduces the probability of failure. The statistical properties of the Θ-set must be provided meaningfully. It is important to avoid an arbitrary increase in safety index by specifying the whole distribution without justification. The subject is clearly relevant to the Bayesian approach, e.g. a priori knowledge and measure of its effects.

2.2.3 Independence Assumption

Another feature is to assume the variables X_i to be independent. We discuss the effect of this assumption on the safety measure. Let P be a probability measure which satisfies the new Θ- set. From the independence property and Chebyshev inequality we have,

$$P(\Xi_{n,\beta}) = \prod_{i=1}^{n} P(|X_i| < \beta)$$

$$\ge (1 - \frac{1}{\beta^2})^n$$

Moreover the lower bound is reached by considering the following probability of a vector with independent components. Each component X_i with support on the respective i - axis of $\mathbf{R}^n$ takes the values $-\beta$, O, β with the associated masses $\dfrac{1}{2\beta^2}, 1 - \dfrac{1}{\beta^2}, \dfrac{1}{2\beta^2}$. As a direct consequence it follows that,

$$P_{upper} = 1 - (1 - \frac{1}{\beta^2})^n \qquad (8)$$

Since $1-(1-\dfrac{1}{\beta^2})^n \le \dfrac{n}{\beta^2}$, independence of the basic variables makes the system safer than under zero-correlation. Equation (8) shows that P_{upper} depends on $(1-1/\beta^2)$ raised to the power n, not n/β^2 as in (6).

2.2.4   Typology of Distributions

Let X be a random variable with zero mean and standard deviation equal to 1. The safety domain is taken in the form of an interval $(-\infty, \beta[$.

We consider the three typical classes of distributions : normal (standard case), double exponential (asymptotic extreme model or Fisher Tippett I), exponential (tail behavior).

The associated one sided Chebychev inequality gives $P_{upper} = \dfrac{1}{1+\beta^2}$.

An upper bound under the additional assumption of continuous unimodal distribution, takes the form $P_{upper} = \dfrac{4}{9(1+\beta^2)}$; it is derived from Gauss inequality [27].

Table 1 shows that different assumptions lead to significantly different safety measures. This simple but significant example leads to the following requirement : there is a need to specify distribution functions for the basic variables. This can be a difficult task in practice, but it is necessary for finding the class in which the system lies.

TABLE 1. A comparison of distribution assumptions on the safety measure

Pupper	β		
	2	3	4
Normal	0,0228	0,0014	0,00003
Double exponential	0,0422	0,012	0,0033
Exponential	0,0498	0,018	0,0067
Gauss inequality	0,089	0,044	0,026
One sided Chebychev inequality	0,2	0,1	0,059

In particular the Pearson and Johnson distribution types cover a large area. They allow us to take into account smoothness conditions by considering continuous and bounded n^{th} derivatives. This question is related to the so-called extremal distributions investigated in the past by statisticians trying to find generalizations of Chebychev's inequalities.

2.3 CHOICE OF AN APPROXIMATION FUNCTION

The most popular assumption is classically that the response can be modelled by a polynomial expression. Mathematically, assuming that f is a continuous function and the set X varies in a finite range, the Weierstrass approximation theorem ensures that one can approximate f by a polynomial function to any desired accuracy. This result is an asymptotic one. It means that generally the corresponding polynomial order will be very large. Polynomials of low order can be used as local approximations, and they can then be pieced together as splines.

2.3.1 Algebraic Forms of Response Functions

A safety domain is a subset of $\mathbf{R}^n$. Its boundary is defined by the response surface. Table 2 presents the main approximation functions for such representations.

The exact form of f is generally unknown , it must be chosen to meet several conflicting requirements :
- the function should describe the data with reduction in storage,
- the function is required to be meaningful, in the sense that it is based as far as possible on some understanding of the underlying mechanism,
- the function is to be used for inference purposes,
- the function can be fitted simply and accurately,
- ...

TABLE 2. Families of response hyper-surfaces

1. Polynomials	$\Sigma a_{t_1,\ldots,\,t_n} X_1^{t_1} \ldots X_n^{t_n}$
- hyperplane	$\sum\limits_{o \le i \le n} a_i X_i$
- quadratic (by convention $X_o=1$)	$\sum\limits_{o \le i,j \le n} a_{i,j} X_i X_j$
2. Exponentials [28]	$\sum\limits_{i=1}^{n} P_i(X) \exp(<\alpha_i, X>)$ α_i frequencies, $P_i(X)$ polynomial $<\theta, X> = \theta_1 X_1 + \ldots + \theta_n X_n$
3. Spline Interpolations [29]	(piecewise polynomial)
4. $\phi(f(X))$ or $f(\phi(X_1), \ldots, \phi(X_n))$	f interior to 1., 2. or 3. ϕ scale function (logarithmic, inverse,...)

A problem is said to be well posed in the sense defined by Hadamard [30] when its solution
i. exists,
ii. is unique,
iii. depends continuously on the initial data and with respect to small perturbations.

For the problem of finding a response surface, physical and/or practical arguments are used to postulate that a solution must exist. The two other conditions (ii) and (iii) are undoubtedly very questionable. The uniqueness is more the consequence of the choice of a mathematical model. In particular, various inference models can give quite different responses. The condition (iii) is very often critical. Many situations fail to satisfy the required criteria. This is due to the sensitivity of a solution and, in particular, on the high order statistics, to uncertainties in the data.

Part of the ill-posedness of RSM problems may be overcome by introducing a **priori** information, and by using a Bayesian approach.

2.3.2 A specific example

Consider an offshore structure buffeted by waves. The safety margin on a failure mode is determined by a response surface ; the set of basic variables includes the strengths of the structural elements and the loads acting on the structure. To set up a computational model, the loads exerted by the waves must be replaced by equivalent concentrated forces applied at the nodes of the model.

Schematically, these forces are found by following the path

The kinematic field corresponding to the waves can be represented through a random linear combination of deterministic vectors. From these we calculate the Morison forces, and from these we compute the concentrated nodal forces by using energy considerations in a finite element model [31].

The total external force F_N at a a node N is obtained as a linear combination of the form,

$$\sum_{i=1}^{n} \lambda_i \ \overrightarrow{F_N^{(i)}} \quad n \approx 10$$

Here λ_i are random multipliers which depend on the basic variables (e.g. : $\lambda_1 = \theta_{mg}^2 \ C_M \ H^2 / T$ where θ_{mg} marine growth screen effect, C_M inertia coefficient, H extreme wave height, T associated wave period), and $\overrightarrow{F}_N^{(i)}$ are deterministic vectors which depend mainly on the structure topology and on the wave location on the structure.

This example illustrates a procedure for constructing response surfaces which combines the stochastic modelling of the basic variables with a deterministic approach in which the successive transfer functions are derived from physical considerations.

3. Applications

3.1 STOKES MODELS IN WAVE MODELLING

Stochastic process modelling has proved to be efficient for introducing different time scales for wind generated waves [32]. The latter are normally the dominant loads for an offshore structure. There are single wave models for the mechanical push over analysis, sea-state descriptions relevant for the resonant response due to second order sum-or difference- frequency wave loads or for the fatigue behavior. The so-called sea-states are well identified as stationary components of a piecewise second order stationary ergodic and regular enough random process [33].

One way of describing a stochastic process is to specify its n-dimensional joint probability laws, for all values of n=1,2,3,.... The basic role which the Gaussian process plays in stochastic modelling arises from the fact that :
- many physical systems can be approximated by Gaussian processes,
- many questions can be answered in a closed form for Gaussian processes more easily than for other processes.

Alternatively, one may give an explicit formula for the value of the process at each index point in terms of a family of random variables whose probability law is known.

There exist several mathematical models (ex : Stokes, Boussinesq, Miche, ...) for predicting the time evolution of the wave propagation. Physical reasoning and data observations allow a classification based on deterministic criteria. But these theoretical models have to be introduced in a reliability analysis, and must be regarded as stochastic models due to their sensitivity to random or uncertain parameters. This has significant effect on the safety domain topology and its probability measurement. The following formal geometrical representation of the wave kinematics leads to a new approach.

Take axes O(U,Z,V), where the origin O is taken at the mean sea level, the axis OZ is vertical and upwards, and OU is the wave directional axis. Let $\overrightarrow{OU}$, $\overrightarrow{OZ}$, $\overrightarrow{OV}$ be unit vectors along the three axes.

Suppose t=time, and x, z are the dimensionless variables,

$$x = U/l, \quad z = 1 + Z/d$$

where l is the width of the structure (diameter of a cylinder containing the platform), and d is water depth.

The small amplitude plane harmonic progressive waves known as Airy waves are derived from a velocity potential,

$$\phi(x, z, t) = \frac{H}{2} \frac{g}{\omega} \frac{\cosh(kdz)}{\cosh(kd)} \sin(klx - \omega t + \varphi)$$

$$(9)$$

where H is the wave height, k is the wave number, ω is the pulsation and φ is the phase angle. The variables k and ω are linked together by the one to one dispersive relation $\omega^2 = gk \ \tanh(kd)$.

The associated velocity vector for Airy Waves is in the wave plane ; it can be written in the form $\lambda(\cos \alpha \ \overrightarrow{OU} + \sin \alpha \ \overrightarrow{OZ})\cdot$ Denote by $H(O,\lambda)$ the similarity with center O and ratio λ, and by $R(\overrightarrow{OV},\alpha)$ the rotation about the OV -axis through the angle α.

A formal geometrical expression of the velocity vector follows,

$$\vec{V}_{airy} = H(O,\lambda) \circ R(\overrightarrow{OV},\alpha) \ \overrightarrow{OU}$$

$$\tan \alpha = \tanh(kdz) \ \tan\left(klx - \omega t + \varphi\right) \qquad (10)$$

$$\lambda = \frac{H\omega \cosh(kdz)}{2\cosh(kd)} \sqrt{1 - \frac{\sin^2\left(klx - \omega t + \varphi\right)}{\cosh^2(kdz)}}$$

The similarity ratio λ and the rotation angle α are random functions indexed by (x,z,t). Their stochastic fluctuations depend on the couple of random variables (H,k).

The range and statistics of k are well adapted to approximate the vector $\overrightarrow{W} = R(\overrightarrow{OV},\alpha) \ \overrightarrow{OU}$ by its normalized first order Taylor expansion $= \overrightarrow{W}^{(1)}$ around the mean wave number $\bar{k}$. We obtain after some algebraic operations and differentiations,

$$\overrightarrow{W}(k) = \overrightarrow{W}(\bar{k}) + \left(k - \bar{k}\right) \frac{\delta}{\delta k} \ \overrightarrow{W}(\bar{k}) + o(k)$$

$$\overrightarrow{W}^{(1)} = H(O, \frac{1}{\sqrt{1+\gamma^2}}) \circ \left[Id + H(O,\gamma) \circ R(\overrightarrow{OV}, \frac{\pi}{2}) \right] \circ R(\overrightarrow{OV}, \underline{\alpha}) \ \overrightarrow{OU} \qquad (11)$$

where,

$$\underline{\alpha} = \alpha\left(\bar{k}\right), \ \gamma = \left(k - \bar{k}\right) \frac{\partial \alpha}{\partial k}\left(\bar{k}\right)$$

The goodness of fit of the approximation can be measured by the inner product of the two vectors. It is given by considering in the wave plane, the distance between the point on the unit circle with angle $\alpha - \underline{\alpha}$, and the straight line of equation $\left[U + \gamma\, Z = 0 \right]$.

Introducing into eq.(10) the approximation $\vec{W}^{(1)}$ instead of $\vec{W}$, we see that the velocity field corresponding to the Airy model has the following response surface,

$$\vec{V}_{airy} = \mu\vec{A} + v\vec{B} \tag{12}$$

where $\vec{A}$ and $\vec{B}$ are deterministic orthonormal vectors defined by,

$$\vec{A} = R(\vec{OV}, \underline{\alpha})\ \vec{OU}, \text{ and } \vec{B} = R(\vec{OV}, \frac{\pi}{2})\ \vec{A}$$

and the coefficients μ and v are random functions.

$$\mu = \frac{\lambda}{\sqrt{1+\gamma^2}}\ ,\ v = \gamma\,\mu$$

4. Remarks :

- Due to the statistics of k, the random variable γ has zero mean and narrow range. It follows that v is small compared to μ. Consequently $\vec{A}$ represents the main axis for the velocity vector. The component following $\vec{B}$ leads to fluctuations of the velocity vector inside a narrow sector around $\vec{A}$.
- The scalar velocity intensity λ contains the factor,

$$a(x, z, t) = \sqrt{1 - \frac{\sin^2(klx - \omega t + \varphi)}{\cosh^2(kdz)}}$$

It is a function of the random variable k, which varies weakly at points interior to the support of k. There it may be concentrated at its value for $k = \bar{k}$. Equation (10) shows that there is a deterministic linear relationship between λ and the random function $\dfrac{H\omega\,\cosh(kdz)}{2\cosh(kd)}$ which depends only on z. This means that the coefficient of variation and the normalized moments of the intensity vector vary with the profile index z only, and not at all with the indices x and t.

- Note that the non linear term $\vec{V}\lVert\vec{V}\rVert$ in the drag component of the force per unit span given by the Morison equation has the form $\lambda\mu(\vec{A} + \gamma\,\vec{B})$; it is a random linear combination of deterministic vectors.

The corresponding velocity field for n^{th} order Stokes waves can be expressed as a combination of the different orders,

$$\vec{V}_{n^{th}} = \sum_{i=1}^{n} \vec{V}_i = \sum_{i=1}^{n} H(O, \lambda_i) \circ R(\vec{OV}, \alpha_i)\, \vec{OU}$$

For each component $\vec{V}_i$, we may apply the same formal geometrical representation as defined for the Airy kinematic field. The issue of an extension of the Airy model can be addressed by considering the projection of each component on the two vectors $\vec{A}$ and $\vec{B}$. It can be shown that [23],

$$\vec{V}_{n^{th}} = \mu_{(n)}\, \vec{A} + v_{(n)}\, \vec{B}$$

$$\mu_{(n)} = \sum_{i=1}^{n} \left(\mu_i \cos \underline{\Delta_i} - v_i \sin \underline{\Delta_i} \right) \tag{13}$$

$$v_{(n)} = \sum_{i=1}^{n} \left(\mu_i \sin \underline{\Delta_i} + v_i \cos \underline{\Delta_i} \right)$$

where we have $\underline{\Delta_i} = \alpha_i - \underline{\alpha}\left(\approx (i-1)\underline{\alpha} \right)$.

Then wave velocity models can be presented with their associated response curves. They are basic inputs, due to causality considerations, for the reliability analysis of offshore structures. The proposed formal geometrical approach allows us to specify the stochastic properties following space/time indices to be considered. It also gives us the opportunity to introduce complexity in wave modelling only when necessary.

4.1 NON LINEAR MECHANICAL BEHAVIOUR

There are softwares which will perform a structural analysis under the assumptions of non linear elastoplastic behaviour and large displacements. The approach is essentially deterministic. The connection with reliability analysis is not direct. A safety measure needs explicit limit state functions and their derivatives in the space of the basic random variables.

The pseudo-random sampling approach and typically the Monte Carlo analysis, can be used with efficiency in such a case, but the computer time increases quickly with the problem dimension [34].

An alternative procedure uses quadratic response surfaces [35]. Quadratic polynomial approximations for response surfaces are simple to apply in finite element methods, and lead to numerical stable computation. Let $f(X)$ be a limit state function and $X = (X_1,...,X_n)$ be the vector of stimuli. An iterative least square method combined with an adaptive mesh of the safety domain uses at each step (k) a quadratic approximation of $f(X)$ of the form,

$$Q^{(k)} (X) = \sum_{0 \le i,j \le n} a_{i,j}^{(k)} X_i X_j.$$

One example of this formulation concerns the codified design reliability of the buckling curves of a steel tubular cross section under compressive load [36]. It is used to compare different codes such as API RP2A-LRFD and EUROCODE3.

4.2 STOCHASTIC DESIGN OF FIBROUS COMPOSITE LAMINATES

The reliability of unidirectional composite laminated materials considered by Tsai/Wu [37], is based on a simplified fibre failure mode under tensile or compressive load. The limit state functions are represented by quadratic polynomials. Safety indices can be computed from the well known FORM/SORM methods. A numerical analysis of graphite epoxy material has been performed [38]. It is shown how the nominal safety factors are very sensitive to the design problem which arises from the lack of dimensional invariance in defining safety margins. This concerns the different types of safety factors such as ultimate strength (I), in plane - load (II), dimensional factor (III). This question is important when the safety measure is highly sensitive to the parameters of the basic variables.

A more complete criterion introduced by Hashin [39], considers in addition, the failure of the matrix material between the fibres due to transverse or shear stresses. This interfibre failure mode can cause large scale collapse. Their analysis illustrates how to modify a preliminary response surface in order to extend its physical meaning. Friction due to compression on the crack surface increases the shear strength. Thus the quadratic polynomial limit state is modified by introducing some friction coefficients [40]. The latter are determined by fitting experimental data of a test specimen under transverse compression. They show that the use of too classical criteria can lead to incompletely designed structures.

5. Combination of Sensitivity Analysis Techniques

The extensive review [41] on sensitivity analysis techniques, states in particular the respective limits for use in radioactive waste disposal. We make some suggestions based on lessons drawn from structural reliability analyses of marine structures.

The Fourier Amplitude Sensitivity Test (FAST) [42], is basically a second order uncertainty and sensitivity technique. The Fourier series representation allows us to obtain the ratios of the contributions of the individual input variables to the variance of the model response. It gives a second order measure of importance for ranking the variables.

Perturbation techniques usually use Taylor's series to approximate models. It is a local approach which gives valuable results when the input variables have small ranges, and the relationship between the input and output variables is relatively smooth. The computation of derivatives needs specialized techniques, such that Green's functions or kernel non-parametric methods [43].

Monte Carlo analysis starts from a pseudo-random sampling approach to represent the system inputs. The well known Latin Hypercube Sampling provides a full coverage of the range of each input variable, but it remains questionable how far the selected samples are representative of the whole underlying distribution. Otherwise the Monte Carlo techniques are particularly appropriate for analysing problems in which large uncertainties occur and where the transfer functions are non linear.

The Response Surface Methodology has been detailed in the previous sections. As a consequence the use of such techniques in structural reliability analysis needs to extend the conventional form which is too concentrated on polynomial approximations. Often, we can combine different techniques : Taylor's expansions around the mean wave number ; responses curves based on a geometrical visualisation of the vector kinematic field ; Monte Carlo simulations ; all these items can be combined to make the total investigation more meaningful.

6. References

1. Wishart, J. (1938). Growth rate determination in nutrition studies with the bacon pig, and their analysis. *Biometrika*, **30**, 16-28

2. Bliss, C.I. (1938). The determination of dosage-mortality curves from small numbers. *Quart.* J. Pharm **11**, 192-216

3. Growther, E.M. and Yates, F. (1941). Fertilizer policy in war-time. *Empire J.Exp. Agric 9,77-97*

4. Box, G.E.P. (1954). The exploration and exploitation of response surfaces : some general mixed considerations and examples. *Biometrics*, **10**, 16-60

5. Box, G.E.P. and Hunter, H.S. (1957). Multifactor experimental designs for exploring response surfaces. *Ann. Math. Statist.* **28**, 195-241

6. Box, G.E.P. and Draper, N.R. (1959). A basis for the selection of a response curve design., J. Amer. *Statist. Ass.* **54**, 622-54

7. Box, G.E.P. and Draper, N.R. (1965). The Bayesian estimation of common parameters from several responses. *Biometrika* **52**, 355-65

8. Robbins, H. and Monro, S. (1951).A stochastic approximation method, *Ann. Math. Statist.* **22**,400-7

9. Rao, C.R. (1958). Some statistical methods for the comparison of growth curves. *Biometrics* **14**, 1-17

10. Kiefer, J. and Wolfowitz, J. (1959). Optimum designs in regression problems *Ann. Math. Statist.* **30**, 271-94

11. Mead, R. and Pike, D.J. (1975). A review of response surface methodology from a biometric viewpoint, *Biometrics* **31**, 803-851.

12. Labeyrie, J., Schoefs, F. (1994). Discussion on response surface approximations for use in structural reliability, *Reliability and Optimization of structural systems, proc. 6th IFIP WG 7.5, Chapman and Hall*, 161-168

13 Worden, K. Stansby, P.K. and Tomlinson, G.R. (1994). Identification of non linear wave forces, *J. fluids and struct.* **8**, 19-71

14. Ditlevsen, O. (1994). Codified reliability of structures, *Reliability and Optimization of Structural Systems, proc. 6th IFIP WG 7.5, Chapman and Hall*, 25-44

15. Iman, R.L., Helton, J.C. and Campbell, J.E. (1981) An approach to sensitivity analysis of computer models : part II. Ranking of input variables, response surface validation, distribution effect and technique synopsis, *J. of Qual. Techn.* **13**, 232-240

16. Bouyssy, V., Rackwitz, R. (1995). Polynomial approximation of Morison wave loading, *Safety and Reliability*, **II**, proc. 14th OMAE, ASME (eds), 91-98

17. Bier, V.M. (1983) A measure of uncertainty importance for components in fault trees. *ANS Trans.* **45** (1), 384-5

18. Iman, R.L (1987) A matrix-based approach to uncertainty and sensitivity analysis for fault trees. *Risk analysis.* **7** (1), 21-33

19. Khatib-Rahbar, M. & al. (1989) A probabilistic approach to quantifying uncertainties in the progression of severe accidents. *Nucl. Sci. Engng.*, **10** (2), 219-59

20. Park, C.K and Anh, K-II (1994) A new approach for measuring uncertainty importance and distributional sensitivity in probabilistic safety assessment, *Rel. Engng. and Syst. Saf.*, **46**, 253-261

21. Scholasmacher, E.J. (1973). An iterative technique for absolute deviations curve fitting, *J. of the Amer. Statist.*, **68** (344), 857-859

22. Lions, J.L. and Magenes, E. (1968) *Problèmes Aux Limites Non Homogènes et applications*, Dunod éditeur, Paris

23. Labeyrie, J., Schoefs, F. (1995). A formal geometrical modelling of wave actions for structural reliability, *Safety and Reliability*, **II**, proc. 14th OMAE, ASME(eds), 85-90

24. Kemperman, J.H.B. (1968) The general moment problem, a geometric approach. *Ann. of Math. Stat,* **39** (1), 93-122

25. Labeyrie, J. (1989) Sharp measures of reliability for components, in VDI Berichte 771, *Zuverlässigkeit von komponenten technischer systeme, München*, pp.237-246

26. Whittle, P. (1958) A multivariate generalization of Tchebychev's inequality, *Quat. J. Math. Oxford.* (2), **9**, 232-40

27. Mallows, C.L. (1956) Generalizations of Tchebycheff's inequalities, *J. Roy. Stat. Soc.*, **18** (2), 139-169

28. Labeyrie, J. (1982) Factorisation de Ritt pour les polynômes exponentiels formels. *Ann. Fac. Sc.*, Toulouse, **4**, 281-289

29. De Boor, C. (1978) *A Practical Guide to Splines*, Springer Verlag (eds), New York

30. Hadamard, J. (1993*). Lectures on the Cauchy Problem in Linear Partial Differential Equations,* Yale University Press, New Haven, CT

31. Labeyrie, J. (1990), Stochastic load models for marine structure reliability, *3 rd. IFIP WG-7.5, Berkeley.*

32. Labeyrie, J. (1991) Time scales and statistical uncertainties in the prediction of extreme environmental conditions, *Rel. Engng. and Syst. Saf.*, **32**, 243-266

33. Labeyrie, J. (1990), Stationary and transient states in random seas, *J. of Marine Struct.*, **3**, 43-58

34. Muzeau, J.P. (1989). Reliability of steel columns subjected to buckling. Comparison of international standard codes of design, *5th Int. Conference on Structural Safety and Reliability, San Francisco.*

35. Muzeau, J.P., Lemaire, M., and El-Tawil, K. (1992). Méthode fiabiliste des surfaces de réponse quadratiques (SRQ) et évaluation des règlements, *Construction Métallique*, **3**, 41-52

36. Muzeau, J.P., Lemaire, M., Besse, P., Locci, J.M. (1995). Evaluation of reliability in case of complex mechanical behaviour, *Safety and Reliability*, **II**, *proc. 14th OMAE, ASME(eds)*, 47-56

37. Tsai, S.W. (1988). *Composite Design* (4th), Thinck Composites (eds.), Tokyo.

38. Nakayasu, H. (1994). Relation between sensitivities and dimensional invariance on stochastic materials design of fibrous composite laminates, *Reliability and Optimization of Structural Systems, prc. 6th IFIP WG 7.5, Chapman and Hall*, 209-216

39. Hashin, Z. (1980). Failure criteria for unidirectional fibre composites, *J. of applied Mechanics*, **47**, 329-334

40. Plica, S., Rackwitz, R. (1994). Reliability of laminated structures by an energy failure criterion, *Reliability and Optimization of Structural Systems, proc. 6th IFIP WG 7.5*, Chapmam and Hall, 233-240

41. Helton, J.C. (1993) Uncertainty and sensitivity analysis techniques for use in performance assessment for radioactive

42. Rae, MC. et al. (1981) Global sensitivity analysis- A computational implementation of the Fourier Amplitude Sensitivity Test (FAST), *Comp. and chem. Engng.*, **5**, 15-25

43. Hwang, J.T. et al (1978) The Green's function method of sensitivity analysis in chemical kinetics, *J. of chem. Phy.*, **69**, 5180-91

STOCHASTIC MODELING OF FATIGUE CRACK GROWTH AND INSPECTION

HENRIK O. MADSEN
Det Norske Veritas AS,
Tuborg Parkvej 8,
DK-2900 Hellerup,
DENMARK.

1. Introduction

In many metallic structures, flaws are inherent due to, e.g., notches, welding defects and voids. Macro cracks can originate from these flaws, and under time varying loading grow to a critical size causing catastrophic failure. The conditions governing the fatigue crack growth are the geometry of the structure and crack initiation site, the material characteristics, the environmental conditions and the loading. In general, these conditions are of random nature. The appropriate analysis and design methodologies should therefore be based on probabilistic methods.

In recent years, considerable research efforts have been reported on probabilistic modeling of fatigue crack growth based on a fracture mechanics approach, see, e.g., (Arone, 1983; ASCE Committee, 1982; Bolotin, 1981; Ditlevesen, 1986; Madsen, 1983; Kozin and Bogdanoff, 1981; Lin and Yang, 1983; Ortiz and Kiremidjian, 1986). In particular, stable crack growth has been studied. This Chapter presents a stochastic model for the stable crack growth phase for which linear elastic fracture mechanics is applicable. A common model is formulated for constant and variable amplitude loading. The model is developed for a semi-elliptical surface crack and for a through-the-thickness crack. Uncertainties in the loading conditions, in the computation of the stress intensity factor, in the initial crack geometry, and in the material properties are included.

The probability that the crack size exceeds a critical size during some time period is of interest. It is demonstrated how this event can be formulated in terms of a limit state function with a corresponding safety margin, and how the probability of failure can be calculated by a first- or second-

59

C. Guedes Soares (ed.), Probabilistic Methods for Structural Design, 59–83.
© 1997 *Kluwer Academic Publishers.*

order reliability method (FORM or SORM). The critical crack size may refer to growth through the thickness or to a size where a brittle fracture or plastic collapse occur. The critical crack size can be modeled as a deterministic or as a random quantity.

Inspections are frequently made for structures in service. Some inspections result in the detection of a crack, while others give no detection. The size of a detected crack is measured either directly or indirectly through a non destructive inspection method, where the measured signal is interpreted as a crack size. Neither the measurement nor the interpretation are possible in an exact way, and the resulting inspection result is consequently of random nature. When the inspection does not reveal a crack, this does not necessarily mean that no crack is present. A detectable crack is only detected by a certain probability depending on the size of the crack and on the inspection method. Whether or not a crack is detected, the inspection provides additional information which can be used to update the reliability and/or the distribution of the basic variables. This can lead to, e.g., modifications of inspection plans, change in inspection method, or a decision on repair or replacement. This Chapter describes inspection results in terms of event margins and formulates the updating in terms of such event margins and the safety margin. The use of first-order reliability methods to perform the calculations is demonstrated. A similar formulation and calculation is introduced to evaluate the reliability after a repair.

Reliability against fatigue damage caused by fatigue crack growth is obtained through a combination of design requirements, inspections and repair or replacement strategies. Each of these efforts introduce cost, and it is of considerable interest to select the solution leading to the smallest expected life time cost including the expected cost of failure. This problem can be formulated as an optimization problem, where the optimization variables for a given strategy are the design parameters, the inspection times and the inspection intervals. The optimal solution can be updated as information about inspection results and repair becomes available.

2. Fatigue crack growth model

In a linear elastic fracture mechanics approach, the increment in crack size, Δa, during a load cycle is related to the range of the stress intensity factor, ΔK, for the load cycle. A simple relation which is sufficient for most purposes was proposed by Paris and Erdogan, (Paris and Erdogan, 1963)

$$\Delta a = C(\Delta K)^m , \quad \Delta K > 0 \tag{1}$$

The crack growth equation is formulated without a positive lower threshold on ΔK below which no crack growth occurs. The equation was based on experimental results, but is also the result of various mechanical and energy based models, see, e.g., (Irving and McCartney, 1977; Paris and Erdogan, 1963). C and m are material constants. A possible dependence of C on the average stress in one load cycle is not included here.

The crack increment in one cycle is generally very small compared to the crack size, and Eq. (1) is consequently written in a "kinetic" form as

$$\frac{da}{dN} = C(\Delta K)^m , \quad \Delta K > 0 \tag{2}$$

where N is the number of stress cycles. The stress intensity factor K is computed by linear elastic fracture mechanics and is expressed as

$$K = \sigma Y(a)\sqrt{\pi a} \tag{3}$$

where σ is the far-field stress and $Y(a)$ is the geometry function. The geometry function depends on the overall geometry, including the geometry of the crack and the geometry of a possible weld. To explicitly account for uncertainties in the calculation of K, the geometry function is written as $Y(a) = Y(a, \boldsymbol{Y})$, where $\boldsymbol{Y}$ is a vector of random parameters. Inserting Eq. (3) in Eq. (2) and separating the variables leads to the differential equation

$$\frac{da}{Y(a, \boldsymbol{Y})^m (\sqrt{\pi a})^m} = C(\Delta\sigma)^m dN , \quad a(0) = a_0 \tag{4}$$

where a_0 is the initial crack size. The equation is applied both for constant and for variable amplitude loading, thus ignoring possible sequence effects.

Eqs. (1) to (4) describe the crack size as a scalar a, which for a through-the-thickness crack is the crack length. For a surface crack, a description of the crack depth, crack length and crack shape is necessary. It is common practice to assume a semi-elliptical initial shape, and to assume that the shape remains semi-elliptical during the crack growth. In that case the crack depth a and the length $2c$ describe the crack. The differential equation Eq. (2) is replaced by a pair of coupled equations, (Shang-Xian, 1985)

$$\frac{da}{dN} = C_a(\Delta\sigma Y_a(a, c, \boldsymbol{Y})\sqrt{\pi a})^m , \quad a(0) = a_0 \tag{5}$$

$$\frac{dc}{dN} = C_c(\Delta\sigma Y_c(a, c, \boldsymbol{Y})\sqrt{\pi c})^m , \quad c(0) = c_0 \tag{6}$$

where Y_a and Y_c are the geometry functions for the deepest point and for the end point of the crack at the surface, respectively. The material constants C_a and C_c may differ due to variation in stress field tri-axiality. The failure

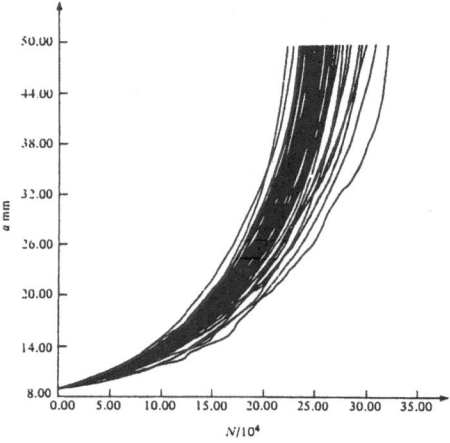

Figure 1. Experimental results.

criterion can refer to a critical value of either a or c individually, or to a function of a and c. The equations are conveniently rewritten as

$$\frac{dc}{da} = \frac{C_c}{C_a}\left(\frac{Y_c(a,c,\mathbf{Y})\sqrt{c}}{Y_a(a,c,\mathbf{Y})\sqrt{a}}\right)^m \tag{7}$$

$$\frac{dN}{da} = [C_a(\Delta\sigma Y_a(a,c,\mathbf{Y})\sqrt{\pi a})^m]^{-1} \tag{8}$$

which are solved simultaneously. The first equation gives c as a function of a and the initial values a_0 and c_0, but independent of the loading and number of stress cycles. The solution for c may be inserted in Eq. (8) which is then of the same form as Eq. (2). For reasons of simplicity in the presentation, the following is limited to a through-the-thickness crack of size a.

Numerous experimental results exist for crack growth under constant amplitude loading. Fig. 1 from (Kozin and Bogdanoff, 1981) shows experimental results reported in (Virkler *et al.*, 1979) for 64 center cracked specimens made of 2024-T3 aluminum. The experiments were highly controlled and performed by the same laboratory using the same equipment and the same personnel.

To capture the essential stochastic behaviour demonstrated by the experimental results we introduce the material parameter C as a random variable. To also capture the irregularity and intermingling of the sample curves in Fig. 1 we modell C as a spatial random process, see e.g. (Ditlevesen, 1986; Ortiz and Kiremidjian, 1986).

A damage function $\Psi(a)$ is introduced from Eq. (4) as

$$\Psi(a) = \int_{a_0}^{a} \frac{dx}{Y(x, Y)^m (\sqrt{\pi x})^m} \tag{9}$$

The stress ranges are denoted $S_i = \Delta\sigma_i$, and solution of Eq. (4) gives

$$\Psi(a) = C \int_{0}^{N} S^m dN = \begin{cases} CS^m N, & \text{constant amplitude loading} \\ C \displaystyle\sum_{i=1}^{N} S_i^m, & \text{variable amplitude loading} \end{cases} \tag{10}$$

The crack growth equation Eq. (1) has here been directly extrapolated to variable amplitude loading where the appropriate value of S is inserted for each stress cycle. It must be emphasized that this is an extrapolation beyond experimental experience, and possible sequence effects are neglected. It is observed that the only difference between the two cases of constant and variable amplitude loading concerns the loading statistics. The crack length after N stress cycles, a_N, is obtained by solving Eq. (10) with respect to a.

It follows from Eq. (10) that if failure is defined by crack growth beyond a critical size a_C, the following equation is valid at failure under constant amplitude loading

$$NS^m = \frac{\Psi(a_C)}{C} = K \tag{11}$$

where K is a constant independent of the loading. This relation is in agreement with the S-N curves generally applied in fatigue calculations.

One way to define a damage index D in terms of the crack size is

$$D = \frac{\Psi(a)}{\Psi(a_C)} \tag{12}$$

From this definition and Eq. (10) it follows that damage increases linearly from zero to one with the number of stress cycles. It can further be shown that the damage increment in one stress cycle of range S_i is $1/N(S_i)$, where $N(S_i)$ is the number of cycles to failure under constant amplitude loading. Damage accumulation is thus in agreement with Miner's rule, and the S-N approach and fracture mechanics approach are very similar.

Let the failure criterion be taken as exceedence of a critical crack size a_C in a time period with N stress cycles,

$$a_C - a_N \leq 0 \tag{13}$$

$\Psi(a)$ is monotonically increasing and the failure criterion can be written as

$$\Psi(a_C) - \Psi(a_N) = \int_{a_0}^{a_C} \frac{dx}{Y(x, Y)^m (\sqrt{\pi x})^m} - CS^m N \leq 0 \tag{14}$$

for constant amplitude loading. The safety margin M is defined as

$$M = \int_{a_0}^{a_C} \frac{dx}{Y(x,Y)^m (\sqrt{\pi x})^m} - CS^m N \tag{15}$$

and failure takes place when $M < 0$.

For variable amplitude loading, the safety margin becomes similarly

$$M = \int_{a_0}^{a_C} \frac{dx}{Y(x,Y)^m (\sqrt{\pi x})^m} - CNE[S_i^m] \tag{16}$$

where the sum of the m'th power of the stress ranges has been replaced by its expected value. This is a good approximation in most practical applications.

The stress range distribution is often chosen as Weibull for long or short term stress response to environmental loading. For a Weibull distribution of stress ranges, the distribution function is

$$F_S(s) = 1 - \exp\left[-\left(\frac{s}{A}\right)^B\right] , \quad s > 0 \tag{17}$$

and the safety margin becomes

$$M = \int_{a_0}^{a_C} \frac{dx}{Y(x,Y)^m (\sqrt{\pi x})^m} - CNA^m \Gamma\left(1 + \frac{m}{B}\right) \tag{18}$$

where $\Gamma(\cdot)$ denotes the Gamma function. For $B = 2$ the stress range distribution becomes of Rayleigh type, which is generally used for stress response modelled as a fairly narrow-band Gaussian process.

3. Failure criteria and basic variables

The previous section has shown an important case where a limit state formulation can be applied for reliability analysis against fatigue crack growth beyond a critical size. In this section some generalizations are presented. Two separate types of failure criteria are envisaged, (ASCE Committee, 1982)

$$a_C - a_N \leq 0 \tag{19}$$

$$K_{IC} - K(a_N) \leq 0 \tag{20}$$

In the first case, a critical crack size a_C is selected perhaps based on serviceability considerations. In the second case, failure occurs when the stress intensity factor K exceeds the fracture toughness K_{IC}; then the crack growth becomes unstable and rapid failure occurs. Four cases are considered, corresponding to the two failure criteria and constant or variable amplitude loading.

Case 1:   Crack growth beyond critical size under constant amplitude loading

Case 2:   Brittle fracture under constant amplitude loading

Case 3:   Crack growth beyond critical size under variable amplitude loading

Case 4:   Brittle fracture under variable amplitude loading

The safety margin for case 1 was given in Eq. (14) as

$$M = \int_{a_0}^{a_C} \frac{dx}{Y(x,Y)^m (\sqrt{\pi x})^m} - CS^m N \ , \quad \textit{case 1} \tag{21}$$

For case 2, the safety margin is

$$M = K_{IC} - Y(a_N, Y)(\sigma_m + \frac{S}{2})\sqrt{\pi a_N} \ , \quad \textit{case 2} \tag{22}$$

where σ_m is there average far-field stress including possible residual stresses. For case 3, the safety margin was given in Eq. (16) as

$$M = \int_{a_0}^{a_C} \frac{dx}{Y(x,Y)^m (\sqrt{\pi x})^m} - CNE[S_i^m] \ , \quad \textit{case 3} \tag{23}$$

For case 4, failure occurs if

$$\sigma > \frac{K_{IC}}{Y(a)\sqrt{\pi a}} \ , \quad \textit{case 4} \tag{24}$$

where $\sigma = \sigma(t)$ is the far-field stress. This is illustrated in Fig. 2. Failure does not occur in the time period $[0,T]$ if the stress process $\sigma(t)$ is below the time varying threshold $\xi(t) = K_{IC}/\{Y(a(t))\sqrt{\pi a(t)}\}$ in $[0,T]$. This probability is approximated by, see e.g. (Madsen *et al.*, 1986)

$$F_{T_f}(T) \approx F_{\sigma(0)} \left(\frac{K_{IC}}{Y(a_0)\sqrt{\pi a_0}} \right) \exp \left[- \frac{\int_0^T \nu_\sigma^+(\xi(t))dt}{F_{\sigma(0)} \left(\frac{K_{IC}}{Y(a_0)\sqrt{\pi a_0}} \right)} \right] \tag{25}$$

T_f is the random life time and $\nu_\sigma^+(\xi(t))$ is the mean upcrossing rate of the level $\xi(t)$ by the process $\sigma(t)$ at time t. This mean upcrossing rate is computed by Rice's formula, (Rice, 1954)

$$\nu_\sigma^+(\xi(t)) = \int_{\dot\xi}^\infty (\dot\sigma - \dot\xi) f_{\sigma,\dot\sigma}(\xi(t), \dot\sigma) d\dot\sigma \tag{26}$$

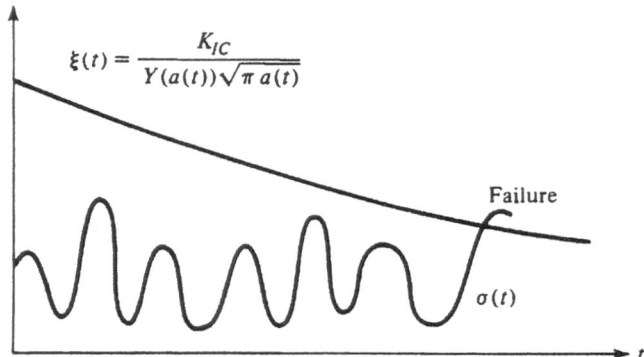

Figure 2. Illustration of failure event for brittle fracture under variable amplitude loading.

where a "dot" denotes a time derivative. In this application, the time derivative $\dot{\xi}$ can be neglected; Eq. (26) therefore reduces to

$$\nu_\sigma^+(\xi(t)) = \int_0^\infty \dot{\sigma} f_{\sigma\dot{\sigma}}(\xi(t), \dot{\sigma})d\dot{\sigma} \tag{27}$$

For a life time T, the failure criterion can be stated as

$$T_f - T \leq 0 \tag{28}$$

The distribution function for T_f is given in Eq. (25) as a function of material, loading and crack size parameters.

4. Parameter estimation for material properties

The value of m in the crack growth equation Eq. (1) is predicted from theoretical models as $m = 2$ or $m = 4$. Statistical estimation of m from experimental results generally results in other values and m should be treated as a random variable in addition to C.

Several studies, e.g. (Gurney, 1978), report a high negative correlation between m and $\ln C$. This is also demonstrated in the study (Tanaka *et al.*, 1981) where crack propagation data for 25 identical specimens under identical loading conditions have been collected. The least square estimates $\hat{m}$ and $\hat{C}$ for m and C have been computed for each specimen, and a joint distribution for m and C has been estimated. The 25 tests are for plane bending of a 0.04% Carbon steel and the following sample statistics are obtained

$$\overline{m} = 2.85 \quad s_m = 0.284 \quad V_m = 0.100 \tag{29}$$

$$\overline{\ln C} = -20.164 \quad s_{\ln C} = 1.067 \quad r_{m,\ln C} = -0.971 \tag{30}$$

An "over bar" denotes a sample mean, s denotes a sample standard deviation, V denotes a sample coefficient of variation and r denotes a sample correlation coefficient. A standard test does not reject an assumption of binormality for $(m, \ln C)$. Based on these results, m and $\ln C$ are expressed in terms of two independent and standardized normal random variables U_1 and U_2 as

$$m = \hat{m} + 0.100 \hat{m} U_1 \tag{31}$$

$$\ln C = \widehat{\ln C} + 1.067(-0.971 U_1 + 0.239 U_2) \tag{32}$$

thus giving a coefficient of variation of 10% for m, a standard deviation of 1.067 for $\ln C$ and a correlation coefficient of -0.971 between m and $\ln C$. The statistical uncertainty in the estimates in Eqs. (29) and (30) is thus ignored. In (Madsen, 1984) a fatigue reliability analysis which includes this statistical uncertainty, is reported.

The negative correlation between m and $\ln C$ is not a physical property but follows form the mathematical form of the crack growth equation. An alternative form is

$$\frac{da}{dN} = C_0 \left(\frac{\Delta K}{K_0} \right)^m \tag{33}$$

where K_0 is a fixed reference value of the same dimension as K. C_0 then has the same dimension as da/dN. The constant C in Eq. (1) is

$$C = C_0 K_0^{-m} \; ; \quad \ln C = \ln C_0 - m \ln K_0 \tag{34}$$

It follows that if the scatter in C_0 is negligible, then $\ln C$ and m are linearly related. Otherwise $\ln C$ and m are expected to be negatively correlated. m and $\ln C_0$ can be made uncorrelated by a suitable choice for K_0. A choice of $K_0 = 38.4$ MPa together with the results in Eqs. (29) and (30) imply that m and $\ln C_0$ are uncorrelated, and the variance of $\ln C_0$ is 0.065, corresponding to a standard deviation of 0.26. In (Ditlevsen and Olesen, 1986) a similar analysis was performed for the data shown in Fig. 1. The coefficient of variation of m was found as 6% and the same value was found for the standard deviation of $\ln C_0$. Both uncertainties in m and in C are thus important and roughly of the same magnitude.

5. Reliability updating based on inspection results

Structures in service are often inspected to detect cracks before they become critical. Let a crack be detected after n_j stress cycles and its length measured as

$$a(n_j) = A_j \tag{35}$$

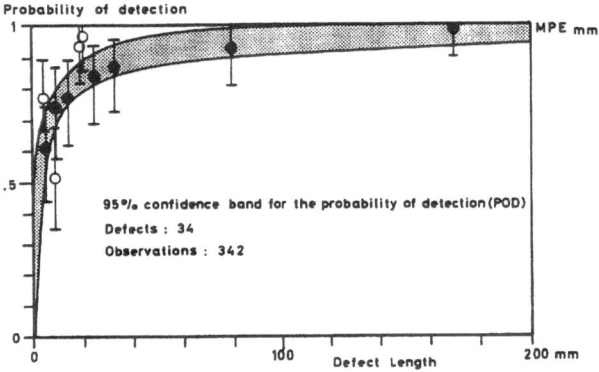

Figure 3. Inspection reliability for MPI.

A_j is generally random due to measurement error and/or due to uncertainties in the interpretation of a measured signal as a crack length. Measurements of the type Eq. (35) can be envisaged for several times corresponding to different values of n_j.

For each measurement Eq. (35), an event margin can be defined as

$$H_j(x) = CS_1^m n_j - \int_{a_0}^{A_j+x} \frac{da}{Y(a,Y)^m(\sqrt{\pi a})^m} \, , \quad j = 1, 2, \ldots, s \qquad (36)$$

These event margins are zero for $x = 0$ due to Eq. (35).

A second type of inspection result is that no crack is detected. For an inspection at a time corresponding to n_i stress cycles, this implies

$$a(n_i) \leq A_{di} \qquad (37)$$

expressing that the crack length is smaller than the smallest detectable crack length A_{di}. A_{di} is generally random since a detectable crack is only detected with a certain probability, depending on the crack length and on the inspection method. The distribution of A_{di} is the distribution of the length of undetected cracks. This distribution is provided through the probability of detection curves (pod curves) for which experimental results exist for various inspection methods. Fig. 3 shows experimental data and a pod curve for magnetic particle inspection (MPI). Information of the type Eq. (37) can also be envisaged for several times. If A_{di} is deterministic, however, and the same for all inspections, the information in the latest observation contains all the information of the previous ones. For each measurement Eq. (37) an event margin M_i can be defined as, (Madsen, 1985a; Madsen, 1985b)

$$H_i = CS^m n_i - \int_{a_0}^{A_{di}} \frac{da}{Y(a,Y)^m(\sqrt{\pi a})^m} \leq 0 \, , \quad i = 1, 2, \ldots, r \qquad (38)$$

These event margins are negative due to Eq. (37).

With one inspection result of the type Eq. (37), the updated failure probability is

$$P[M \leq 0 \mid H \leq 0] = \frac{P[M \leq 0 \cap H \leq 0]}{P[H \leq 0]} \qquad (39)$$

Evaluation of the reliability of a parallel system (numerator) and a component (denominator) are thus required. A FORM or SORM analysis can be directly applied, (Madsen, 1987). With one inspection result of the type Eq. (35), the updated failure probability is

$$P[M \leq 0 \mid H(0) = 0] = \frac{\frac{\partial}{\partial x} P[M \leq 0 \cap H(x) \leq 0]}{\frac{\partial}{\partial x} P[H(x) \leq 0]} \qquad (40)$$

where the derivatives are computed for $x = 0$. An evaluation of the sensitivity factor for a parallel system (numerator) and a component (denominator) are thus required, and a FORM or SORM analysis can be directly applied. The analysis is easily generalized to simultaneous consideration of several inspection results (Madsen, 1987).

The interest is now on updating after repair, and it is assumed that a repair takes place after n_{rep} stress cycles when a crack length a_{rep} is observed. An event margin H_{rep} is defined as

$$H_{rep}(x) = CS^m n_{rep} - \int_{a_0}^{a_{rep}+x} \frac{da}{Y(a,Y)^m(\sqrt{\pi a})^m} \qquad (41)$$

which equals zero for $x = 0$. The crack length present after repair and a possible inspection is a random variable a_{new}, and the material properties after repair are m_{new} and C_{new}. The safety margin after repair is M_{new}

$$M_{new} = \int_{a_{new}}^{a_C} \frac{da}{Y(a,Y)^{m_{new}}(\sqrt{\pi x})^{m_{new}}} - C_{new}S^{m_{new}}(n - n_{rep}) \qquad (42)$$

and the updated failure probability is

$$P[M_{new} \leq 0 \mid H_{rep}(0) = 0] = \frac{\frac{\partial}{\partial x} P[M_{new} \leq 0 \cap H_{rep}(x) \leq 0]}{\frac{\partial}{\partial x} P[H_{rep}(x) \leq 0)]} \qquad (43)$$

where the derivatives are computed for $x = 0$.

Example: Center cracked panel.

Consider a panel with a center crack as in the experiments of (Virkler *et al.*, 1979), see Fig. 4. The loading is a constant amplitude loading, leading

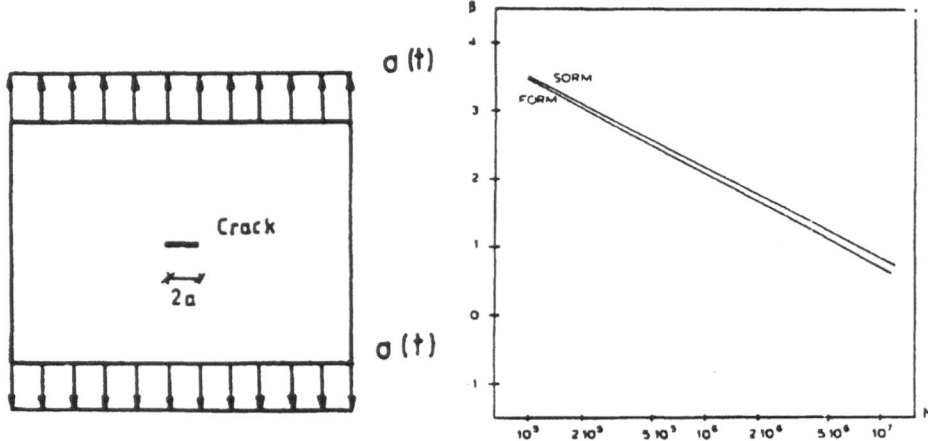

Figure 4.   Center cracked panel.

Figure 5.   FORM and SORM reliability index design calculation.

to a far-field stress range S. The geometry function is modeled as

$$Y(a, \mathbf{Y}) = \exp\left[Y_1 \left(\frac{a}{50} \right)^{Y_2} \right]$$

The geometry function takes the value one for $a = 0$. Lengths are measured in mm, and stresses in N/mm^2. The distribution of the basic variables is rather arbitrarily taken as

$$S \in N(60, 10^2)$$
$$Y_1 \in LN(1, 0.2^2)$$
$$Y_2 \in LN(2, 0.1^2)$$
$$a_0 \in EX(1)$$
$$a_C \in N(50, 10^2)$$
$$(\ln C, m) \in N_2(-33.00, 0.47^2, 3.5, 0.3^2; -0.9)$$

$N(\mu, \sigma^2)$ denotes a normal distribution with mean value μ and variance σ^2. Similarly $LN(\mu, \sigma^2)$ denotes a log-normal distribution with mean value μ and variance σ^2. $N_2(\mu_1, \sigma_1^2, \mu_2, \sigma_2^2; \rho)$ denotes a bi-normal distribution with mean values μ_1 and μ_2, variances σ_1^2 and σ_2^2 and correlation coefficient ρ. $EX(\mu)$ denotes an exponential distribution with mean value μ. The example has seven basic variables which are transformed into standardized and independent normal variables.

TABLE 1. Sensitivity Factors. $N = 1.5 \cdot 10^6$, $\beta = 1.817$

Variable	α_i	α_i^2
S	0.3577	13%
Y_1	0.0085	0%
Y_2	-0.0060	0%
a_0	0.5514	30%
a_C	-0.0001	0%
m	-0.6141	38%
$C_1 \mid m$	0.4362	19%

The FORM and SORM approximations to the reliability index are shown in Fig. 5 for various life times expressed in terms of the number of stress cycles N. The two approximations are close, implying that the curvatures of the limit state surface are moderate at the design point. Statistics for the distribution of life time T can be directly approximated from the results of Fig. 6. For the mean life times, the approximation is

$$E[T] = \int_0^\infty (1 - P(T \le t))dt \approx \int_0^\infty \Phi(\beta(t))dt$$

For $N = 1.5 \cdot 10^6$ stress cycles, the reliability index is $\beta = 1.817$ and the α's are shown in Table 1.

α_i^2 is interpreted as the fraction of the total uncertainty due to uncertainty arising from the ith basic variable. The major contribution to the overall uncertainty arises from the uncertainty in the material parameters. The critical crack size uncertainty is of little relative importance in this case, and the same is concluded in almost all cases where the critical crack size is significantly larger than the initial crack size. The uncertainty in the geometry function contributes very little to the total uncertainty in this case. This is because the value for $a = 0$ is completely known. When this initial value is not known, the uncertainty is comparable to the uncertainty in the loading. The uncertainty contribution from the uncertainty in the change in the geometry function from the initial value is generally found to be low. For tubular joints, where the geometry function is approximately proportional to $a^{-1/2}$ for large values of a, this statement may not always be true.

Reliability updating based on inspection results is concidered. First, the situation where a crack is found in the first inspection is considered. It is envisaged that the inspection is carried out after $N_1 = 10^5$ stress

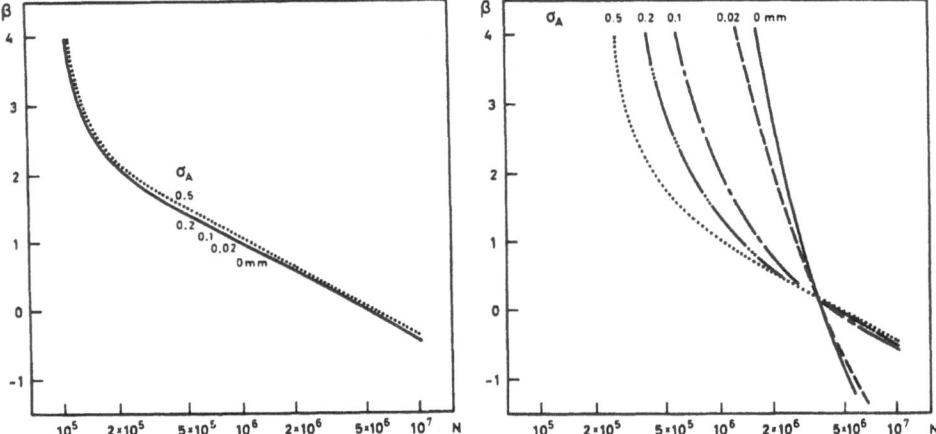

Figure 6. Updated first-order reliability index after first inspection with measurement 3.9 mm.

Figure 7. Updated first-order reliability index after second inspection with crack measurements 3.9 mm and 4.0 mm.

cycles, and a crack length of 3.9 mm is measured. The measurement error is assumed to be normally distributed with standard deviation σ_A. Fig. 6 shows the updated reliability index as a function of σ_A. The result is almost independent of σ_A in this example, as the uncertainty in the initial crack size dominates the uncertainty in A_1.

When the crack is detected, a decision has to be made, and two options are present. It may be decided to repair the crack immediately, or to leave the crack as it is, and base a decision on repair on more inspection results. With just one inspection, it is not possible to determine if the crack was initially large but is growing slowly enough that repair is not needed, or the crack was initially fairly small, but is growing fast and must be repaired. If a requirement on the reliability index in a period without inspections is formulated, e.g., $\beta \geq 2$, the latest time of the next inspection is determined from Fig. 6.

Assume that the crack is not repaired, but a second inspection at $N = 2 \cdot 10^5$ stress cycles is required. Let the inspection method be the same as in the first inspection, and let the measured crack size be 4.0 mm. The measurement error is again assumed to be normally distributed with standard deviation σ_A, and the two measurement errors are assumed to be statistically independent. Fig. 7 shows the updated reliability index after this second inspection. Different inspection qualities now lead to very different results. With $\sigma_A = 0$, the negative slope of the reliability index curve becomes very large, demonstrating that the crack growth behaviour is basically determined by two combinations of the basic variables. With

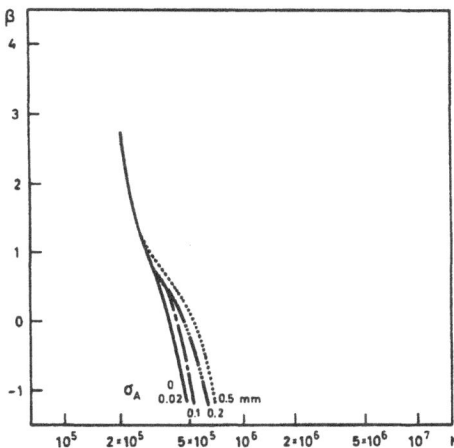

Figure 8.   Updated first-order reliability index after second inspection with crack measurements 3.9 mm and 5.0 mm.

a large measurement uncertainty, there is an immediate and large increase in reliability, but after some time the curve becomes almost identical to the curve resulting after the first inspection. Due to large uncertainty in both inspections, only little information is gained on the crack growth rate. If the inspection quality is very high, it may be possible to state that the crack does not grow to a critical size within the design life time. Repair and further inspections are then unnecessary. For a poorer inspection quality, a time period until the next inspection can be determined, and the decision on repair be further delayed.

Fig. 8 presents results similar to those in Fig. 7, but for the case where a crack size of 5 mm is reported in the second inspection. Together, the two inspection results now indicate that a large and fast growing crack is present. Repair is therefore necessary within a short period.

Consider now different situations where the inspections show no cracks. An attempt is made to illustrate possible means to achieve a required reliability. Let the reliability requirement be $\beta \geq 3.0$, and let the design life time correspond to $1.5 \cdot 10^6$ stress cycles. Fig. 9 shows the reliability index as a function of number of stress cycles, for two plate thicknesses. With a plate thickness t, the reliability requirement is fulfilled for the design life time, and no inspections are needed. With a plate thickness of only 60% of t, the reliability requirement is fulfilled for the period until $N = 2 \cdot 10^5$ stress cycles, where an inspection is needed. The quality of the inspection is reflected in the distribution of the smallest detectable crack size. An exponential distribution is assumed, with a mean value λ. Cracks initially present are cracks which have passed the inspection at the production site,

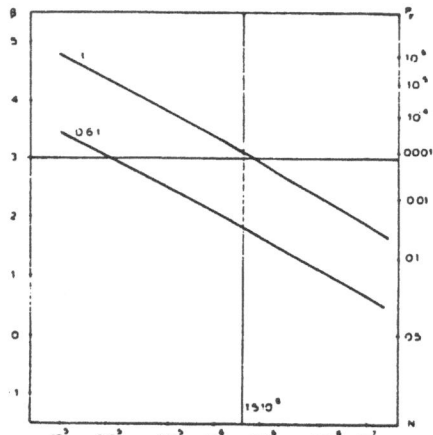 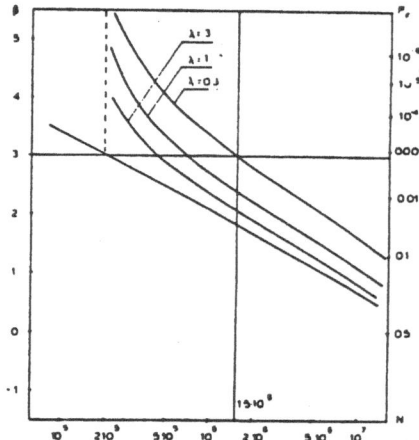

Figure 9. First order reliability index for two plate thicknesses.

Figure 10. Updated first-order reliability index after first inspection with no crack detection.

either because they were not detected or because they were below the acceptance level. If no cracks were accepted in fabrication, the fabrication inspection therefore corresponds to $\lambda = 1$.

Fig. 10 shows the initial reliability index and updated reliability indices for three inspection qualities. The best inspection quality $\lambda = 0.3$ is better than the fabrication inspection quality, and if no crack is found with this method, the increase in reliability is sufficient to make further inspections unnecessary. For the two other inspection qualities, periods are determined until the next inspection.

Fig. 11 shows the total inspection requirement for $\lambda = 1$ when no crack is detected in any inspection. For this case, two inspections are needed. Finally, Fig. 12 shows the total inspection requirement for $\lambda = 3$ when no crack is detected in any inspection; for this case five inspections are needed. It is thus demonstrated that different strategies on design and inspection planning can be used to achieve a required reliability. Based on costs of each strategy, including expected failure costs, a cost optimal solution can be determined.

The results of a reliability analysis following a repair of a detected crack is illustrated in Fig. 13. It is assumed that a crack of size $a_{rep} = 8$mm is repaired after $N_{rep} = 2 \cdot 10^5$ stress cycles. The distribution of the initial crack size after repair a_{new} is taken as an exponential distribution, with a mean value of 1 mm, i.e., as the same initial distribution as after fabrication. Two situations are considered, with either identical or independent material properties before and after repair. When independent properties

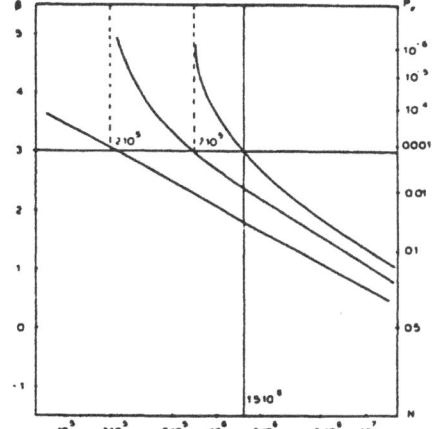

Figure 11. Updated first-order reliability index after inspections with no crack detection, mean size of non-detected cracks 1 mm.

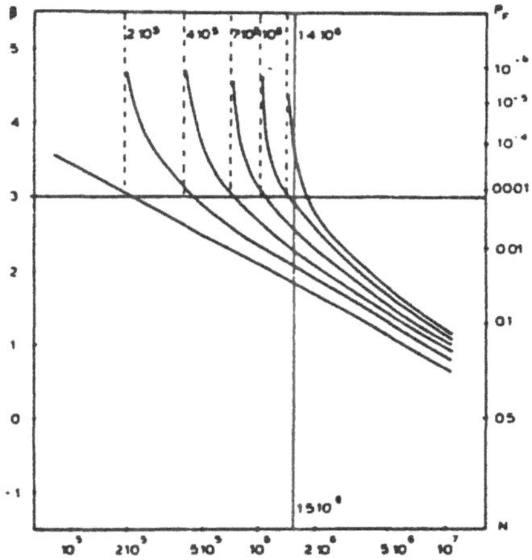

Figure 12. Updated first-order reliability index after inspections with no crack detection, mean size of non-detected cracks 3 mm.

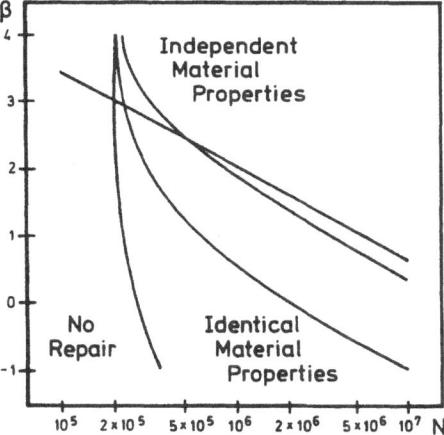

Figure 13. Updated first-order reliability index after repair of an 8 mm crack at $N = 2 \cdot 10^5$ stress cycles.

are assumed, the same distribution is used for the properties before and after repair. It follows from the results that there is an immediate increase in reliability after repair, but the reliability quickly drops to a level below the level obtained for the calculations before repair. This reflects the possibility that the cause for the large repaired crack size is a larger than anticipated loading of the crack tip, which is also acting after the repair.

6. Probability based optimization of design, inspection, and maintenance

Reliability against fatigue crack growth is achieved through efforts in design, inspection, and repair or replacement. These efforts all introduce cost, and the minimum total expected cost solution is of interest. An optimization problem can be formulated which can further include a constraint on the smallest allowable reliability, e.g. as specified by a regulatory body. Contributions to the formulation and solution of this optimization problem can be found in e.g. (Skjong, 1985; Thoft-Christensen and Sørensen, 1987; Madsen *et al.*, 1989; Fujita, Schall and Rackwitz, 1989).

Following (Madsen *et al.*, 1989) this section presents a consistent formulation of cost optimal fatigue design, inspection and repair. Reliability and sensitivity calculations are performed by a first-order reliability method, and the optimization is carried out by a general non-linear optimization algorithm.

A one-dimensional description of the crack size a is employed. A Weibull distribution with random distribution parameters A and B is used for the distribution of the stress ranges S. This is a relevant choice for the long term distribution for offshore jacket structures. The stress ranges depend on the structural optimization parameters. The number of stress cycles per unit time is ν, and the safety margin M for failure — defined as crack growth to a critical size a_C — before time t is, see Eq. (16)

$$M(t) = \int_{a_0}^{a_C} \frac{dx}{Y(x)^m (\sqrt{\pi x})^m} - C\nu t A^m \Gamma \left(1 + \frac{m}{B}\right) \tag{44}$$

The first inspection at time T_1 leads to a crack detection or no crack detection. An event margin H is defined as

$$H(t) = \int_{a_0}^{a_{d1}} \frac{dx}{Y(x)^m (\sqrt{\pi x})^m} - C\nu T_1 A^m \Gamma \left(1 + \frac{m}{B}\right) \tag{45}$$

The event margin is negative when a crack is detected, and is otherwise positive. Values for the smallest detectable crack size a_d for different inspections are assumed to be mutually independent. When a crack is detected and repaired at time T_1, the safety margin after repair is

$$M^1(t) = \int_{a_R}^{a_C} \frac{dx}{Y(x)^m (\sqrt{\pi x})^m} - C\nu(t - T_1) A^m \Gamma \left(1 + \frac{m}{B}\right) \quad ; \quad t > T_1 \tag{46}$$

The geometry function is identical before and after repair. The material parameter C, before and after repair, is either assumed fully dependent or completely independent. Crack sizes after repair a_R are assumed mutually independent. A notation is introduced to describe the sequence of repair/no

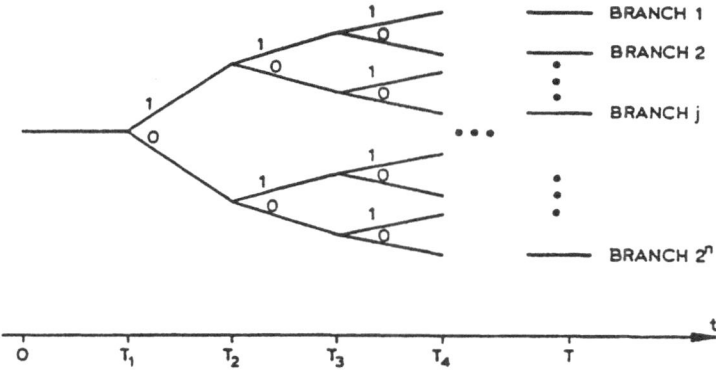

Figure 14.   Repair realizations. 0 denotes no repair, while 1 denotes repair.

repair events. For example, with repair at times T_1 and T_2 and no repair at T_3, the safety margin for $T_3 \leq T_4$ is

$$M^{110}(t) = \int_{a_R}^{a_C} \frac{dx}{Y(x)^m (\sqrt{\pi x})^m} - C\nu(t - T_2)A^m\Gamma\left(1 + \frac{m}{B}\right) \; ; \; T_3 < t \leq T_4$$
(47)

and the event margin for crack detection at time T_4 is similarly

$$H^{110}(t) = \int_{a_R}^{a_{d4}} \frac{dx}{Y(x)^m (\sqrt{\pi x})^m} - C\nu(T_4 - T_2)A^m\Gamma\left(1 + \frac{m}{B}\right) \qquad (48)$$

The crack growth formulation and subsequent optimization has in (Madsen, 1988) been extended to include a possible positive threshold value for the stress intensity factor range in Eq. (2), a constant corrosion rate, and a crack initiation period.

The following strategy for repair is selected:

All detected cracks are repaired

In (Madsen, 1988) also other repair and replacement strategies are included. These are

  − Detected cracks smaller than a threshold value are repaired by grinding, while larger detected cracks are repaired by welding
  − Only detected cracks larger than a reference size are repaired (by welding)
  − All detected cracks are "repaired" by replacement of the element.

With n inspections performed at times $T_1, \ldots, T_n$, the total number of different repair courses is 2^n, see Fig. 14.

The failure probability before time t is $P_F(t)$. The corresponding reliability index is

$$\beta(t) = -\Phi^{-1}(P_F(t)) \qquad (49)$$

In terms of the safety and event margins, failure before time t is as an example

$$P_F(t) = \mathrm{P}\left[M(T_1) \leq 0\right] + \mathrm{P}\left[M(T_1) > 0 \cap H > 0 \cap M^0(t) \leq 0\right]$$
$$+ \mathrm{P}\left[M(T_1) > 0 \cap H \leq 0 \cap M^1(t) \leq 0\right] ; \quad T_1 < t \leq T_2 \tag{50}$$

and similarly for other inspection time intervals and the life time T. With n inspections between 0 and T, $2^{n+1} - 1$ parallel systems are analysed to compute the failure probabilities.

The expected number of repairs $\mathrm{E}[R_i]$ at time T_i is identical to the probability of repair at time T_i. It is at time T_2

$$\mathrm{E}[R_2] = \mathrm{P}\left[M(T_1) > 0 \cap H > 0 \cap M^0(T_2) < 0 \cap H^0 \leq 0\right]$$
$$+ \mathrm{P}\left[M(T_1) > 0 \cap H \leq 0 \cap M^1(T_2) > 0 \cap H^1 \leq 0\right] \tag{51}$$

and similarly for other inspection times. With n inspections between 0 and T, 2^{n-1} parallel systems are analysed to compute the repair probabilities.

The inspection quality is defined by the pod (probability of detection) curve $p(a)$ for which an exponential form may be chosen for illustration.

$$p(a) = F_{a_d}(a) = 1 - \exp\left[-\frac{a}{\lambda}\right] ; \quad a > 0 \tag{52}$$

The pod curve is identical to the distribution function of the smallest detectable crack size a_d. The inspection quality is thus characterized by the parameter λ, which can take values between 0 and ∞. In the optimization, an auxiliary measure of inspection quality q is introduced.

$$q = \frac{1}{\lambda} \tag{53}$$

q can take values in the interval $[0; \infty[$. $q = 0$ corresponds to no inspection while $q \to \infty$ corresponds to a perfect inspection where infinitely small cracks are found.

The number of inspections n during the life time T is selected beforehand. This is done to avoid an optimization with a mixture of integer and real valued optimization variables. The analysis is repeated for several values of n and the resulting optimal costs are compared. The n-value with the smallest total expected cost is the optimal value. Inspection times and qualities are optimization variables together with the structural design parameter vector z.

The following cost items are included: initial cost, $C_I = C_I(z)$, inspection cost, $C_{IN} = C_{IN}(q)$, cost of repair, C_R, and cost of failure, $C_F = C_F(t)$.

Inspection and repair costs are assumed to increase with the rate of inflation. The difference between the desired rate of return and the rate of inflation is assumed to be constant, r. The cost of failure may be a function of time.

The optimization is formulated as a minimization of the total expected cost, with a constraint on the reliability index for the life time, and simple constraints on the optimization parameters:

$$\min_{t,q,z} \quad C_I + \sum_{i=1}^{n} \{C_{IN}(q_i)(1 - P_F(T_i)) + C_R \mathrm{E}\,[R_i]\} \frac{1}{(1+r)^{T_i}}$$

$$+ \sum_{i=1}^{n+1} C_F(T_i)\{P_F(T_i) - P_F(T_{i-1})\} \frac{1}{(1+r)^{T_i}}$$

$$\text{s.t} \quad \beta(T) \geq \beta^{\min} \tag{54}$$

$$t^{\min} \leq t_i = T_i - T_{i-1} \leq t^{\max}, \quad i = 1, 2 \ldots, n$$

$$t^{\min} \leq T - \sum_{i=1}^{n} t_i \leq t^{\max}$$

$$q^{\min} \leq q_i \leq q^{\max}, \quad i = 1, 2 \ldots, n$$

$$z_i^{\min} \leq z_i \leq z_i^{\max}, \quad i = 1, 2, \ldots, k$$

The possibility of predetermining one or more of the inspection times and qualities as well as elements in z is available.

The optimization problem is solved for each n using the NLPQL algorithms (Schittkowski, 1986). The value of the objective function and of the constraints are computed in a separate routine. This routine calls upon the reliability analysis program PROBAN (Tvedt, 1986) for analysis of $2^{n+1} - 1$ parallel systems for calculation of failure probabilities and $2^n - 1$ parallel systems for calculation of expected repair cost. PROBAN provides a reliability index calculated by a first-order reliability method (Madsen, 1992) for each parallel system, together with exact partial derivatives of the reliability index with respect to λ_i, T_i and z_j. From these partial derivatives, the partial derivatives with respect to q_i, T_i, and z_j are easily derived. Possibilities are included to perform the optimization with only the most important branches in the event tree in Fig. 14 included.

A more general formulation than presented in Fig. 14 has different inspection times and qualities in different branches. The number of optimization variables is thereby increased drastically. To overcome this problem, a procedure is here chosen in which the inspection plan is first optimized at the design stage. When the result of the first inspection is known, a new optimal inspection plan is determined by applying this information

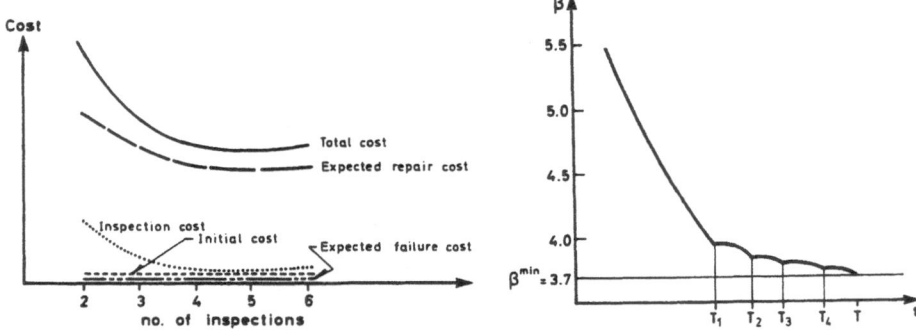

Figure 15.   Cost functions at optimal solution; reliability index with time.

in addition to the information available at the design stage. The various failure probabilities and probabilities of repair are then conditional probabilities, conditioned upon the result of the first inspection. Actual crack measurement results can be considered at this stage. As each inspection result becomes available, the tree of possibilities in Fig. 14 is reduced to one half of its size as the actual branch at the inspection time is known. With inspection results available at times $T_1, \ldots, T_{j-1}$, the optimization problem is formulated as

$$\min_{t,q} \quad \sum_{i=j}^{n} \{C_{IN}(q_i)(1 - P_F(T_i)) + C_R \mathrm{E}\,[R_i]\} \frac{1}{(1+r)^{T_i}}$$

$$+ \sum_{i=j}^{n+1} C_F(T_i)\{P_F(T_i) - P_F(T_{i-1})\} \frac{1}{(1+r)^{T_i}}$$

s.t $\beta(T) \geq \beta^{\min}$ (55)

$t^{\min} \leq t_i = T_i - T_{i-1} \leq t^{\max}, \quad i = j, \ldots, n$

$t^{\min} \leq T - \sum_{i=1}^{n} t_i \leq t^{\max}$

$q^{\min} \leq q_i \leq q^{\max}, \quad i = j, \ldots, n$

where failure and repair probabilities are computed conditioned upon the results of the first $j - 1$ inspections.

Example: Design and Maintenance Optimization for Tubular Joint.

An analysis of a tubular joint in an offshore jacket structure is considered. The selected input data are described in detail in (Madsen, 1988). Fig. 15

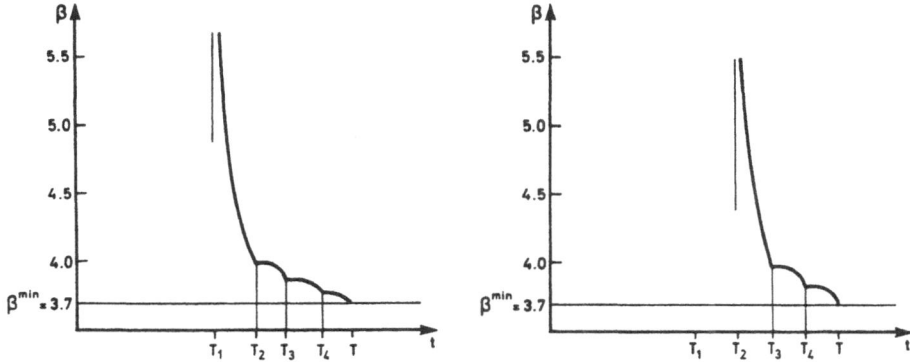

Figure 16. Reliability index with time in optimization after inspections.

shows the different cost items as a function of the number of inspections, and it shows the change in reliability index with time for the case of $n = 4$ inspections. If the inspection at time T_1 does not result in a crack detection, a new optimization is performed as described in the previous section. Fig. 16 shows the change in reliability index with time for the new optimal solution and for the optimal solution determined with no crack detection in the first two inspections.

7. Conclusions

The following conclusions can be stated:

1. A stochastic model for fatigue crack growth has been formulated which accounts for uncertainties in loading, initial defects, critical crack size, material parameters, and in the computation of the stress intensity factor. Both constant and variable amplitude loading is considered.
2. Different failure criteria are considered, and limit state functions and corresponding safety margins are formulated. Basic uncertain variables are described, and statistics for the material parameters and the loading are analyzed.
3. Two types of inspection results have been considered, and the inspection uncertainty has been modeled. Event margins have been defined for both types of inspection results. Updated reliabilities have been expressed in terms of the safety margin and the inspection event margins. Updated reliabilities after repair have been described in a similar manner.
4. Reliability calculations and reliability updating are conveniently and accurately done by first-and second-order reliability methods (FORM and SORM). These methods in addition provide a set of useful impor-

tance and sensitivity factors.

5. The analysis has been presented for an example panel with a center crack. The reliability index has been based on information at the design stage and has been updated based on inspection results both resulting in crack detecting and in no detection. Different inspection qualities have been considered, resulting in different effects on the updated reliability index.

6. Optimal allocation of resources in design and for inspection and repair to achieve a minimum total life cycle cost is described as an optimization problem and a solution method is presented.

References

R. Arone. On Reliability Assessment for a Structure with a System of Cracks. In *DIALOG 6-82*, Danish Engineering Academy, Lyngby Denmark, 1983.

ASCE Committee. Series of Articles by the ASCE Committee on Fatigue and Fracture Reliability. *Journal of the Structural Division*, ASCE, 108, 1982, pages 3–88.

V.V. Bolotin. Wahrscheinlichkeitsmetoden für Berechnung von Konstruktionen. VEB Verlag für Bauwesen, Berlin, 1981.

O. Ditlevsen. Random Fatigue Crack Growth – A First Passage Problem. *Engineering Fracture Mechanics*, 23, 1986, pages 467–477.

O. Ditlevsen and R. Olesen. Statistical Analysis of the Virkler Data for Fatigue Crack Growth. DCAMM Report No. 318, Technical University of Denmark, Lyngby, Denmark, 1986.

M. Fujita, G. Schall and R. Rackwitz. Adaptive Reliability-Based Inspection Strategies for Structures Subject to Fatigue. Proceedings of *ICOSSAR'89*. San Francisco, USA, 1989.

T.R. Gurney. An Analysis of Some Fatigue Crack Propagation Data for Steel Subjected to Pulsating Tension Loading. The Welding Institute Report 59/1978/E, 1978.

P.E. Irving and L.N. McCartney. Prediction of Fatigue Crack Growth Rates: Theory, Mechanisms and Experimental Results. In *Metal Science*, Aug./Sept. 1977, pages 351–361. Fatigue 77 Conference, University of Cambridge.

F. Kozin and J.L. Bogdanoff. A Critical Analysis of Some Probabilistic Models for Fatigue Crack Growth. *Engineering Fracture Mechanics*, 14, 1981, pages 59–89.

Y.K. Lin and J.N. Yang. On Statistical Moments of Fatigue Crack propagation. *Engineering Fracture Mechanics*, 18, 1983, pages 243–256.

H.O. Madsen. Probabilistic and Deterministic Models for Predicting Damage Accumulation due to Time Varying Loading. *DIALOG 5-82*, Danish Engineering Academy, Lyngby, Denmark, 1983.

H.O. Madsen. Bayesian Fatigue Life Prediction. Paper presented at *IUTAM Symposium on Probabilistic Methods in the Mechanics of Solids and Structures*. Stockholm, Sweden, June 19-21, 1984.

H.O. Madsen. Random Fatigue Crack Growth and Inspection. In *Structural Safety and Reliability*, Proceedings of ICOSSAR. Kobe, Japan, IASSAR, 1985 1, pages 475–484.

H.O. Madsen. Model Updating in First-Order Reliability Theory with Application to Fatigue Crack Growth. In proceedings *Second International Workshop on Stochastic Methods in Structural Mechanics*. University of Pavia. Pavia, Italy, 1985.

H.O. Madsen. Model Updating in Reliability Theory. *Proceedings ICASP-5*, Vancouver, 1987, 1, pages 564–577.

H.O. Madsen. Theoretical Manual PRODIM - Probability-based Design, Inspection and Maintenance. A.S. Veritas Research Report No. 88-2019. Høvik, Norway, 1988.

H.O. Madsen. Sensitivity factors for parallel systems. In G. Mohr, editor, *Miscellaneous Papers in Civil Engineering*. Danish Engineering Academy, Lyngby, Denmark, 1992. DIA 35'th Aniversary '92.

H.O. Madsen, S. Krenk and N.C. Lind. *Methods of Structural Safety*. Prentice-Hall, Inc., Englewood Cliffs, New Jersey, 1986.

H.O. Madsen, J.D. Sørensen and R. Olesen. Optimal Inspection Planning for Fatigue Damage of Offshore Structures. Proceedings of *ICOSSAR'89*. San Francisco, USA 1989.

K. Ortiz and A.S. Kiremidjian. Time Series Analysis of Fatigue Crack Growth Data. *Engineering Fracture Mechanics*, 1986.

P. Paris and F. Erdogan. A Critical Analysis of Crack Propagation Laws. *Journal of Basic Engineering*, Trans. ASME, 85, 1963, pages 528–534.

S.O. Rice. Mathematical Analysis of Random Noise. In N. Wax, editor, Selected Papers on Noise and Stochastic Processes, Dove, 1954.

K. Schittkowski. NLPQL: A FORTRAN Subroutine Solving constrained Non-Linear Programming Problems. Annals of Operations Research, 1986.

W. Shang-Xian. Shape Change of Surface Crack During Fatigue Growth. *Engineering Fracture Mechanics*, 22, 1985, pages 897–913.

R. Skjong. Reliability-Based Optimization of Inspection Strategies. In *Structural Safety and Reliability*. Proceedings of ICOSSAR'85, Japan, IASSAR, 1985, III, pages 614–618.

S. Tanaka, M. Ichikawa and S. Akita. Variability of m and C in the Crack Propagation Law $da/dN = C(\Delta K)^m$. *International Journal of Fracture*, 17, 1981, pages 121–124.

P. Thoft-Christensen and J.D. Sørensen. Optimal Strategies for Inspection and Repair of Structural Systems. *Civil Engineering Systems*, 4, 1987, pages 94–100.

L. Tvedt. User's Manual PROBAN - PROBabilistic ANalysis. A.S. Veritas Research Report No. 86-2037. Høvik, Norway, 1986.

D.A. Virkler, B.M. Hilberry and P.K. Goel. The Statistical Nature of Fatigue Crack Propagation. *Journal of Materials and Technology*, 101, 1979, pages 148–153.

PROBABILISTIC FATIGUE ASSESSMENT OF WELDED JOINTS

Navil K Shetty
WS Atkins Science & Technology
Woodcote Grove, Ashley Road
Epsom, Surrey KT18 5BW
U.K.

Abstract

A methodology for probabilistic fatigue assessment of welded joints using the S-N and fracture mechanics approaches is proposed. A fatigue crack propagation model is presented for a semi-elliptical crack in a welded plate which accounts for the effects of weld geometry, residual stresses, stress ratio, fatigue threshold and variable amplitude loading. A simple lognormal format and a rigorous FORM/SORM approach is used for evaluating the reliability of a joint against failure by fatigue. The model accounts for the uncertainties in fatigue loading, stress analysis, stress intensity factors, initial defect size and crack growth material properties. Examples involving reliability analysis of tubular joints of offshore structures are presented.

1. Introduction

Conventional design against fatigue has been based on the S-N approach, in which all parameters are taken at their expected values except for a conservative choice of fatigue strength (S-N curve). The computed fatigue life of a joint is thus a single valued quantity arrived at in a deterministic manner. However, many of the parameters affecting the fatigue life of a joint, such as weld defect size, magnitude of stress range, and resistance of the material against crack propagation, are random in nature, and should be treated probabilistically. Modern methods of structural reliability analysis provide a rational basis for the probabilistic fatigue assessment of welded joints.

In this chapter, a brief review of the S-N and fracture mechanics methods for fatigue life estimation of welded joints is presented. A number of uncertainties associated with fatigue analysis, for example in the load modelling, structural modelling, stress analysis and fatigue damage modelling are discussed. The use of First Order (FORM) and Second Order (SORM) reliability methods for computing the probability of joint failure under fatigue are explained with practical examples.

C. Guedes Soares (ed.), Probabilistic Methods for Structural Design, 85–111.
© 1997 *Kluwer Academic Publishers.*

2. Factors Influencing Fatigue of Welded Joints

In welded joints, fatigue cracks develop from weld defects because of their stress raising effect. Since defects of significant size are invariably present, the crack initiation phase is relatively short for welded joints, and much of the fatigue life is characterised by crack propagation. Fatigue damage in welded joints of metallic structures is influenced by many factors which may be grouped as below.

(a) *Loading:* Fatigue is a cumulative, time-dependent phenomenon. Each cycle of load will cause some fatigue and thus the magnitudes of all load cycles over the service life of the structure characterize the fatigue loading.
(b) *Joint Geometry:* Since fatigue crack advance depends on the cyclic stress range at the very tip of the crack, the stress analysis of joints should account for the stress raising effects of the overall geometry of the joint, and the local geometry of the weld at the crack site.
(c) *Weld Defects:* Fatigue cracks in welded joints originate from weld defects such as inclusions, undercuts, lack of fusion etc. which are to some extent unavoidable. Though there may be many defects in the joint, only those defects which are present in regions of high stress concentration develop into fatigue cracks.
(d) *Material Property:* Crack propagation depends on the cyclic elasto-plastic response of the material near the crack tip. Material properties describing crack growth behaviour are obtained from experiments, and generally a large scatter in test results is observed.
(e) *Others:* The presence of residual stresses, mean stress due to static loads, and the sequence of load cycles in a variable amplitude loading are observed to have some influence on fatigue life.

3. Fatigue Analysis Using S-N Approach

In this method, the fatigue strength of a component is characterized in terms of an S-N curve which is a plot of stress range (as ordinate) versus number of stress cycles to failure (as abscissa), both plotted on a logarithmic scale. Most design codes specify S-N curves for the fatigue design of various types of welded joints which are derived from fatigue tests on a large number of real scale joints.

3.1 THE S-N CURVE

The basic S-N curves for various joint types are often established for a reference plate thickness and for air conditions. Additional corrections to the basic curve are applied to account for the effect of other factors. A log-linear S-N curve can be expressed in the form of

$$\log(N) = \log(K) - m \log(S_B) \tag{1}$$

where N is the number of stress cycles to failure at a constant amplitude stress range S_B,

K and m are the intercept and inverse slope of the S-N curve.

The mean values of m and log(K) are obtained from regression of test data. The design S-N curve is then derived by shifting the mean curve downwards by typically two standard deviations of log(N) of the data-set. However, for the purposes of reliability analysis, only the mean value S-N curve is relevant.

The effects of weld geometry, residual stresses and through-thickness stress variation are implicitly included in the S-N curve. The effect of other factors such as, plate thickness, sea water environment, weld toe grinding and post-weld heat treatment, etc. are accounted through appropriate corrections to the basic S-N curve.

3.2 COMPUTATION OF FATIGUE DAMAGE

In an S-N approach, fatigue damage is quantified in terms of Miner's damage summation. Miner's rule assumes that every stress cycle causes some fatigue damage, and that the damages caused by various stress cycles are linearly additive. Miner's rule neglects all stress-cycle interaction effects, and thus damage due to $n(S)$ cycles of constant amplitude loading of stress-range S can be expressed as

$$D_n = \frac{n(S)}{N(S)} \qquad (2)$$

where $N(S)$ is the total number of stress cycles to failure at the same stress-range S and D is the damage indicator. Eqn. (2) shows that D should be equal to unity for failure. An S-N curve of constant slope m can be expressed as

$$NS^m = K \qquad (3)$$

Combining the two equations, we can express the damage indicator as

$$D_n = \frac{1}{K} n S^m \qquad (4)$$

In the above, it is assumed that all the corrections for plate thickness, environment, weld improvement etc. have been incorporated through an appropriate value of K. If D_{1y} is the damage computed for all stress cycles in one year, then $1/D_{1y}$ gives the fatigue life or the number of years to failure of the joint.

3.3 COMMENTS ON THE USE OF S-N APPROACH

The S-N method of fatigue analysis is very simple to apply and is widely used for design purposes. Since an S-N curve is developed from a direct observation of fatigue lives of a set of joints, it incorporates the effect of all known and unknown factors on fatigue life which were present in the test data. Consequently, when used with joint types and loading conditions which are similar to those from which the S-N curve was developed, a high degree of confidence can be assigned to the predicted fatigue lives. However, in practical applications, extrapolations outside the original data-sets have to be made and, in general, the method suffers from the following disadvantages:

(a) Fatigue tests on welded joints required for developing S-N curves are very expensive, and every additional factor, such as thickness effect, weld toe grinding etc., that is to be accounted necessitates new tests.
(b) S-N curves cannot be easily extended to cases other than those covered by the original data-base used for developing the curves. Thus a change of material, or major changes in joint configuration and sizes requires new S-N curves to be developed.
(c) To develop S-N curves, joints of different types, different weld geometries and different loading modes are normally combined into one data-set, and consequently large scatter in fatigue lives is observed over the data-set. The design S-N curve can thus be over conservative for some joint types.
(d) If during in-service inspection a crack of significant size is found in a joint, the S-N method cannot be used to estimate the remaining life of the joint.

For these reasons the emphasis of research in recent years has turned towards the development of fracture mechanics methods for the fatigue analysis of welded joints, and this is discussed in the next section.

4. Fatigue Analysis Using Fracture Mechanics Approach

For welded joints, the crack initiation phase is short and most of the fatigue life is consumed in crack propagation from small weld defects. Because of this, fracture mechanics principles can be used advantageously for the fatigue assessment of welded joints. Fracture mechanics provides an accurate description of the stress-strain field around the crack, and its propensity to extend. It explicitly accounts for the influence of weld geometry, residual stresses, stress ratio, fatigue threshold, etc. It is observed that, because the fatigue process involves typically low stress levels, the crack tip plastic zone is usually small compared to the crack dimensions, and hence linear elastic fracture mechanics is applicable. The main elements involved in a fracture mechanics fatigue assessment procedure are discussed below.

4.1 BASIC CRACK GROWTH LAW

Analogous to the S-N approach, the fracture mechanics approach uses the stress intensity factor range ΔK to correlate the crack propagation rate da/dn. A crack propagation model typically gives a functional relation of the form

$$\frac{da}{dn} = f(\Delta K) \tag{5}$$

A number of attempts have been made to describe the form of the function f, either in a deterministic or probabilistic manner, and accordingly the various crack propagation models can be broadly grouped into deterministic and probabilistic models, see Shetty (1992) for a detailed review. In this chapter, only the deterministic models are discussed, while the stochastic models are described in the next chapter by Madsen. The simplest and the most widely used model is the one given by Paris and Erdogan (1963), commonly known as the Paris law, which can be written as

$$\frac{da}{dn} = C(\Delta K)^m \qquad for \quad \Delta K > 0 \qquad (6)$$

where C and m are material constants which can be determined from experiments on simple specimens. A typical plot of da/dn versus ΔK on a logarithmic scale obtained from experimental data is shown in Fig.1. The crack propagation rate curve exhibits a sigmoidal behaviour which can be considered to fall in three regimes based on the value of ΔK as shown in the figure. It can be seen that the Paris law correlates well with the observed crack propagation behaviour in regime II where $\log(da/dn)$ is more or less linearly related to $\log(\Delta K)$. In practical applications the crack propagation behaviour is often idealized as shown by the dotted lines in the figure, and the Paris law is assumed to be valid even in regimes I and III.

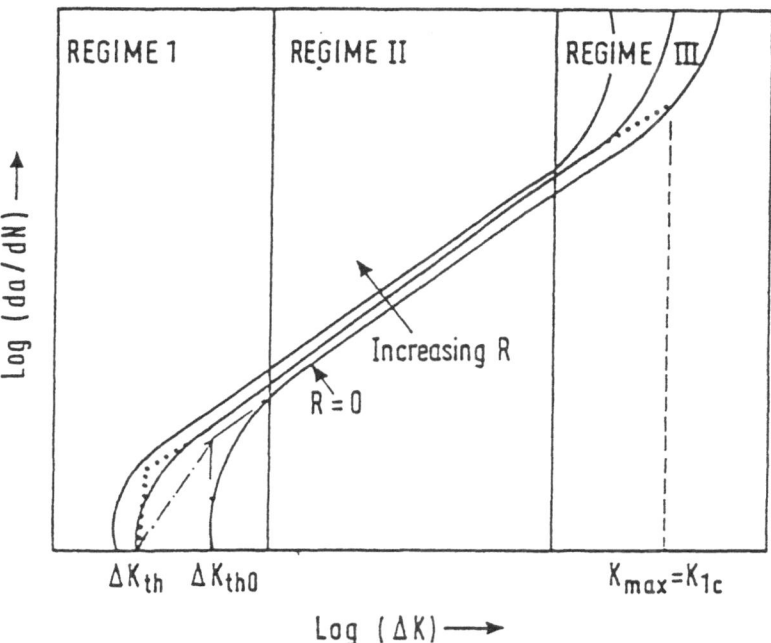

Figure 1: Typical variation of crack propagation rate with stress intensity factor range

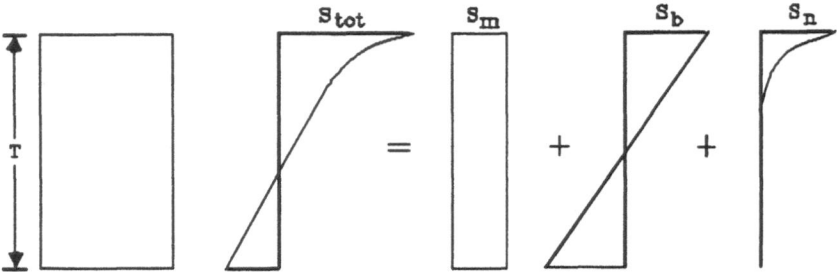

Figure 2: Through-thickness stress distribution at a welded joint

4.2 STRESS DISTRIBUTION AROUND THE CRACK

In order to calculate the stress intensity factor range ΔK for use in the crack growth law, a detailed description of the stress system in the vicinity of the crack (but of an uncracked section) is required. For welded joints, in addition to the stresses due to the applied loading, there are residual stresses, often reaching yield stress in magnitude. The stress distribution due to the applied loading is influenced by the overall geometry of the joint and the local geometry of the weld. The total stress system causing the propagation of a crack through the thickness of a plate can be represented typically as in Fig.2. The stress distribution at the joint due to the applied loading can be determined by finite element analysis. The distribution of residual stresses is usually obtained from experimental measurements on typical joint and weld details.

The through-thickness distribution of geometric stress is often assumed to be linear, consisting of *membrane* and *bending* components as shown in Fig.2, and is characterised in terms of the ratio of the bending stress to the total geometric stress, called *degree-of-bending*.

The geometric stress is amplified very close to the weld toe due to the effect of weld notch. The notch stress distribution is highly non-linear, as shown in Fig.2.

4.3 STRESS INTENSITY FACTORS

The stress distribution of an un-cracked body is further modified when a crack is introduced into it. The stress field close to the crack can be described by the linear-elastic stress intensity factor (SIF) K given as

$$K = YS_a\sqrt{\pi a} \qquad\qquad (7)$$

where a is the crack size, S_a the applied stress and Y is called the *stress intensity geometry correction* factor or *compliance function*. The Y-factor depends on the geometry of the joint, the nature of stress distribution and the crack size. Solutions for

Y-factor are available for a number of standard stress distributions and joint/crack geometries, see for example Tada et al (1973) and Rooke and Cartwright (1976).

For welded joints, it is observed that micro-cracks initiate from surface-breaking defects at the toe of the weld. These micro-cracks coalesce to form a single, dominant fatigue crack of roughly semi-elliptical shape. Semi-elliptical cracks in plated structures are of interest in many practical applications, and the SIF solutions for this case are presented below.

Stress Intensity Factors for a Semi-Elliptical Crack:
Empirical stress intensity factor solutions for a semi-elliptical surface crack in a plate subjected to membrane and bending stresses through the thickness of the plate have been developed by Newman and Raju (1981) based on results from 3-D finite element analyses. For a semi-elliptical crack of depth a and semi-length c, the stress intensity factor K can be expressed in the form

$$K = (H\rho + 1 - \rho)S_g Y_g (\pi a/q)^{1/2} \tag{8}$$

where S_g is the geometric stress at the crack site, ρ is the degree-of-bending factor for the stress distribution through the plate thickness and q is the elliptical integral of the second kind. The factors H and Y_g are functions of the relative crack depth a/T, crack aspect ratio a/c, relative crack length c/W and angle ϕ to the point on the crack front, for which expressions have been given in the above reference. In the above T is the plate thickness and W is the plate width. Using this equation, we can determine the stress intensity factor at any point on the crack front, but the interest is normally on the deepest point of the crack in the thickness direction ($\phi=90°$) and the surface tips ($\phi=0°$).

Effect of Weld Geometry:
The presence of a weld at the plate surface gives rise to a non-linear notch stress distribution as shown in Fig.2. To account for this notch stress a weld toe correction factor is used. The calculation of stress intensity factors for a semi-elliptical crack under a varying stress field is generally very difficult. For practical applications, simplified parametric expressions have been developed which have the general form

$$Y_w = Y_1\left(\frac{a}{T}, \frac{l_w}{T}\right) Y_2\left(\frac{a}{T}, \phi_w\right) Y_3\left(\frac{a}{T}, \frac{\rho_w}{T}\right) \tag{9}$$

where l_w is the weld leg length, ϕ_w is the weld toe angle and ρ_w is the weld toe radius. Empirical expressions for function Y_1 as a function of relative crack depth and relative weld leg length for $\phi_w=45°$ and $\rho_w=0$ are given in BS PD6493 (1991) separately for membrane and bending components of the through thickness geometric stress distribution. Similar expressions for functions Y_2 and Y_3 can be obtained from Dijkstra et al (1990). However, it is to be noted that these expressions are developed from parametric studies using 2-D finite element analysis on edge cracks in weldments, and can be conservative for semi-elliptical cracks, Dijkstra et al (1990). Solutions are not available for semi-elliptical cracks at present.

Effect of Stress Gradient:
The stress intensity factors calculated from Newman- Raju solutions are valid for a plate subjected to uniform stress across the width of the plate. An additional correction is required to account for stress gradient. Mattheck et al (1983) give a method for calculating stress intensity factors for semi-elliptical cracks in plates subjected to a varying stress field using the weight function technique. This can be used if the stress distribution around the intersection is known accurately. Alternatively, the effect of stress gradient can be modelled approximately by an *average stress factor* defined as

$$\gamma_s = \frac{average\ stress\ over\ the\ crack\ length}{geometric\ hot-spot\ stress} \tag{10}$$

This stress gradient correction is applied for the crack growth at the surface tips only, and can be conservatively neglected for the growth at the deepest point of the crack.

4.4 EFFECT OF STRESS RATIO

For each loading cycle, two stress ratios can be defined by

$$R_n = \frac{S_{min}}{S_{max}}$$

$$R = \frac{K_{min}}{K_{max}} \tag{11}$$

where R_n is called a *nominal stress ratio* which is calculated from the minimum and maximum values of the applied cyclic stress alone, while R is a *stress intensity factor ratio* (simply called *stress ratio*) which is based on the total stress which includes applied cyclic stress, applied static stress, residual stress etc. The influence of the stress ratio can be seen from Fig.1 which shows that crack propagation rate increases as positive R ratio becomes larger. However, the influence of R is more pronounced in regimes I and III, and is highly dependent on the microstructural properties of the material. The effect of stress ratio in these regimes can be accounted through the modifications suggested by Klesnil and Lucas (1972) and Forman (1964). In structural steels, the crack propagation in regime II is found to be almost independent of stress ratio. In addition to this influence on crack propagation rate, the stress ratio also influences threshold stress intensity factor range and effective stress range as discussed in the following.

4.5 EFFECT OF RESIDUAL STRESSES

The effect of tensile residual stresses is to alter the stress ratio and stress intensity factor range experienced at the crack tip during load cycling, and these are calculated as

$$
\begin{aligned}
K_{min} &= K_{app.min} + K_{res} \\
K_{max} &= K_{app.max} + K_{res} \\
\Delta K &= K_{max} - K_{min} \\
&= K_{app.max} - K_{app.min} \qquad for \quad K_{min} > 0 \\
\Delta K &= K_{app.max} + K_{res} \qquad for \quad K_{min} \leq 0 \\
R &= K_{min} / K_{max} \qquad for \quad K_{min} > 0 \\
R &= 0 \qquad for \quad K_{min} \leq 0
\end{aligned} \tag{12}
$$

where K_{app} is the stress intensity factor due to the applied loading and K_{res} is the stress intensity factor due to residual stress. The suffixes *min* and *max* refer to the minimum and maximum values within a stress cycle. Here it is implicitly assumed that for $K_{min} \leq 0$ (i.e. under net compressive stress), the crack will be closed and cannot propagate. Only that part of the loading cycle for which the crack remains open is considered to be effective in crack propagation. Equations (12) show that residual stresses do not have any influence on ΔK as long as $K_{min} > 0$. For $K_{min} \leq 0$, i.e. when the applied stress cycle has a very high compressive component, residual stresses will have a significant influence. However, the stress ratio is altered in both cases, which in turn has an influence on crack propagation, as mentioned above. The presence of other stresses, such as those due to static loading, which contribute to the mean stress, will have a similar influence, and should be considered in calculating ΔK.

The calculation of K_{res} should be based on the through-thickness distribution of residual stresses which is idealized as a combination of bending and membrane components as in Fig.2. The residual stresses are believed to dissipate as the crack propagates through the thickness, the exact nature of this is however not known. This dissipation can be approximated as

$$
\begin{aligned}
S_{res}(a) &= S_{b\,res}(a) + S_{m\,res} \\
&= \left[\rho_r \zeta(a) + 1 - \rho_r \right] S_{res.0} \\
\zeta(a) &= 1 - \left(\frac{a}{T} \right)
\end{aligned} \tag{13}
$$

where $S_{res.0}$ is the initial value of residual stress at the surface, $S_{b\,res}$ is the bending component and $S_{m\,res}$ is the membrane component of the residual stress, ρ_r is the degree of bending factor for residual stress and $\zeta(a)$ is a *residual stress relaxation factor*. This relaxation is assumed to be effective for the deepest point of the crack only and no relaxation is used at the surface tips.

The effect of residual stresses and stress ratio R can be conveniently introduced into the crack growth law through an effective stress intensity factor ratio U defined as

$$U = \frac{\Delta K_{eff}}{\Delta K} = \frac{K_{max} - K_{op}}{K_{max} - K_{min}} \qquad (14)$$

where ΔK_{eff} is the effective stress intensity factor range, and K_{op} is the crack opening stress intensity factor which is regarded as a material property to be determined from experiments. This reduces the crack growth law to a Paris type with ΔK in Eq.6 now replaced by $U \Delta K$. Combining Eqs. (12) and (14), we can express U in terms of R as

$$\begin{aligned} U &= 1 & for \quad R \geq 0 \\ &= 1/(1-R) & for \quad R < 0 \end{aligned} \qquad (15)$$

4.6 EFFECT OF FATIGUE THRESHOLD

Fig.2 shows that for values of applied stress intensity factor range ΔK less than a threshold value, called the *threshold stress intensity factor range* ΔK_{th}, the crack does not propagate. The exact mechanism of fatigue threshold is not completely understood, but it is regarded as a material property which can be determined from experiments. The threshold value is seen to be significantly influenced by stress ratio and can be explicitly modelled as suggested in BS PD6493 (1991) using

$$\Delta K_{th} = f_R \, \Delta K_{th0}$$
$$\qquad (16)$$
$$f_R = (190 - 144\,R)/190 = 1 - 0.757\,R$$

where ΔK_{th0} is the threshold value determined from experiments at $R=0$.

4.7 COMPUTATION OF FATIGUE DAMAGE

Based on the crack propagation model and the stress intensity factor solution presented above, a procedure for the computation of fatigue damage is summarized here. For the sake of simplicity the following derivation assumes constant amplitude fatigue. The extension of the procedure to variable amplitude loading is discussed in Section 5. Fatigue damage at any instant, after an exposure period of time t, or an average number of stress cycles $n(t)$, can be characterised in terms of the crack dimensions $a(t)$ and $c(t)$. Corresponding to the two principal directions of crack growth of a semi-elliptical crack

$$\begin{aligned} \frac{da}{dn} &= C \, \Delta K_{eff.a}^{m} \, G(a) \\ \frac{dc}{dn} &= C \, \Delta K_{eff.c}^{m} \, G(c) \end{aligned} \qquad (17)$$

where $G(a)$ and $G(c)$ are threshold correction factors which are discussed in Section 5.

The effective stress intensity factor ranges are

$$\Delta K_{eff.a} = U_a S Y_a [\pi a/q]^{1/2}$$
$$\Delta K_{eff.c} = U_c S Y_c [\pi a/q]^{1/2} \qquad (18)$$

where S is the stress range and U_a and U_c are the effective stress intensity factor ratios, which incorporate the effects of residual stresses and stress ratio, to be determined based on Eq.15. The geometry correction factors Y_a and Y_c are determined from the stress intensity factor solutions discussed in Section 4.3 as

$$Y_a = Y_{wa}(H\rho + 1 - \rho) Y_{ga}$$
$$Y_c = Y_{wc}(H\rho + 1 - \rho) Y_{gc} Y_s \qquad (19)$$

where Y_s is the average stress factor obtained from Eq.11. The coefficients H and Y_g are functions of relative crack depth a/t and crack aspect ratio a/c and are obtained from the stress intensity factor solutions for a semi-elliptical crack given by Newman and Raju (1981), Y_w is the weld geometry correction factor obtained as a function of weld geometry from BS PD6493 and Dijkstra et al (1990). The correction factor Y_{wc} is calculated corresponding to a constant crack depth of $a=1.5$mm. The crack propagation equations in the two directions are coupled, as the stress intensity factors at the deepest point and at the surface tip are functions of the parameters a/t and a/c.

Integration of Eq.17 in the thickness direction from an initial defect size a_0 to a crack size $a(t)$ after time t, and average number of stress cycles $n(t)$, gives

$$\int_{a_0}^{a(t)} \frac{dx}{C G(x) U_a^{\ m} Y_a^{\ m} [\pi x/q]^{m/2}} = n(t) S^m \qquad (20)$$

Equation 20 needs to be integrated numerically using a computer program. From an initial crack depth a_0 and initial semi-crack length c_0, crack depth is incremented in steps of, typically, 10-20% of its current value until the final crack length is reached. Within each increment, the fatigue damage is calculated using an 8-point Gauss quadrature integration. For each increment of crack depth Δa, the increment of crack length Δc, and the crack aspect ratio a/c for the jth increment, are

$$\Delta c = \frac{G(c)}{G(a)} \left[\frac{U_c Y_c}{U_a Y_a} \right]^m \Delta a$$
$$\left(\frac{a}{c} \right)_j = \frac{a_{j-1} + \Delta a}{c_{j-1} + \Delta c} \qquad (21)$$

Alternatively, when the fatigue life of a joint is required, Eq.20 is integrated from the initial defect size a_0 to a final crack size equal to the plate thickness, and the corresponding time t for failure is determined.

The effect of crack coalescence and short crack effect can also be incorporated, as discussed in Shetty and Baker (1990b).

5. Fatigue Under Variable Amplitude Loading

In Sections 3 and 4, the fatigue analysis method using S-N and fracture mechanics approaches have been developed assuming constant amplitude fatigue stress cycles. However, in many practical applications, the fatigue loading is of variable amplitude type. Since each load cycle causes some fatigue damage, in principle a knowledge of all stress cycles expected at a joint over the life-time of a structure is required. This is often developed in the form of a probability distribution of fatigue stress cycles. The main elements involved in a variable amplitude fatigue analysis are summarized below.

5.1 PROBABILITY DISTRIBUTION OF STRESS RANGES

The stress process under variable amplitude loading can be characterised in a time domain as a stochastic process $S(t)$, or equivalently in a frequency domain using a *spectral density function* $G_{ss}(\omega)$. The nth moment of a spectrum is computed as

$$m_n = \int_0^\infty \omega^n G_{ss}(\omega)\, d\omega \tag{22}$$

where ω is the angular frequency. Usually the zeroth, first, second, and fourth moments, corresponding to $n=0$, 1, 2 and 4 are of interest in developing a probability distribution of stress cycles, as discussed next.

From the moments of the spectrum of the stress response, the following statistical properties of the stress process can be obtained

$$E[S^2] = \sigma_s^2 + \mu_s^2 = m_0$$
$$T_0 = \frac{1}{\nu_0} = 2\pi \sqrt{m_0/m_2}$$
$$T_p = \frac{1}{\nu_p} = 2\pi \sqrt{m_2/m_4} \tag{23}$$
$$\alpha = \frac{T_p}{T_0} = \frac{\nu_0}{\nu_p} = \frac{m_2}{\sqrt{m_0 m_4}}$$
$$\varepsilon = \sqrt{1 - \alpha^2}$$

where E[.] is the Expectation operator, σ^2 is the variance, μ is mean, T_0 mean time between successive upcrossings of the mean level, T_p the mean time between successive peaks, ν_0 and ν_p are the average frequency of zero-crossings and average frequency of peaks of the stress response process. The ratio α is called the *spectral irregularity factor* and ε as the *spectral bandwidth parameter*. When $\varepsilon \rightarrow 0$ the spectrum is called *narrow-banded* and for $\varepsilon \rightarrow 1$ it is called *broad-banded*.

If the stress spectrum is Gaussian and narrow-banded it can be shown that the peaks of the stress process, and hence the stress ranges, follow a Rayleigh distribution. When the stress process is broad-banded, the stress cycles cannot be easily distinguished, and a convention is required for defining them. One of the approaches is to simulate the time history of the stress process and use one of the peak, range or rainflow counting schemes to count the stress cycles, see Dowling (1972). The rainflow counting method is seen to give the best correlation with experimental results of fatigue tests under variable amplitude loading. However, this approach is computationally very expensive for practical applications.

In recent years, a number of attempts have been made to obtain analytical distributions for stress ranges of a wide-banded stress process, see Shetty and Baker (1990a) for a detailed review. These methods either attempt to fit empirical distributions to the results obtained from rainflow simulations, or modify one of the standard distributions to derive a density function for stress ranges. In Shetty and Baker (1990a) a distribution for stress ranges is derived based on Rice's distribution for peaks of a broad-banded process. Zhao and Baker (1990) gives the following empirical 5-parameter mixed Weibull distribution which is seen to give very good comparison with the rainflow counting method.

$$
\begin{aligned}
f_Y(y) &= \gamma \frac{b}{a} (\frac{y}{a})^{b-1} \exp\left[-(\frac{y}{a})^b\right] \\
&+ (1-\gamma)\frac{d}{c} (\frac{y}{c})^{d-1} \exp\left[-(\frac{y}{c})^d\right] \\
a &= (8-7\alpha)^{-1/b} \\
b &= 1.1 \qquad\qquad\qquad for \quad \alpha \leq 0.9 \qquad\qquad (24) \\
&= 1.1 + 9(\alpha - 0.9) \qquad for \quad \alpha > 0.9 \\
c &= \sqrt{2} \\
d &= 2 \\
\gamma &= (1-\alpha)/\left[1-\sqrt{2/\pi}\ a\ \Gamma(\frac{1}{b}+1)\right]
\end{aligned}
$$

where $y = s/2\sigma$ is the normalised stress range and σ is the *root mean squared (rms)* value of the process.

5.2 COMPUTATION OF FATIGUE DAMAGE

The fatigue damage procedures using the S-N and fracture mechanics methods developed earlier can be extended to variable amplitude loading by summing the fatigue damage caused by each stress cycle. Since the stress cycles under variable amplitude loading are described using a probability density function, the cumulative loading sum can be replaced by its expected value. The Eq.4 for S-N approach and Eq.20 for the fracture mechanics approach can be generalised to variable amplitude loading as below.

S-N Approach:

$$D_n = \frac{1}{K} \sum_{i=1}^{n} S_i^m = \frac{1}{K} E[n(t)] E[S^m] \qquad (25)$$
$$= \frac{1}{K} \psi_L(t)$$

Fracture Mechanics Approach:

$$\int_{a_0}^{a(t)} \frac{dx}{C\,G(x)\,U_a^m\,Y_a^m\,[\pi x/q]^{m/2}} \qquad (26)$$
$$= E[n(t)]\,E[S^m] = \psi_L(t)$$

where $\psi_L(t)$ is sometimes referred to as the *fatigue loading function.*
    The mth expected value of stress range $E[S^m]$ can be calculated from the probability density function of stress ranges. Often a partial expectation considering stress ranges above a threshold level is also required. For the stress range density function given in Eq.25, the following expressions can be derived.

$$E[Y^m]_{y_{th}}^{\infty} = \gamma\,a^m\Gamma\!\left(\frac{m+b}{b};y_{th}\right) + (1-\gamma)c^m\Gamma\!\left(\frac{m+d}{d};y_{th}\right) \qquad (27)$$

where $\Gamma(.;.)$ is the incomplete Gamma function. When the expectation is over all stress cycles, this can be replaced by the complete Gamma function in the above expressions.

6. Quantification of Uncertainties

In the fatigue analysis procedure described in the previous sections, considerable uncertainties are involved which influence the fatigue life estimates. These uncertainties can be broadly grouped as (i) uncertainties in fatigue load estimation, (ii) uncertainties in stress calculations (iii) uncertainties in strength models, and (iv) uncertainties due to the natural randomness of fatigue material properties. In a probabilistic analysis these sources of uncertainty should be identified, quantified and accounted for. The first two categories of uncertainties depend on the type of the component considered and the type of loading to which it is subjected, and cannot be discussed independently of the type of structure being considered. In this chapter attention is focused on the modelling of uncertainties in fatigue material properties and strength models.

6.1 MINER'S SUM

Miner's hypothesis of linear damage accumulation, used in S-N fatigue analysis, states that the component will fail by fatigue when the accumulated damage D equals unity. However, a number of tests under variable amplitude loading indicate a scatter in the value of the accumulated damage at failure. This could be attributed as a model uncertainty in Miner's hypothesis. This uncertainty can be modelled by treating Miner's damage sum at failure (Δ) as a random variable. Wirsching (1984) suggest that Δ be modelled as a lognormal variable with a mean of 1.0 and coefficient of variation of 0.3.

6.2 FATIGUE CRACK PROPAGATION MODEL

The fatigue crack propagation model presented in Section 4 involves a number of parameters which are subject to considerable uncertainty. These uncertainties arise from the use of simplified empirical formulations for stress intensity factors, unknown distribution of residual stresses, modelling of weld geometry effect, crack coalescence and relaxation of residual stresses. It is difficult to quantify the uncertainty for each of these sources separately due to scarcity of data. It is convenient to express all the above sources of uncertainty through a single basic variable B_{sif} which gives the ratio of stress intensity geometry correction factor obtained by experiment, Y_{exp}, to that computed using the proposed model Y_{mod}. The statistics for this variable can be developed by comparison with experimental compliance function curves. The factor B_{sif} can be modelled as a lognormal variable with a mean of unity and coefficient of variation typically in the range of 0.15-0.25. Lower variability may be used if the stress intensity factors are computed using finite element models such as line-spring and weight function models.

6.3 S-N CURVE PARAMETERS

S-N curves are developed from fatigue tests on welded joints, and generally large scatter is observed in the number of cycles to failure. The uncertainty in the S-N curve can be expressed through the parameters m and K. Because of the mathematical form of the equation, the parameters m and K are expected to be highly correlated, and it is common to consider only one of the parameters as random and the other as fixed. Considering K as random, the statistic for this variable can be obtained from the test data on number of cycles to failure. The variable is often modelled as lognormal, see for example Wirsching (1984) with coefficient of variation in the range of 0.4-0.6.

6.4 CRACK GROWTH PARAMETERS

Crack propagation rate data are usually obtained from tests on simple specimens, and exhibit considerable scatter. The uncertainty modelling for crack growth depends on the crack propagation model used. A number of probabilistic crack propagation models attempt to describe the scatter in test results by treating crack propagation as a stochastic process, for example as a lognormal process or a Markov process (see next chapter by Madsen). However, when a simple Paris type crack propagation model is used, as presented in Section 4, a random variable description is usually adequate. In this approach, the crack growth parameters, C and m, in the Paris law can be used to describe the scatter in crack propagation data.

The crack growth parameters C and m usually exhibit a high negative correlation, and it is therefore common to use a fixed value of m, and express all the uncertainty through C. The variable C is well described by a lognormal model with coefficient of variation typically in the range of 0.3-0.4.

6.5 INITIAL DEFECT SIZE

The weld defects, which act as crack initiation points, are of several types and defect size, and occurrence rates are found to be highly random and are influenced by such factors as fabrication yard, welding procedure, welding position, type of joint etc, Baker et al (1988). Some of the reported studies on plated joints have been reviewed by Kirkemo (1988). Statistical analysis of a large amount of weld defect data obtained from the Conoco Hutton TLP structure has been carried out by Kountouris and Baker (1989). For use in a reliability analysis, the distribution of defect sizes should be based on defects existing in a structure entering into service, considered acceptable according to quality control standards, as well as those remaining undetected during fabrication. Thus defect occurrence rate, and the amount and quality of NDE used, should be taken into account in developing a distribution for weld defect size. From the limited data available, the mean value and standard deviation of weld defect depth a_0 can be estimated to be 0.15mm and 0.10mm for "sound" quality welds. Lognormal, exponential and Weibull distributions have been used by researchers to fit weld defect data. Similarly, initial defect aspect ratio, defined as the ratio of initial defect depth to defect semi-length (a_0/c_0) may also be modelled as a lognormal variable with a mean of 0.62 and coefficient of variation of 0.4, Kountouris and Baker (1989).

7. Probabilistic Fatigue Assessment

In a deterministic approach for design against fatigue, all the loading parameters are taken at their expected values, while the resistance parameters are taken at values corresponding typically to mean minus two standard deviations. Additional safety is ensured by keeping the computed fatigue lives by a factor of 2-10 times higher than the planned service life of the structure. However, in view of the considerably large sources of uncertainty, both in the loading and fatigue resistance parameters, a fully probabilistic approach to fatigue design is considered appropriate. In this section reliability analysis methodology for the fatigue limit-state is presented, using both an S-N approach and a fracture mechanics approach. Two types of reliability analysis methods, namely, (i) the Lognormal formulation and (ii) the FORM/SORM approach are described. The sensitivity of the probability of failure to uncertainties in various parameters is studied using numerical examples.

7.1 RELIABILITY ANALYSIS USING LOGNORMAL FORMULATION

This formulation has been extensively used in code calibration studies based on a reliability approach, see for example Wirsching (1984). The method is often used with the S-N approach, in which only Δ, K, B_{gtf} and B_{scf} are treated as random variables, and all of which are assumed to follow a lognormal distribution. The variables B_{gtf} and B_{scf} are introduced to account for the uncertainty in the global stress analysis of the structure and local stress analysis of the welded component, respectively.

Using the expression for fatigue damage, Eq.25, from Section 5, we can compute the damage due to n_t cycles of loading in a service life of t as

$$D_L = \frac{\omega t}{K} B_{gtf}^m B_{scf}^m E[S^m] \qquad (28)$$

where ω is the average frequency of stress cycles and the mth expected value of stress-range can be computed as in Eq.27, using the stress range density function given in Eq.24.

In addition, considering the value of Miner's damage sum at failure Δ as a random variable, the probability of failure due to fatigue can be expressed as

$$p_f = P[\Delta \leq D_L] \qquad (29)$$

Using the Lognormal format, we can write the reliability index β as

$$\beta = \frac{\ln\left(\tilde{\Delta}/\tilde{D}_L\right)}{\sigma_{lnM}} \qquad (30)$$

$$\sigma_{lnM}^2 = \ln[(1+V_\Delta^2)(1+V_K^2)(1+V_{B_{gtf}}^2)^{m^2}(1+V_{B_{scf}}^2)^{m^2}]$$

where V denotes the coefficient of variation, and the tilde denotes the median value of a variable. The median fatigue damage can be calculated from Eq.28 using median values for the random variables K, B_{gtf} and B_{scf}. The probability of failure can be obtained from the reliability index using $p_f = \Phi(-\beta)$, where $\Phi(.)$ is the standard normal distribution.

The main advantage of this approach is that an exact closed form expression of the reliability index can be obtained which makes reliability analysis simple and efficient. However, the main disadvantage is that all the variables have to be assumed to follow a lognormal distribution. Moreover, the method cannot be used if an explicit expression for damage in terms of all the basic variables is not available. The Lognormal formulation cannot be easily used with a fracture mechanics approach, as this method does not provide a closed form expression for fatigue damage.

The sensitivity of β with respect to the COV of each of the variables can be derived as

$$\frac{\partial \beta}{\partial V_\Delta} = \frac{-\beta}{\sigma_{lnM}^2} \frac{V_\Delta}{1+V_\Delta^2}$$

$$\frac{\partial \beta}{\partial V_K} = \frac{-\beta}{\sigma_{lnM}^2} \frac{V_K}{1+V_K^2}$$

$$\frac{\partial \beta}{\partial V_{B_{gtf}}} = \frac{-\beta}{\sigma_{lnM}^2} \frac{m^2 V_{B_{gtf}}}{1+V_{B_{gtf}}^2}$$

$$\frac{\partial \beta}{\partial V_{B_{scf}}} = \frac{-\beta}{\sigma_{lnM}^2} \frac{m^2 V_{B_{scf}}}{1+V_{B_{scf}}^2}$$

(31)

7.2 RELIABILITY ANALYSIS USING FORM/SORM

The first step in a reliability analysis using FORM/SORM is to identify a set of basic random variables which influence the failure mode or the limit-state under consideration. Let $X = (X_1, X_2,....., X_n)$ represent a vector of n basic variables.

Next, a *limit-state function* or a *safety margin equation* is formulated in terms of the n basic variables, Thoft-Christensen and Baker (1982),

$$M = g(\bar{X}) = g(X_1, X_2,..., X_n) \tag{32}$$

in such a way that $g(\cdot)$ satisfies the following:

$$\begin{aligned} g(\bar{x}) &> 0 \qquad when \ \bar{x} \in \omega_s \\ g(\bar{x}) &\leq 0 \qquad when \ \bar{x} \in \omega_f \end{aligned} \tag{33}$$

$$g(\bar{x}) = g(x_1, x_2,..., x_n) = 0 \tag{34}$$

Eq.34 defines an *(n-1)* dimensional surface in the space of n basic variables. This surface is called a *failure surface* or a *limit-state surface* and divides the basic variable space into a *safe region* ω_s and an *unsafe* region ω_f. Note that the limit-state function $g(\cdot)$ is entirely deterministic, and any of the existing deterministic models for strength prediction could be used. The random variable M is called a *safety margin*.

The reliability or the probability that the limit-state will not be reached is then expressed as

$$R = 1-p_f = 1-P[M \leq 0] = 1 - \int f_{\bar{X}}(\bar{x})d\bar{x} \tag{35}$$

where $f_{\bar{X}}(\cdot)$ is the joint probability density function of the n basic variables and p_f

denotes probability of failure. The n-dimensional integral is defined over the failure region ω_f.

In practical applications, the reliability cannot be evaluated in the exact manner as given by Eq.35. This is so, first, because enough statistical data is usually not available to develop the n-dimensional joint density function of the basic variables. Secondly, even when the joint density function is available, analytical or numerical integration is possible only for a few simple cases. The FORM/SORM methods provide a way of evaluating the reliability efficiently with reasonably good accuracy which is adequate for practical applications.

In a FORM/SORM reliability method, see Madsen et al (1986), the set of basic variables $\bar{X}$ is first transformed to a new set, $\bar{U}=(U_1,U_2,...,U_n)$ using a one-to-one transformation

$$T: \quad \bar{X} = (X_1,X_2,...,X_n) \rightarrow \bar{U} = (U_1,U_2,...,U_n) \tag{36}$$

such that the new set of variables are independent, standardized and normally distributed. A number of transformation methods are available depending on the initial distribution of basic variables. The equation for the limit-state surface in u-space becomes

$$g(\bar{x}) = g\left(T^{-1}(\bar{u})\right) = g_u(\bar{u}) = 0 \tag{37}$$

In the transformed space, a point on the failure surface closest from the origin, $\vec{u}^*$ is determined by a minimization solution with one constraint such that

$$\bar{u}*: \quad min \, |\,\bar{u}\,| \,; \quad with \quad g_u(\bar{u}) = 0 \tag{38}$$

The point $\vec{u}^*$ is called a *design point* or a *most likely failure point* as it represents a point of highest probability density in u-space on the failure surface. At the design point the unit normal vector or the direction cosines to the failure surface with respect to each of the variables are

$$\bar{\alpha} = (\alpha_1,\alpha_2,...,\alpha_n) = \frac{\nabla g_u(u^*)}{|\,\nabla g_u(u^*)\,|} \tag{39}$$

These direction cosines are often referred to as *sensitivity factors* in reliability literature, see Thoft-Christensen and Baker (1982).

In a First Order Reliability Method (FORM) the failure surface at the design point is approximated by a tangent hyperplane defined by

$$\sum_{i=1}^{n} \frac{\partial g_u}{\partial u_i}\Big|_{\bar{u}=\bar{u}*}(u_i - u_i^*) = 0 \tag{40}$$

and the shortest distance to this hyperplane from the origin is determined by

$$\beta = -\bar{\alpha}^T \bar{u}* = \left[\sum_{i=1}^{n} u_i*^2\right]^{\frac{1}{2}} \qquad (41)$$

Then the first order approximation to the failure probability is computed as

$$p_f = \Phi(-\beta) \qquad (42)$$

and β is termed as a *first order reliability index*. In principle any optimization algorithm can be used to search for the design point in u-space, and a number of computer programs are now available for this purpose. In the RASOS (Reliability Analysis System for Offshore Structures) software, the NLPQL algorithm given by Schittkowski (1986) is used.

A number of studies examining the accuracy of the first order reliability method have been reported in literature, see for example Dolinski (1983). In most cases of practical applications, FORM is seen to give a reasonably accurate estimate of the probability of failure. The accuracy of the FORM method derives from the rotational symmetry property of the standard normal space, and from the fact that the probability density function of the variables decreases very quickly, namely as $\exp(-r^2/2)$, with the distance r from the origin. The area of integration giving maximum contribution to the failure probability is therefore located in the region of the design point, and the failure surface is well approximated by a tangent hyperplane around this point. However, when the principal curvatures of the failure surface at the design point are large (i.e. highly curved failure surface) a tangent hyperplane approximation is not satisfactory. For convex failure surfaces (with respect to the origin) FORM overpredicts the probability of failure while for concave failure surfaces it underpredicts p_f. Unfortunately FORM procedure does not provide a means for estimating the level of error involved.

The accuracy of the result can be improved by approximating the failure surface by a quadratic surface, with the same principal curvatures as the true failure surface at the design point. This method is called a Second Order Reliability Method (SORM). Unfortunately, results for the calculation of the probability content of the failure region bounded by a general quadratic surface are not available. Good results are available for hyperparabolic failure boundaries, Madsen et al (1986). For a general failure surface Breitung (1984) gives the following asymptotic result for the probability of failure

$$p_f \approx \Phi(-\beta) \prod_{i=1}^{n-1} (1-\beta \kappa_i)^{-1/2} \qquad \beta \to \infty \qquad (43)$$

where β is the first order reliability index and κ_i are the principal curvatures of the failure surface at the design point. The result is asymptotic in the sense $\beta \to \infty$ with $\beta\kappa_i$ fixed. This second order correction to the first order result gives a very good estimate of the probability of failure for large values of β.

The computational effort required for FORM/SORM methods is very small compared to numerical integration or simulation methods.

7.2.1   FORM/SORM Reliability Analysis Using S-N Approach:

Using a FORM/SORM approach and the S-N method for fatigue damage, we can write an expression for safety margin in a more general form as

$$M = \Delta - D_L$$
$$= \Delta - \frac{\omega t}{K}\left(\frac{T_p}{T_B}\right)^{m\xi} B_{scf}^{m}\left[\gamma a^{m}\Gamma\left(1+\frac{m}{b}\right)+(1-\gamma)c^{m}\Gamma\left(1+\frac{m}{d}\right)\right] \tag{44}$$

where ξ is the thickness correction exponent on stress, T_p is the thickness of the plate and T_B is the reference plate thickness. In this approach Δ, K, ξ, B_{scf} and the five Weibull model parameters γ, a, b, c, and d, describing the fatigue loading, can be modelled as random variables using appropriate probability distributions.

Example 1: Fatigue Reliability of a Tubular Joint Using S-N Approach

This reliability analysis approach is demonstrated with an example. The tubular joint considered here is part of a 6-legged jacket structure in 60m water depth. A bi-linear S-N curve is used and the loading uncertainties are modelled through the 5 parameters of the mixed Weibull model. The correlations used for the Weibull parameters are: $\rho_{ab}=\rho_{cd}=-0.6$, $\rho_{ac}=\rho_{bd}=0.4$, and $\rho_{ad}=\rho_{cb}=-0.3$, while γ was modelled as independent of others. In addition, the thickness correction exponent ξ and the stress level S_0 at which a change in slope of the S-N curve occurs are modelled as random. The slope m_2 is taken as 5, and K_2 is calculated as $K_2=K_1 S_0^{m2-m1}$. The results are given in Table 1.

A deterministic assessment of the joint, including thickness correction, gives a mean fatigue life of 270 years, and a nominal life of 80 years based on the D.En. 'T' curve. The reliability analysis, relating again to a service life of 20 years, incorporates a more realistic representation of uncertainties in fatigue assessments, and the reliability index of 1.847 is typical of joints with estimated lives of 80 years based on a design S-N curve. A study of the sensitivity factors show that the uncertainty in fatigue loading and calculation of stress concentration factors are the dominant variables. The high sensitivity of reliability index to uncertainties in Weibull parameters a and b in comparison with c and d, derives from the fact that the weighting factor is 0.63, and therefore the first part of the mixed-Weibull density function contributes most to the stress-range density function. However, when the stress spectra are bi-modal, as in most deep-water jacket platforms, the contribution of the second part of the density function will be significant, and in such cases the uncertainty in parameters c and d will also be important. The reliability index is seen to be less sensitive to the uncertainty in total number of stress cycles, thickness correction and stress level S_0 in comparison with other resistance variables.

Table 1:   Results of reliability analysis for the limit-state of fatigue using an S-N
             approach and FORM/SORM formulation.

Reliability index β = 1.847
Probability of failure p_f = 3.24E-02

Variable	Distribution	Mean	COV	α
Δ	lognormal	1.00	0.30	0.228
K	lognormal	1.012E+13	0.58	0.366
ξ	normal	0.30	0.15	-0.147
S_0	normal	90.0	0.30	-0.001
B_{scf}	lognormal	1.0	0.25	-0.502
A	lognormal	1.0^*	0.20	-0.644
B	lognormal	1.0^*	0.15	0.354
C	lognormal	1.0^*	0.20	-0.042
D	lognormal	1.0^*	0.15	0.009
W	lognormal	0.63	0.15	-0.074
B_{cyc}	lognormal	1.0	0.10	-0.068

* bias   (Units in N, mm)

7.2.2   FORM/SORM Reliability Analysis Using a Fracture Mechanics Approach

The fracture mechanics approach gives a safety margin equation for the limit-state of
fatigue, see also Shetty and Baker (1990c)

$$M = a_c - a(t) \tag{45}$$

where a_c is the limiting crack depth (for example plate thickness) and $a(t)$ is the crack
depth after a service exposure of time t. Starting from an initial crack depth of a_0, we
can calculate the crack depth $a(t)$ after time t using the fracture mechanics crack
propagation model presented in Section 4. In this case, $a(t)$ is a function of the random
variables such as initial defect size, fatigue material properties, uncertainties in service
loading, etc. Alternatively, in terms of a fatigue resistance function and a fatigue
loading function, the safety margin for fatigue failure can also be expressed as

$$M = \psi_R(a_c) - \psi_L(t)$$

$$= \int_{a_0}^{a_c} \frac{dx}{C\,G_a\,B_{sif}^{\,m}\,Y_a^{\,m}\,U_a^{\,m}\,[\pi\,x/q]^{m/2}} - \omega t\,B_{gtf}^{\,m}\,B_{scf}^{\,m}\,E[S^{\,m}] \tag{46}$$

where G_a, Y_a and U_a are all functions of the instantaneous crack depth as defined earlier.

In the above formulation for safety margin, the parameters a_0, C, B_{sif}, B_{scf} and B_{gtf} can be modelled as random variables to account for the uncertainties in initial defect size, fatigue material properties, stress intensity factor model, stress concentration factors and the response analysis under fatigue loading, respectively. In addition, the uncertainty in threshold stress intensity factor range ΔK_{th} can be introduced through G_a and the uncertainty in residual stresses through U_a. The safety margin is thus implicitly defined in terms of the basic variables and a simple lognormal format cannot be used for reliability analysis of this limit-state. However, the FORM iterative method can be used efficiently, as demonstrated in the following example.

Example 2:   *Fatigue Reliability Analysis Using a Fracture Mechanics Approach.*

The tubular joint considered in Example 1 is re-analyzed here using a fracture mechanics approach. Maintaining the same service load history and using the crack propagation model developed in Section 4 gives a mean fatigue life of 287 years, which is very close to that obtained using the S-N approach. Properties have been slightly adjusted for this joint to obtain nearly equal mean lives from the two approaches. A reliability analysis is carried out by considering the plate thickness as the limiting crack depth, and the results are given in Table 2.

All the variables in this example are considered as statistically independent of each other, except for the Weibull parameters a and b which are assigned a correlation of $\rho_{ab}=-0.6$. The other parameters c, d and γ are not treated as random variables as they are not important for the joint considered, see Table 2.

The reliability index for the joint is obtained as 1.747, which is significantly lower than that obtained using the S-N approach, although the mean lives obtained by the two approaches are nearly equal. This can be attributed to the additional uncertainty involved in the calculation of degree-of-bending factors and stress intensity factors required in a fracture mechanics approach, and the fact that the reliability index is highly sensitive to these variables. If the variability in B_{dob} and B_{sif} is ignored (or reduced to zero!) the reliability index can be increased to 2.536. This value is slightly higher than that obtained using the S-N approach, as the variability in material parameter C is smaller than that of the S-N parameter K, other variabilities being the same.

From the direction cosine sensitivity factors given in Table 2, it can be seen that the reliability index is highly sensitive to the model uncertainty in the response calculations, and to the SCFs, SIFs and DoBs, and less sensitive to the variability in material parameter C. The uncertainties in initial defect size, threshold factor and residual stresses do not have a significant influence on the probability of joint failure by fatigue.

Table 2:   Results of the reliability analysis for the limit-state of fatigue using a
            fracture mechanics approach.

Reliability index   = 1.747
Probability of failure   = 4.03E-02

Variable	Distribution	Mean	COV	α
Weib-A	lognormal	1.0	0.20	-0.640
Weib-B	lognormal	1.0	0.15	+0.364
B_{scf}	lognormal	1.00	0.25	-0.474
B_{sif}	lognormal	1.00	0.20	-0.379
B_{dob}	lognormal	1.00	0.15	0.298
Paris-C	lognormal	1.0E-11	0.40	-0.247
a_0	lognormal	0.15	0.66	-0.088
$(a/c)_0$	lognormal	0.62	0.66	-0.006
ΔK_{th}	normal	7.60	0.20	-0.001
S_{res}	lognormal	300	0.20	-0.000

Table 3: Improvement in reliability with reduced variability.

Reliability index = 1.954   (2.128)   [2.318]

Variable	COV			α^2 X 100		
B_{gtf}	0.30	(0.30)	[0.25]	55	(64)	[55]
B_{scf}	0.15	(0.15)	[0.15]	14	(17)	[20]
B_{sif}	0.20	(0.15)	[0.15]	18	(12)	[15]
B_{dob}	0.10	(0.00)	[0.00]	6	(0)	[0]
C	0.40	(0.40)	[0.40]	7	(7)	[10]

The statistics used in Table 2 correspond to a simplified design situation, where parametric equations are used for the calculation of stress concentration factors, degree of bending factors and stress intensity factors. Next, we study the improvement in reliability that can be obtained by successively refining the analysis techniques and thereby reducing the variability in model uncertainty parameters. First, the variability in B_{scf} and B_{dob} is reduced to correspond to a situation where the stress analysis of the intact joint is carried out using finite element methods, while the stress intensity factors are calculated using parametric equations. This increases the reliability index to 1.954 and the order of importance of variables changes, as shown in Table 3. In the next step, the variability in B_{sif} is also reduced to correspond to a situation where the stress intensity factors are calculated using finite element methods, such as line springs or 3-D f.e.m, in which case degree of bending factors are not used. This gives a further increase in β to 2.128, and the sensitivity factors for this case are shown in braces in Table 3. Finally, if a more accurate environmental description is used and a rigorous stochastic response analysis is carried out, the variability in B_{gtf} is slightly reduced. This, however, gives a significant improvement in reliability ($\beta=2.318$) as B_{gtf} happens to be the most important variable. The results for this case are enclosed in square brackets.

In conclusion, it can be said that when a fracture mechanics analysis is used for fatigue assessment, it is important to have a reasonably good estimate of the stress distribution of the intact joint. This can be obtained by a marginal increase in computational effort, but it will give a significant increase in the computed reliability.

The uncertainties in the calculation of response transfer functions to determine nominal member stresses, are of paramount importance in fatigue assessment, and effort should be made to reduce these uncertainties by improving the quality of analysis.

In addition to the value of the reliability index for a joint, which can be used to assess the adequacy of the design, what is of more interest is to generate information about the variation of reliability through the service life of the structure. This can be used for judiciously planning the inspection times for various joints.

The variation of reliability index with service exposure for the joint is shown in Fig.3. The reliability index is as high as 3.50 at the end of 2 years of service exposure, but drops to a value of 1.747 at the end of 20 years, considered to be the design life of the structure. On this plot, possible target level for inspection is indicated, which more appropriately should be recommended by regulatory authorities. From this plot, the time for first inspection can be obtained as 8 years. After the inspection, and depending on whether a crack is detected or not the reliability of the joint can be updated using Bayesian updating methods which are discussed in the chapter by Madsen.

8. Summary and Conclusions

Fatigue is a dominant mode of failure for many metallic structures. Fatigue in welded joints is characterised by crack growth. In recent years, the emphasis of research has been on the application of fracture mechanics methods for the fatigue assessment of welded joints.

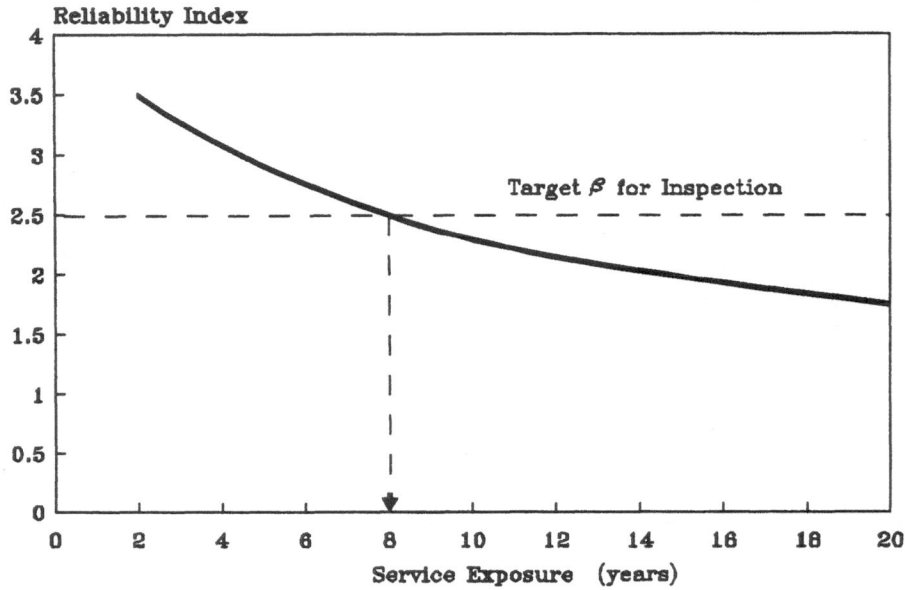

Figure 3: Variation of reliability index with service exposure

In this chapter, S-N and fracture mechanics methods of fatigue assessment of welded joints are discussed. A simplified model for predicting the propagation of semi-elliptical cracks in plated joints has been presented. The model predicts growth of the crack in both the thickness and the longitudinal directions of the joint, and explicitly accounts for factors such as weld geometry, fatigue threshold, residual stresses, etc.

A methodology for probabilistic assessment of welded joints has been presented which accounts for the uncertainties involved in the modelling of the fatigue loading, stress analysis, stress intensity factors, fatigue material properties, initial defect size and residual stresses.

Examples have been presented of reliability analysis of tubular joints of offshore structures which show that the probability of joint failure by fatigue is highly sensitive to the uncertainty in crack growth parameter C, initial defect depth a_0, model uncertainty in SCF calculation B_{scf}, model uncertainty in DoB calculation B_{dob}, model uncertainty in SIF calculation B_{sif} and the uncertainty in the calculation of stress response B_{gtf}.

References

Baker, M.J. (1985): The reliability concept as an aid to decision making in offshore structures, *Proc. Behaviour of Offshore Structures Conference*, BOSS-85, Amsterdam, The Netherlands.

Baker, M.J., Kountoris, I.S. and Ohmart, R.D. (1988): Weld defects in an offshore structure -A detailed study, *Proc. Behaviour of Offshore Structures Conference*, BOSS-88, Trondheim, Norway.

Breitung, K. (1984). Asymptotic approximate for multinormal inregrals, ASCE, J. of Eng. Mech., Vol. 110, No.EM3.

British Standards Institution, (1991): Published document for guidance on some methods for the derivation 6 acceptance levels for flaws in fusion welded joints, (Revision of PD6493), BSI, U.K.

Dijkstra, O.D., Snijder, H.H. and Van Rongen, H.J.M. (1990): Assessment of the remaining fatigue life of defective welded joints, *Proc. Conf. Int. Association of Bridge and Structural Engineers*, IABSE-90.

Dolinski, K. (1983): First-order second-moment approximation in reliability of structural systems: critical review and alternative approach, *Structural Safety*, Vol.1.

Dowling, N.E. (1972): Fatigue prediction for comlicated stress-strain histories, *J. of Materials, JMASA*, 7.

Forman, R.G. et al (1964): Numerical analysis of crack propagation in cyclic loaded structures, *ASME*, Paper No. 66-WA/Met 4.

Kirkemo, F. (1988): Application of probabilistic fracture mechanics to offshore structures, *Applied Mechanics Reviews*, Vol. 41, No. 2. Also in *Fracture Mechanics in Offshore Industry*, ASME Book No. AMR032.

Klesnil, M. and Lucas, P. (1972): Influence of strength and stress history on growth and stabilisation of fatigue cracks, *Engineering Fracture Mechanics*, Vol. 4.

Kountoris, I.S., and Baker, M.J. (1989): Defect Assessment - Analysis of defects detected by MPI in an offshore structure, *CESLIC Report No. OR6, Dept. of Civil Engineering*, Imperial College, London, U.K.

Madsen, H.O., Krenk, S. and Lind, N.C. (1986): *Methods of Structural Safety*, Prentice-Hall Inc., Englewood Cliffs, N.J.

Madsen, H.O. (1987): Model updating in reliability theory, *Proc. ICASP-5*, Vancover, Canada.

Mattheck, C., Morawietz, P. and Munz, D. (1983): Stress intensity factor at the surface and at the deepest point of semi-elliptical surface crack in plates under stress gradients, *Int. J. of Fracture*, Vol.23.

Newman, J.C. and Raju, I.S. (1981): An empirical stress intensity factor equation for the surface crack, *Engineering Fracture Mechanics*, Vol.15.

Paris, P.C. andd Erdogan, F. (1963): A Critical analysis of crack propagation laws, *J. of Basic Engineering*, ASME, Vol. 85.

Rooke, D.P. and Cartwright, D.J. (1976): *Compedium of Stress Intensity Factors*, HMSO, London, U.K.

Schittkowski, K. (1986): NLPQL: A fortran subroutine solving constrained nonlinear programming problems, *Annals of Operations Research*, Vol.5.

Shetty, N.K. and Baker, M.J. (1990a): Fatigue reliability of tubular joints in offshore structures: Fatigue loading, *Proc. 9th Offshore Mechanics and Arctic Engineering Conf.*, Houston, Texas.

Shetty, N.K. and Baker, M.J. (1990b): Fatigue reliability of tubular joints in offshore structures: Crack propagation model, *Proc. 9th Offshore Mechanics and Arctic Engineering Conf.*, Texas.

Shetty, N.K. and Baker, M.J. (1990c): Fatigue reliability of tubular joints in offshore structures: Reliability analysis, *Proc. 9th Offshore Mechanics and Arctic Engineering Conf.*, Houston, Texas.

Shetty, N.K. (1992): System Reliability of Jacket Type Offshore Structures under Fatigue Deterioration", Ph.D. Thesis, University of London.

Tada, H. et al (1973): *The Stress Analysis of Cracks Handbook*, Dell Research Corporation, Hellertown, Pennysylvania.

Thoft-Christensen, P. and Baker, M.J. (1982): *Structural Reliability Theory and its Applications*, Springer-Verlag, Berlin.

Wirsching, P.H. (1984): Fatigue reliability for offshore structures, *ASCE, J. of Structural Engineering*, Vol. 110, No. 10.

Zhao, W. and Baker, M.J. (1990): A new stress-range distribution model for fatigue analysis under wave loading, in *Environmental Forces on Offshore Structures and their Prediction*, Kluwer.

PROBABILISTIC MODELLING OF THE STRENGTH OF FLAT COMPRESSION MEMBERS

C. GUEDES SOARES
Instituto Superior Técnico
Universidade Técnica de Lisboa
1096 Lisboa, Portugal

1. Introduction

The structural components that are subjected to compressive forces are often critical elements in many structures. This is even more important if one has in mind that compressive failure is followed by a decrease in load carrying capacity which will overload adjacent structural components, being bound to precipitate an overall failure of the structure.

Another important feature of compression member behaviour is their sensitivity to the shape and amplitude of initial imperfections, which in real structures are of a random nature. Thus only probabilistic treatments are appropriate to describe the initial status of the imperfect structural components as well as their collapse strength.

This work deals with flat compression members, that is, with columns and plates, leaving aside the strength of shells. A description is provided of the methods for predicting the strength of these components, describing afterwards the probabilistic approaches that have been used up to now. These range from stochastic descriptions of the spatial variability of initial distortions to a simple statistical analysis of experiments relating to component strength.

2. Probabilistic Modelling of Column Strength

Columns are very imperfection sensitive structural elements, a property that probably explains the early interest in representing their strength probabilistically. Various approaches have been used, ranging from stochastic descriptions of the initial imperfections, to a statistical analysis of experimental strength data, irrespective of initial imperfections. This work provides a review of the main types of approaches used, first introducing the basic concepts about column strength.

C. Guedes Soares (ed.), Probabilistic Methods for Structural Design, 113–140.

2.1 PREDICTION OF COLUMN STRENGTH

In the classical theory, the elastic buckling of an initially straight elastic slender column is an eigenvalue problem. For compressive loads smaller than the Euler critical load, the column remains straight, but when the load reaches the critical value, there is a bifurcation i.e. the column can either remain straight or it will achieve an indeterminate sideway bowing. The Euler critical stress is σ_E given by:

$$\sigma_E = \pi^2 E / (L/r)^2$$

where E is the modulus of elasticity, L is the column length, r is the radius of gyration of the cross-section, and $\lambda=(L/r)$ is the slenderness ratio. This approach is valid while σ_E remains below the elastic limit, which often occurs with slender columns.

For stocky columns, the elastic limit is exceeded before inception of buckling and the structural modulus becomes a function of the stress level i.e., the deformation becomes controlled by the tangent modulus of elasticity. Engesser proposed determining the critical stress by substituting the tangent modulus for the modulus of elasticity in Euler's expression. However, Cosidere realized that since buckling would take place at constant load some material of the column would unload when deflection began. Thus he proposed using a reduced or double modulus of elasticity which lies between the tangent and the Young´s modulus of elasticity.

Although the reduced modulus theory was for a long time considered to be the exact theory for the buckling of perfect columns in the inelastic range, the tangent modulus has been widely used by engineers because of its simplicity, since it is independent of the cross-section shape, and because of its safety, since it predicts a lower value of strength. Furthermore, experimental results seem to lie closer to the tangent modulus results. The reduced modulus load is an upper bound which in practice cannot be attained, while the tangent modulus load is a lower bound, thus being the basis for several design codes. Its principal shortcoming lies in the assumption that the member is initially perfectly straight, although it can account for residual stresses, and for non-linear stress-strain relationships. Thus, the tangent modulus load is a fair representation of the strength of a column, as long as the imperfections are small.

Considering the load-deflection behaviour of the column, one is led to the maximum strength theory. The maximum strength of a column will depend on the stress-strain relation and on the yield stress of the material, on the method of fabrication, on the size of the column, on its cross section and bending axis as well as on the initial out-of-straightness. Only the maximum strength theory can account for the last aspect. In addition to the out-of-straightness, the residual stresses produced by cooling after rolling, by cold-straightening or by welding, constitute the single most important factor for column strength.

The maximum strength theory is based upon the equilibrium equations:

$$P = P_{\text{int}}$$

$$P.\delta = M_{int}$$

where the subscript *int* stands for internal force P and moment M, while δ is the out of straigthtness. The total stress in any element i of the column at any stage of the loading may be expressed non-dimensionally as:

$$\frac{\sigma_i}{\sigma_o} = \frac{\varepsilon_i}{\varepsilon_o} = \frac{\varepsilon_{ri}}{\varepsilon_o} + \frac{\varepsilon_p}{\varepsilon_o} + \theta \frac{\xi_i}{\varepsilon_o} \quad \text{for} \quad |\sigma_i| < \sigma_o$$

where ε_{ri} is the residual strain of element i, ε_p is the axial strain due to the applied load P, θ is the curvature of the column due to the deflection δ caused by the force P, and ξ_i is the distance from the centroid of element i to the bending axis considered.

If the initial and subsequent deflected shapes of the column may be described by a half sine wave, the curvature θ at the column midheight is given by:

$$\theta = \pi^2 \delta / L^2$$

where δ is the lateral deflection of the column. In order to avoid this simplified approach, one may iteratively determine the actual deflected shape of the column at each load increment. This is achieved by incrementing the deflection δ, and finding the corresponding equilibrium load P for every value of the deflection.

The Euler elastic load, the tangent modulus load and the maximum strength load are the three main theories (fig. 1) that have been used in connection with probabilistic modelling of column strength, as will be described in next section.

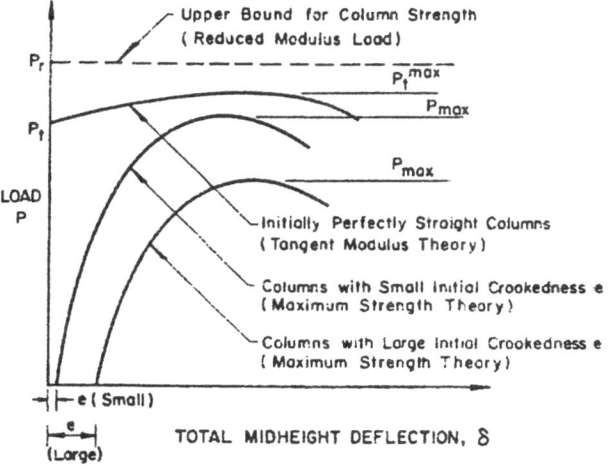

Figure 1 - Predictions of three theories of column strength

2.2 APPROACHES BASED ON STOCHASTIC FORMULATIONS

One of the initial probabilistic treatments of column buckling was based on describing the initial imperfections as random processes. Then, making use of methods of stochastic differential equations, one obtains the relationship between the critical load and the initial imperfection. Imposing the assumption that the random process is ergodic leads to a deterministic critical load which, in the asymptotic case, depends only on the spectral density of the imperfection shape. This type of study dealt only with elastic buckling.

Boyce (1961) derived the relation between the axial load and the mean and variance of the transverse displacement of a simply supported Bernoulli-Euler column, when one component of curvature is initially a stationary random function of the spatial variable. The linear differential equation of equilibrium is:

$$Y''(x) + k^2 Y(x) = -k^2 W(x) \qquad 0 < x < 1$$

and adopting the boundary conditions $Y(0) = Y(1) = 0$, where $k^2 = PL / EI$ and $W(x)$ is the initial displacement of the centerline, assumed to be a random process, it has the solution in terms of Green's functions:

$$G(x,\xi) = \frac{\sin kx . \sin(1-\xi)}{k . \sin k} \qquad 0 < x < \xi$$

$$G(x,\xi) = \frac{\sin k(1-x) \sin k\xi}{k . \sin k} \qquad \xi < x < 1$$

The mean value of displacement becomes:

$$E[Y(x)] = k^2 \int_0^1 G(x,\xi) \, E[W(\xi)] \, d\xi$$

in terms of the mean value of the initial imperfection displacement.

If $W(x)$ is a stationary function, the covariance is independent of x and the autocorrelation ρ depends only of the difference between ξ and η. Thus, the variance of the deflection is:

$$D[Y(x)] = k^4 \sigma^2 \int_0^1 \int_0^1 G(x,\xi) \, G(x,\eta) \, \rho(\xi - \eta) \, d\xi \, d\eta$$

With some assumptions concerning the form of $\rho(\xi-\eta)$, the integral can be solved, giving the variance as a function of the critical load. The result that the critical load is deterministic is somewhat surprising, since one would expect a random imperfection to yield a random critical load. However, this is a result of the ergodicity assumption, which makes the spatial averages of initial imperfections equivalent to their statistical moments.

Bernard and Bogdanoff (1971) generalized Boyce's approach, dealing with a simply supported column that is both randomly bent and twisted. It is described by a set of three non-linear differential equations with nonhomogeneous terms that are random. Instead of following Boyce, who assumed an explicit form for the covariance of the initial displacement, Bernard and Bogdanoff gave the functions describing the initial components of curvature and twist as sums of ordinary functions with a set of random constants.

If the customary linear analysis is applied to the equilibrium equations governing the column behaviour, the reduction of the equations reduces to a trivial form in which the equations governing the transverse displacement components are independent of the twist. In a modified linear analysis, the equations governing the displacements remain coupled by the rate of change of the angle of twist. Describing the initial curvature components and the twist by random functions with known statistical properties, Bernard and Bogdanoff developed a method whereby the mean values, variances and other moments of the displacement components may be found. Specific expressions are worked out for the means, variances, and covariances when the initial curvature components and the twist are taken to be Fourier-series functions in which the coefficients are random variables.

Bernard and Bogdanoff (1971) concluded that the variances of the transverse displacement at the midspan increases as the load increases, becoming very large near the Euler load for bars that are twisted very little, and increasing more and more rapidly as the amount of twist increases. This indicates that the load at which displacements of the type associated with the Euler load may occur, decreases as the amount of twist increases. As additional wave components are added to the twist and bending functions, the variance of the midspan deflection increases, the increase being larger in the case of addition to the twist. The higher the frequencies that are added, the less the increase in the variance.

Fraser and Budiansky (1969) used Boyce's idea of representing the initial deflection by a Gaussian stationary random function of known autocorrelation, to study the buckling of an infinite column resting on non linear elastic foundations. They used an exponential cosine autocorrelation, and concluded that the buckling load depends only on the autocorrelation of the initial deflection functions.

Still in the same line, Jacquot (1972) extended the theory to examine the mean square deflection of a column subjected to an initial deflection which is a non stationary process.

2.3 APPROACHES BASED ON FULL PROBABILISTIC DESCRIPTION OF THE VARIABLES

Another line of early work on elastic buckling of columns, by Bolotin (1965), Thompson (1967) and Roorda (1969), is based on a deterministic transfer function between initial imperfections and critical load. Given a probabilistic description of initial imperfections, they obtain a probabilistic description of critical load. However, the initial imperfections must be of a specified shape with random amplitude or they are represented by the sum of a series of shapes with random amplitudes. The same type of approach was also adopted by Brown and Evans (1972) in the study of the elastic instability of beam-columns.

Augusti and Baratta (1971) followed the same general approach, but they developed a probabilistic theory for the strength of imperfect columns made of elastic-perfectly plastic material, initially bent in a shape of a sinusoidal half-wave with maximum amplitude δ. The mean compressive strength at collapse (critical stress) σ_c is given by:

$$\gamma = 3\left(\frac{\sigma_o}{\sigma_c} - 1\right)\left[1 - \left(\frac{\sigma_c}{\sigma_E}\right)^{1/3}\right]$$

where $\gamma = \delta/r$ is the imperfection parameter. The critical stress is maximum for an ideally perfect structure, and decreases with increasing imperfection, at first rapidly and later more and more slowly. This critical stress is limited by the yield stress in the plastic region and by the Euler stress in the elastic region.

They considered γ, σ_o and the slenderness λ as normally distributed random variables, calculating numerically $F(\sigma_c)$, the cumulative probability distribution function of σ_c, for realistic values of the three governing random variables. The calculated distribution functions were very close to Gaussian distributions, although showing some positive skewness.

Augusti and Baratta (1975) extended their analysis by calculating the probability of failure of a column accounting for the random variability of the applied σ_a. Defining the probability density functions of the applied compressive load by $g(\sigma_a)$, the probability of failure becomes:

$$P_f = \int_0^\infty g(\sigma_a) \, F(\sigma_a) \, d\sigma_a$$

They performed several calculations, assuming the distribution of the load to be Gaussian. Their results indicated that P_f was very sensitive to the relative value of the dispersion of the various random variables, but insensitive to the type of the probability distribution of the applied load. For probabilities of failure on the order of 10^{-5} and smaller, a simplified integration was shown to yield acceptable results. It was based on substituting the exact distribution $F(\sigma_c)$ by an exponential approximation that coincides with $F(\sigma_c)$ and its derivative in the neighbourhood of the average stress $\bar{\sigma}_a$. They concluded that due to the sensitivity of P_f on the design load, the approximate integrations are preferable to the full numerical calculations.

Another aplication of classical methods, is due to Chung and Lee (1971), who have used the tangent modulus approach to determine the weak axis compressive strength of H shaped columns with residual stresses. The critical stress was given by:

$$\sigma_c = \sigma_E = \pi^2 E / (L/r)^2 \qquad\qquad \sigma_a < \sigma_o - \sigma_r$$
$$\sigma_c = [\pi / (L/r)] E(\alpha\sigma_o)^{-3/2}(\sigma_o - \sigma_r) \quad \sigma_o - \sigma_r < \sigma_a < \sigma_o$$
$$\sigma_c = \sigma_o \qquad\qquad\qquad\qquad\qquad \sigma_a = \sigma_o$$

where α is the ratio of maximum residual stress to yield stress. Expressing the critical stress as a function of the random variables E and σ_o gives the probability of failure as:

$$P_f = \int_R f_E(e) f_\sigma(\sigma_o) \, dE \, d\sigma_o$$

where the fs are the normal density functions of E and σ_o, and R is the domain defined in the previous equation. This expression is valid for the elastic and the tangent modulus of elasticity, i.e., it accounts for both elastic and elasto-plastic collapse.

The probability of failure is the probability that the critical stress is less then a specified value. Thus the density function of the critical stress is the derivative of P_f:

$$f_{\sigma_c}(\sigma_c) = \frac{d}{d\sigma_c} P_f$$

The mean and variance of σ_c are determined from:

$$\bar{\sigma}_c = \int_{-\infty}^{\infty} \sigma_c \, f(\sigma_c) \, d\sigma_c$$

$$\sigma_{\sigma_c}^2 = \int_{-\infty}^{\infty} (\sigma - \bar{\sigma}_c)^2 \, f(\sigma_c) \, d\sigma_c$$

Bjorhovde (1972) developed a different approach, based on the improved formulation of the maximum column strength, according to which P_{max} is given by:

$$P_{max} = f(\sigma_o, \sigma_r, E, b, t, d, W, \delta_o, L)$$

where b, t, d and W, are geometric dimensions of the column crossection and δ_o is the amplitude of the initial out-of-straightness. This equation represents a multidimensional probability density function or a response surface since all the parameters involved can be treated as random variables. The modulus of elasticity E and the column length L are treated as constants, the column is subjected to a deterministic load p which remains as such from the onset of loading until the maximum capacity is reached. The cross sectional properties of wide flange and box-shaped columns, as well as the residual stresses, are assumed to be normally distributed. The yield stress and the initial out-of-straightness were considered to be Type I asymptotic extreme distributions.

For each value of the parameters, one load-deflection curve can be determined. Performing the calculations for several input values, will lead to a probabilistic distribution of load-deflection curves. Thus:

$$\frac{\tilde{\sigma}_i}{\tilde{\sigma}_o} = \frac{\tilde{\varepsilon}_i}{\tilde{\varepsilon}_o} = \phi(\varepsilon) = \frac{1}{\tilde{\varepsilon}_o}(\tilde{\varepsilon}_{ri} + \tilde{\varepsilon}_p + \tilde{\theta}\tilde{\xi}_i)$$

where the variables were defined in section 2.1 and ($\tilde{}$) indicates that the variable is random.

The results of this study were presented as confidence intervals on the strength curves of the column (fig.2). It was found that the variability of yield stress, of cross-sectional properties and of variations of residual stresses is relatively small (3% to 7%). The overriding factor contributing to the spreading in strength is the initial out-of-flatness.

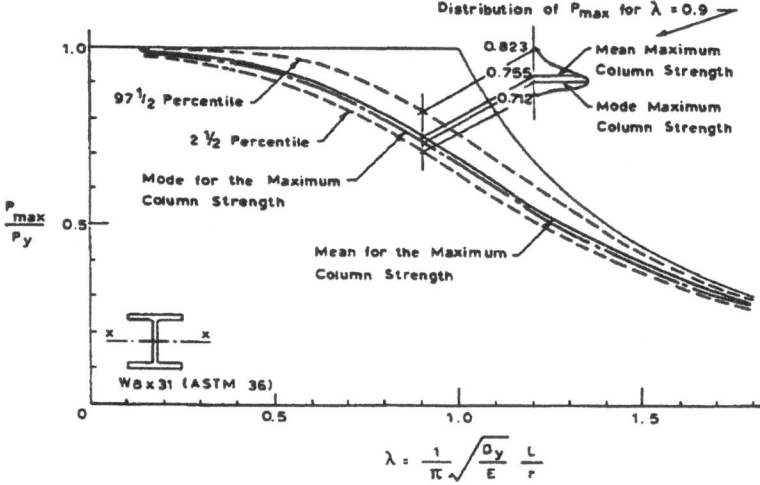

Figure 2 - Confidence intervals of the strength curve for columns (Bjorhorde, 1972).

2.4 FIRST ORDER SECOND MOMENT ANALYSIS

The methods based on stochastic descriptions of the imperfections mainly have an academic interest since the design information that they provide is limited, and because they are based on a detailed description of the initial deformations which in general cannot be supported by data.

The approaches based on a full probabilistic description of the variables can be applied in design and analysis, although they are not practical when the number of variables is large.

For design work, simplified methods based on First Order Second Moment (FOSM) information of the design variables have been commonly adopted (Ang and Cornell, 1974). These methods operate with mean values and variances of the different random variables, and in advanced formulations of first order reliability methods (FORM) (Hasofer and Lind, 1974, Rackwitz and Fiessler, 1978, Ditlevsen, 1979) they are even able to include information about the type of probability distribution functions of the variables. An important application in which first order methods have been particularly useful, is in codified design, that is, in formulating design codes and in determining the values of the safety factors.

In general, for a performance function Z of n variables:

$$Z = g(x_1, x_2, ..., x_n)$$

the FOSM approach predicts the mean and the variance of the function by:

$$\mu_z = g(x_1, x_2, ..., x_n)$$

$$\sigma_z^2 = \sum_{i=1}^{n} \sum_{j=1}^{n} \rho_{ij} \left(\frac{\partial g}{\partial x_i} \right) \left(\frac{\partial g}{\partial x_j} \right) \sigma_i \sigma_j$$

where σ_i and σ_j are the standard deviation of x_i and x_j respectively, and ρ_{ij} is the correlation coefficient between x_i and x_j, which is equal to unity when $i=j$. In the initial version of FOSM methods, the partial derivatives of the performance function are evaluated at the mean value of the variables, while in FORM they are evaluated at the most likely failure point, which is found by an iterative procedure.

First order reliability methods often produce results that are sufficiently accurate for design purposes; thus it may be interesting to compare their results with a full probabilistic analysis. Applying the FORM approach to the full probabilistic analysis of Augusti and Baratta (1971), we find the mean critical stress $\overline{\sigma}_c$ is:

$$\bar{\gamma} = 3\left(\frac{\bar{\sigma}_o}{\bar{\sigma}_c} - 1\right)\left[1 - \left(\frac{\bar{\sigma}_c}{\bar{\sigma}_E}\right)^{1/3}\right]$$

The variance of σ_c is obtained from

$$\sigma_{\sigma_c}^2 = \left(\frac{\partial \sigma_c}{\partial \lambda}\right)^2 \sigma_\lambda^2 + \left(\frac{\partial \sigma_c}{\partial \gamma}\right)^2 \sigma_\gamma^2 + \left(\frac{\partial \sigma_c}{\partial \sigma_o}\right) \sigma_{\sigma_o}^2$$

where the derivatives are calculated at the mean value of the variables, and are multiplied by the variance of λ, γ and σ_o. For very small imperfections, this shows close agreement with the curve obtained by Augusti and Baratta for zero mean imperfections.

Another interesting comparison is with the results of Chung and Lee (1971) for H shaped columns with residual stresses. The results are very similar, the differences being acceptable in design situations. Further examples of application of first order methods for studying the strength of columns, can be found in Frangopol and Hung (1977) and in Djalaly (1977).

An interesting feature of the application of first order methods to codified design, is that they allow the modelling not only of the intrinsic variability of the design variables, but also of the uncertainty in the design models. Codified design must be accomplished with simple expressions which cannot be accurately predicted for all combinations of variables. Thus, a certain degree of error is accepted in each design equation, and it can be represented by a model uncertainty.

Frequently, a multiplicative idealisation is used; this represents the real value of a variable X by the product of the design prediction $\hat{X}$ and a model error B_x, as suggested by Ang and Cornell (1974):

$$X = B_x \hat{X}$$

The mean value of B_x gives the bias or systematic error of the model, and its coefficient of variation (cov) indicates the model uncertainty. Normally, this information is derived from comparisons of predictions with test results or with predictions of more accurate theories. Once the bias is assessed, it can be incorporated in the design equation which becomes then unbiased. However, there is still a model uncertainty associated with it.

An application of this concept is due to Lenz et al (1973) who, based on a first order approach, developed safety factors for use in design codes of columns. In the American Institute of Steel Construction (AISC) specification, the required cross sectional area A of a column is,

$$A = P / 1.7 \sigma_a$$

where σ_a is the allowable stress given by:

$$\sigma_a = \frac{\sigma_y 1 - \dfrac{(kL/r)}{2C_c}}{\dfrac{5}{3} + \dfrac{3}{8}\left(\dfrac{kL/r}{C_c}\right) - \dfrac{1}{8}\left(\dfrac{kL/r}{C_c}\right)^3}$$

and

$$C_c = \left(2\pi^2 E / \sigma_o\right)^{1/2}$$

Lenz et al (1973) used the interesting concept of model error B_x, which was determined from comparisons of the equation predictions with test results. The mean value and the coefficient of variation of B_x were respectively 1.03 and 0.13 in that case; these represented the bias and the uncertainty of the model.

In addition to calibrating a design rule with a set of measurements, one must also calibrate it for the whole family of possible designs that a code can govern. This can be accomplished by using second moment methods to calculate the reliability index over the full range of the design variables. One may be interested in achieving a constant reliability over the full range of a variable, or a reliability that is a function of the variable value. This type of study has been performed by Hawranek and Rackwitz (1976), by Bjorhovde (1978) and by Galambos (1983), among others.

First order methods have also been used to study the strength of beam-columns. Rojiani and Woeste (1982) used FORM method accounting for three possible modes of failure, while Ellingwood and Reinhod (1980) compared a FOSM analysis with an advanced FOSM method, obtaining results that were close enough for design purposes. Kotoguchi, Leonard and Shiomi (1985) determined the model uncertainty of Japanese and American code specifications on the basis of experimental data. Finally, Bjorhovde, Galambos and Ravindra (1978) used FOSM methods to calibrate a design specification in the probabilistically based code of the the AISC.

2.5 APPROACHES BASED ON STATISTICAL ANALYSIS OF DATA

Recognizing that column strength depends on many variables that are random in nature, some authors considered that a convenient approach to predict column strength could be based on a statistical analysis of experimental data of column strength. This data could be obtained directly from experiments or from Monte Carlo simulations which extrapolated experimental or numerical analysis.

The European Convention for Constructional Steelwork (ECCS) has carried out an extensive experimental program on buckling of concentrically loaded hinged columns with imperfections (Beer and Schuls 1970, Sfintesco, 1970). The test series has been statistically designed so that a buckling failure with a certain probability of occurrence could be derived from a statistical analysis of the data.

A similar approach was taken by Perry and Chilver (1976), who conducted experiments on model columns instead of the full-size members of the ECCS study. They found that variation of maximum strength depends on the slenderness ratio, with the greater scatter of maximum strength occurring in the region of the slenderness ratio where there is a transition from elastic to plastic buckling. This is because the variations of E are not important for plastic buckling, and the variations of yield stress are not important for elastic buckling. At the transitional slenderness ratio (L/r=60-100) there is not only a dependence on both Young's modulus and on yield stress, but also a maximum sensitivity to geometrical imperfections.

More recently, Fukumoto and Itoh (1983) conducted another statistical analysis of data which they have compiled in a data bank from experiments conducted by several authors. As in the previous cases, design curves were produced for different probability levels.

Based on the same principles, Stating and Vos (1973) used Monte Carlo simulations to reproduce the scatter of buckling strength. Starting from a probabilistic description of each of the governing parameters from which samples were drawn, and using a deterministic model of column strength, they collected results on the column strength; these were analysed statistically, leading to a probabilistic description of column strength.

3. Probabilistic Modelling of Plate Strength

In comparison with the theory of stability of columns, the problem of stability of plates is complicated by the fact that the critical buckling load may be different from the ultimate load which the plate can carry. While the buckling load is for practical purposes the largest load any column can carry, plates may be able to sustain in the buckled state ultimate loads far exceeding the buckling load. This is specially important for very thin plates and for materials with low modulus of elasticity like aluminium alloys.

There is significantly less work reported on the probabilistic modelling of plate strength than is available for columns. Most of the work referred to here is based on FOSM formulations and on statistical analysis of experimental data, with a particular emphasis on the development of design methods.

3.1 STATISTICAL ANALYSIS OF DATA ON PLATE CHARACTERISTICS

One of the major factors contributing to the uncertainty in plate strength, is the randomness of the initial imperfections. Thus, it is appropriate to refer to some studies that collected data on plate initial distortions. Much less data is available on measurements of weld induced residual stresses.

Faulkner (1975) and Antoniou (1980) reported surveys of plate deflections, although measuring only the maximum central deflection. Kmiecik (1971) concluded that the shape of the imperfections is important, and that the elastic buckling strength is governed mainly by the amplitude of the buckling mode component, as determined from a Fourier analysis. Thus the available statistical analysis of plate imperfections has concentrated on these components (Antoniou, Lavidas and Karvounis, 1984 and Jastrzebski and Kmiecik, 1986). More recently Kmiecik, Jastrzebski and Kuzniar (1995) have analysed the data further by fitting a Weibull distribution to the amplitudes of the buckling components that describe the imperfect surface, and by providing additional regression equations. These studies provided the basic statistical data in terms of the statistical moments, which were used in different studies of plate strength.

A different type of statistical analysis was conducted by Itoh and Fukumoto (1987); they directly analysed the results of tests of plate collapse which had been collected in their data base. They conducted a statistical analysis of the test results and derived design curves. This is a different use of test data than was given by Guedes Soares (1988b), who used it to isolate the effect of the various parameters that govern plate strength, namely, slenderness, residual stresses and initial deflections. A related approach was adopted by Ueda and Yao (1985), who derived design formulas from a regression equation on results calculated by finite-element analysis.

3.2 PREDICTION OF PLATE STRENGTH

The buckling strength of rectangular plate elements can be predicted with various methods of different degrees of sophistication and accuracy. The elastic methods can be linear and non-linear, including large deflections and the effect of initial imperfections; they can be analytical or numerical. Numerical methods allow the inclusion of the effects of plastic deformation as well as of residual stresses and initial distortions that are present in welded plates.

Several systematic studies have been conducted, both of experimental and of numerical nature, which have given good information about the compressive strength of plates and of the effect of parameters such as aspect ratio, residual stresses, initial distortions, boundary conditions and different types of loading. A summary of their influence as well as of their contribution to the compressive strength of plates can be found for example in Guedes Soares (1988b). These results served as the basis for the calibration of simple design formulas which had been proposed in the past. Much of the work on probabilistic analysis of plate strength has been associated with calibration of those formulas.

A plate with simply supported edges will buckle at a stress:

$$\sigma_E = \frac{\pi^2 E \sqrt{\tau}}{12(1-\nu^2)} \left(\frac{t}{b}\right)^2 \left(\frac{a/b}{n\sqrt[4]{\tau}} + \frac{n\sqrt[4]{\tau}}{a/b}\right)^2$$

where n is the number of half waves in which the plate buckles, a/b is the plate aspect ratio, τ is the ratio of the tangent to the elastic modulus of elasticity, and ν is Poisson's ratio.

In the elastic domain, $\tau=1$, and in long plates the length of the half waves approaches b (i.e. $k=4$), yielding:

$$\sigma_E = \frac{\pi^2}{12(1-\nu^2)}\left(\frac{t}{b}\right)^2 k = 3.62\left(\frac{t}{b}\right)^2$$

where k is the effect of aspect ratio.

The plate starts to buckle when the critical load is reached, but as the load increases the rate of increase of the deflections decreases. When the applied load exceeds the critical load, the supported edges parallel to the acting load supply the plate with an additional element of strength which comes into play when the middle part buckles. The redistribution of stresses that take place together with the stabilising membrane stresses, enables the plate to regain stability in the distorted shape. Finally, with increasing load the largest stresses of the plate will approach yield and the plate will reach the ultimate collapse load. The margin between buckling and ultimate strength increases with decreasing critical stress, and approaches zero near the yield stress.

The most important parameter governing plate strength is the reduced slenderness:

$$\lambda = \frac{b}{t}\left(\frac{\sigma_o}{E}\right)^{1/2}$$

In bridges, λ can be in the range 1 to 3, in ships 1 to 5 and in airplanes in 3 to 7. Several design methods that predict the plate's ultimate strength are based on λ.

A method due to Faulkner (1975) has been widely used for strength assessments and also for probabilistic modelling, as we now describe. The strength of a simply supported rectangular plate with residual stresses is given by:

$$\phi = \frac{\sigma_u}{\sigma_o} = \phi_b - \Delta\phi_b$$

where σ_u is the ultimate stress and $\Delta\phi_b$ is the reduction of strength due to the residual stresses:

$$\Delta\phi_b = \frac{\sigma_r}{\sigma_o}\frac{E_t}{E}$$

and

$$\phi_b = \frac{\sigma_u}{\sigma_o} = \frac{a_1}{\lambda} - \frac{a_2}{\lambda^2} \quad \text{for} \quad \lambda > 1.0$$

$$\phi_b = 1.0 \quad\quad\quad\quad\quad \text{for} \quad \lambda \le 1.0$$

is the strength of a plate without residual stresses but with average initial distortions. The constants a_1 and a_2 account for the boundary conditions:

$$a_1 = 2.0 \text{ and } a_2 = 1.0 \text{ for simple supports}$$

$$a_1 = 2.56 \text{ and } a_2 = 1.56 \text{ for clamped supports}$$

The residual stresses σ_r are assumed to be uniformly distributed across the central zone of the plate, being equilibrated by two strips of tensile yield stresses at the edges, each with a breadth of ηt. The magnitude of σ_r is:

$$\frac{\sigma_r}{\sigma_o} = \frac{2\eta}{(b/t) - 2\eta} \quad \text{for} \quad 1 < \lambda < 2.5$$

and the tangent modulus of elasticity can be approximated by (Guedes Soares and Faulkner, 1987):

$$\frac{E_t}{E} = \frac{\lambda - 1}{1.5} \quad \text{for} \quad 1 < \lambda < 2.5$$

$$\frac{E_t}{E} = 1.0 \quad \text{for} \quad \lambda > 2.5$$

This method does not account explicitly for the magnitude of the initial imperfections. The effect of the initial imperfections on plate strength was examined in Guedes Soares (1988b), where Faulkner's expression was generalised to account explicitly for that effect. In that work the model error of the expressions for ϕ_b and $\Delta\phi_b$ was included and assessed from comparisons with experiments. The final form of the design equation proposed was:

$$\phi = 1.08 \, \phi_b \, R_\eta \, R_\delta \, R_{\eta\delta}$$

$$R_\eta = \left[\left(1 - \frac{\Delta\phi_b}{1.08\phi_b} \right) (1 + .0078\eta) \right]$$

$$R_{\delta} = \left[1 - (0.625 - .121\lambda)\frac{\delta_o}{t}\right]$$

$$R_{\eta\delta} = \left[.665 + .006\eta + .36\frac{\delta_o}{t} + .14\lambda\right]$$

where the term $1.08\,\phi_b$ predicts the strength of perfect plates, R_{η} is the reduction factor due to residual stresses, R_{δ} is the reduction factor due to initial defections and the four terms should be used for plates that have both initial defections and residual stresses i.e. $R_{\eta\delta}$ accounts for the interaction between initial deflection and residual stresses. It should be noted that the value of 1.08 in the first square term is a bias term, as is the term $(1+0.0078\eta)$ which affects the component due to the residual stresses. The uncertainty associated with the whole equation is 0.07, which is the standard deviation of the residuals obtained from the regression.

Guedes Soares (1988c) proposed another equation which depends only on plate slenderness and has inbuilt the influence of the average levels of initial distortions and of residual stresses existing in merchant ships:

$$\phi_{GS_m} = \frac{1.6}{\lambda} - \frac{0.8}{\lambda^2} \text{ for simple supports}$$

$$\phi_{GS_m} = \frac{2.0}{\lambda} - \frac{1.25}{\lambda^2} \text{ for clamped supports}$$

or in warships:

$$\phi_{GS_w} = \frac{1.5}{\lambda} - \frac{0.75}{\lambda^2} \text{ for simple supports}$$

This series of equations which are specific for one kind of ship were derived from a probabilistic description of the plate geometry in terms of a/b and b/t ratios in those ship types. Since the strength equations are conditional on those distributions, the mean strength was determined by unconditioning on the initial imperfections of η and δ which are functions of a/b and b/t.

This procedure allowed the calculation of the mean bias that arises from the combined effect of the three reduction factors R_{η} R_{δ} $R_{\eta\delta}$, taking into account the probability distribution of the plate geometry that governs those factors.

The same type of approach was later applied to different types of merchant ships and it was concluded that, despite different geometrical characteristics of the plates in these ships, the resulting design equations were similar (Guedes Soares, 1992).

Thus, while Faulkner proposed a method that accounts explicitly for the levels of residual stresses and implicitly for the degree of initial imperfection, the method of Guedes Soares (1988b) deals explicitly with all parameters that have the same level of contribution to the assessment of plate strength. Based on a probabilistic description of the geometry in specific ship types, Guedes Soares (1988c, 1992) derived simplified design formulations which depend implicitly both on residual stresses and initial distortions, but which reflect appropriately their probability of occurrence.

This systematic approach and the explicit incorporation of the probability of occurrence of initial distortions in different types of structures has not been yet generally adopted.

Ivanov and Rousev (1979) took no account of residual stresses, but considered initial distortions explicitly; they took

$$\phi_1 = \frac{1}{1 + (0.3\lambda + 0.08)\,\delta_0}$$

where $\delta_0 = w_0/t$ and w_0 is the maximum amplitude of initial distortions.

The method proposed by Carlsen (1977) accounts explicitly for both types of initial defects:

$$\phi_C = \phi_b \left(\frac{1}{1 + \sigma_r/\sigma_0}\right)\left(1 - \frac{0.75\delta_0}{\lambda}\right)$$

The coefficients for ϕ_b in this case are $a_1 = 2.1$, $a_2 = 0.9$ for simple supports.

In addition to these equations, a few others have been proposed for marine structures. Ueda and Yao (1985) have made a regression study and proposed for simply supported plates the following:

$$\phi_U = \frac{1.3388\delta_0^2 + 4.380\delta_0 + 2.647}{\lambda + 6.130\delta_0 + 0.720} - 0.271\delta_0 - 0.088$$

Soreide and Czujko (1983) studied plates under biaxial loading, but also proposed an equation for almost perfect plates uniaxially loaded:

$$\phi_{S_0} = \frac{2.74}{\lambda} - \frac{2.56}{\lambda^2} + \frac{0.921}{\lambda^3}$$

To account for initial distortions and residual stresses, the previous equation ϕ_{S_0} should be modified to become:

$$\phi_S = 1.52\phi_{S_0}\left\{1 - 2.528\left(\frac{\delta_0}{b/t}\right)^{0.113}\left(\frac{1.207}{\lambda} - \frac{1.467}{\lambda^2} + \frac{0.59}{\lambda^3}\right)\right\}$$

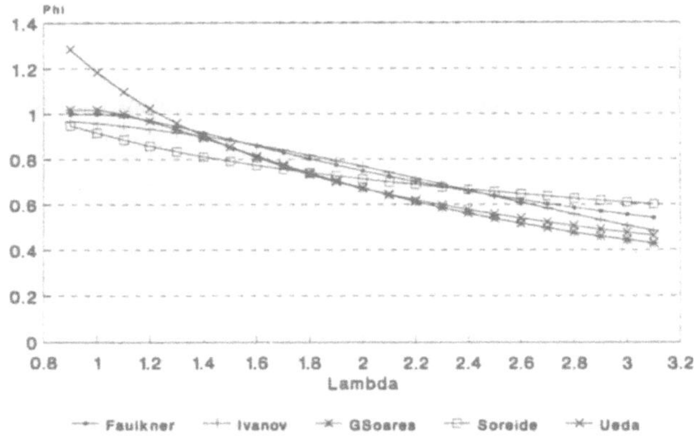

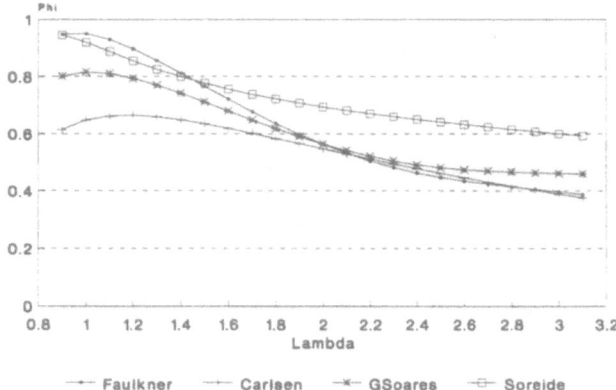

Figure 3 - Non dimensional strength (ϕ) of plates with initial distortions under uniaxial load, as predicted by different methods.

Figure 3 shows a comparison of the prediction of the different methods for an average level of intial distortions, while Fig. 4 shows the predictions of the few methods that also account for residual stresses.

Figure 4 - Non dimensional strength (ϕ) of plates with initial distortions and residual stresses, under uniaxial load, as predicted by different methods.

It can be observed in the figures that the range of the spreading in the model predictions is 10 to 15% in the low slenderness range, but becomes closer to 30% for slenderness around 3.

While each author may claim that the design equation he proposed best fits the data he had available, a designer may want to interpret the differences between the various methods, as shown in figures 3 and 4, as measures of model uncertainty.

3.3 FIRST ORDER RELIABILITY ANALYSIS

The plate slenderness, which governs plate strength, is in reality an uncertain quantity that depends on geometric variables b and t, and on material properties, E and σ_o, as was pointed out by Ivanov and Rousev (1979). A FOSM analysis indicates that the uncertainty of λ of varies between 0.06 and 0.07, most of it being due to the contribution of the yield stress (Guedes Soares 1988a).

Guedes Soares (1988a) adopted Faulkner's method as a basis for a first order-second moment analysis of the compression strength of rectangular plates and distinguished between the laboratory test, the analysis calculations and the design decisions, which can be based on the same equations but involve different uncertainty levels for the basic variables. In a laboratory test, the plate geometry is known, as well as the residual stresses and the initial deflections which will be close to zero unless artificially created. The support conditions involve small uncertainty, and the load distribution and intensity are carefully controlled, thus with small uncertainty also.

Introducing the fundamental uncertainty of λ in the design equation, one is able to calculate the fundamental uncertainty of ϕ_b. Comparing the predictions of ϕ_b with experimental results, one assesses the total uncertainty which, by comparing with the calculated fundamental uncertainty, allows one to derive the model uncertainty of the method. While the fundamental uncertainty was about 0.06, the model uncertainty was found to be 0.08 for the examples in Guedes Soares (1988a).

In the analysis situation, the plate is considered to have weld induced residual stresses, the boundary conditions are not completely known, and the plate may be subjected to complicated load patterns. The model uncertainty in this case is around 0.10 to 0.12 in the mid slenderness range, i.e. for elasto-plastic collapse. It decreases to 0.08 for elastic buckling and for plastic collapse. In the design situation, corrosion is present, bringing an additional uncertainty to the plate thickness; this increases the uncertainty of the predictions from 0.10 to 0.15, although the mean value is almost unchanged. Figure 5 shows the mean plate strength in the analysis (ϕ_a) and design (ϕ_d) situations almost coinciding but their uncertainty, as reflected in the coefficients of variation of the model uncertainty (V_a, V_d), have a fairly constant difference of 5% along the plate slenderness range (Guedes Soares 1988a).

Guedes Soares and Faulkner (1987) used a probabilistic approach to extend Faulkner's equation to assess the strength of plates with aspect ratio smaller than unity.

The new equation was used to predict the strength of long rectangular plates that collapse in a mode higher than the fundamental elastic buckling mode.

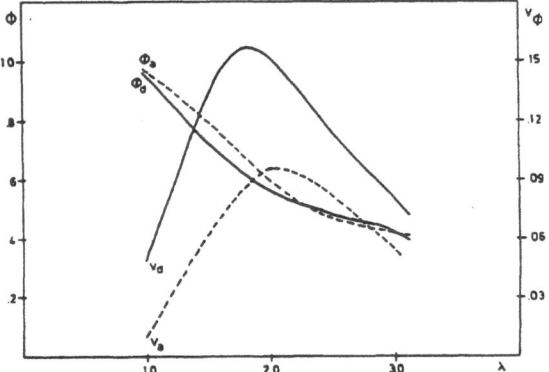

Figure 5 - Mean plate strength in the analysis (ϕ_a) and design (ϕ_d) situation, and its uncertainty as reflected in the coefficient of variation of the model uncertainty (V_a, V_d) (Guedes Soares, 1988a)

Akita (1988) used a FORM approach to predict the strength of square panels. He used an implicit formula to predict plate strength σ_u which took explicit account only of plate slenderness and amplitude of initial distortions:

$$\frac{\lambda^2 \sigma_u}{\pi^2} = \frac{1 - \dfrac{\delta_o}{\xi / \sigma_u}}{3(1 - v^2)} + \frac{\left(\dfrac{\xi^2}{\sigma_u}\right) - \delta_o^2}{8}$$

where

$$\xi = \frac{9\left(1 - \sigma_u^2\right)}{\left(16 - 15\sigma_u^2\right)^{1/2}}$$

Akita obtained the probability distribution of σ_u from the distribution of imperfections δ_o which was assumed to be exponential. This is a simplified model in that only the imperfections were considered to be a random variable and the above expression is an explicit relationship between δ_o and σ_u.

Guedes Soares and Silva (1991) assessed the reliability of rectangular plate elements whose strength was predicted by some of the expressions discussed here. Figure 6 compares the reliability indeces calculated by FOSM and FORM, applied to Faulkner's strength equation. It is apparent that both give almost the same results. The figure also shows the prediction of the strength equation. It is apparent that as the slenderness increases the reliability indices decrease quicker than the strength predictions.

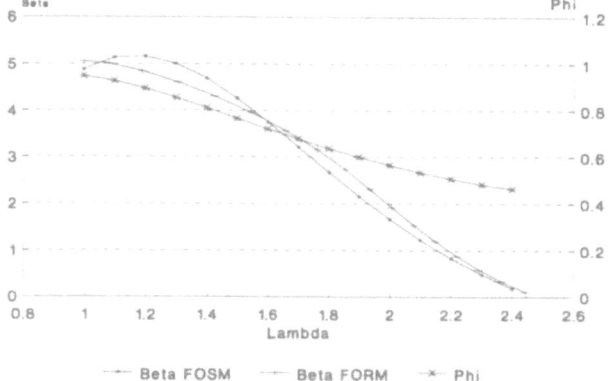

Figure 6 - Reliability index (β) for uniaxially loaded plate element as predicted by FOSM and FORM methods, as a function of plate slenderness (λ). Also indicated is the mean plate strength (ϕ) (Guedes Soares and Silva, 1991).

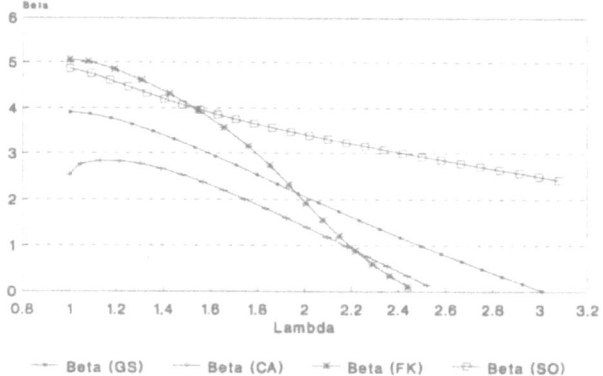

Figure 7 - Reliability index as function of plate slenderness for simply supported plates with initial distortions and residual stresses, with strength predicted by different methods (Guedes Soares and Silva, 1991).

Figure 7 shows the values of the reliability index calculated for different design methods in which the loading was assumed to consist of two components: one Weibull distributed, representing the wave induced stresses; and the other normally distributed, idealising the still-water loads in ships. There is some spread in the results, as one could expect. The figures show that, even though the strength formulations are very similar, as was shown in figure 2, the reliability indices for the different formulations differ by one or two units. The methods of Guedes Soares, and of Carlsen yield reliability indices with the same type of dependence on λ; Faulkner's method leads to a steeper decrease of β with λ; and the one of Soreide and Czujko predicts a small variation of β in the whole range of λ.

The sensitivity analysis showed that the yield stress and the two loading variables were the most important ones contributing to the probability of failure.

Bonello et al. (1991, 1993a) used the concept of system reliability to access the reliability of one plate element; they assumed that, for collapse, the plate must simultaneously have reached the collapse load, and the strain at the edges of the plate must be sufficiently large in the post-collapse region. This is sometimes what happens in large panels: one plate element is in the post-collapse region, but the rest of the structure does not allow it to develop excessive strain.

Guedes Soares and Silva (1991) also studied the system reliability effect; they considered a panel composed of several plates as a parallel system. Each plate could fail individually, and after failure it would continue supporting its post-collapse load (which was considered to be half of the collapse load) while the rest of the load was equally distributed by the intact plates. For two plates the probability of system failure was 3 times smaller than the probability of first plate failure. For a system of 5 plates, it increased to about 10. In this latter case, considering the plates independent and without load redistribution would yield a probability of failure 12 orders of magnitude smaller, which is clearly incorrect.

Guedes Soares and Silva (1991) showed that different design formula, although equally appropriate for predicting the expected design strength, would however lead to different values of the reliability index because of its different relation with the various design parameters. As an alternative to this type of analysis, Guedes Soares and Kmiecik (1993) assessed the variability of the compressive strength of plate elements from the results of a Monte-Carlo simulation of the shape of initial distortions. The strength of each panel with the sampled initial distortions was calculated by a non-linear finite-element code which would yield the "correct" strength for each plate defined by a set of material and geometrical variables as well as by initial imperfections. The method was able to account for the variability of the shape of initial imperfections and of material properties. It led to larger values of the coefficient of variation than the approach of Guedes Soares (1988a), which however did not deal with the effect of the shape of imperfections. However, the dependence of the coefficient of variation on the slenderness was the same, with larger values near a slenderness of 2 where the collapse is elasto-plastic and decreasing towards the stocky and the slender ends.

Although this approach is interesting, in that a more accurate mechanical model of plate collapse is used, it is very time consuming in that one calculation of plate collapse can take several hours of computing time. Kmiecik and Guedes Soares (1994) proposed an alternative formulation: the use of a response surface approach to represent the limit state function describing the effect of different parameters. The situation analysed in that paper involved only four random variables, three of which described the shape of the imperfections and the fourth one was the plate slenderness. In this case only nine computations of plate strength were necessary to define the response surface. A linear response surface was fitted to the results, and the probability distribution of plate strength was constructed using FORM.

4. Probabilistic Modelling of the Strength of Stiffened Plates

Stiffened plates are important structural components in many structures; their structural strengths are predicted by combining the contributions from plates and columns. The usual approach is to associate an effective width of plate with each stiffener; the effective width is less than the original width of the plate, which is meant to account for the strength degradation due to the plate element buckling. The design methods predict reviewed in sections the contribution of the plate to the plate-stiffener assembly, which thus becomes a beam column to be treated by the methods discussed in section 2 of this work.

Design methods for stiffened plates that are common in marine structures are due to Faulkner et al (1973) and to Carlsen (1977); for civil engineering structures, reference can be made to Dwight and Little (1976) or to Horne and Narayanan (1977) for example. The methods proposed by Faulkner et al (1973) and by Carlsen (1977) have been compared against experimental data by Guedes Soares and Soreide (1983); they found that the first one had a bias of 1.04 and an uncertainty of 0.12, while the second had a bias of 0.90 and an uncertainty of 0.11.

Note that Faulkner's method is intended to predict the mean value of the strength, while Carlsen's proposal is a lower bound design method which is intended to provide enough strength for all but 2.5% of the cases. In fact, in the analysis of Guedes Soares and Soreide (1983) the strength predicted by Faulkner's model exceeded the experimental results in 58% of the 74 cases while Carlsen's predictions exceeded it in 12% of the cases. A discussion on the relative merits of both types of approaches can be found in Faulkner, Guedes Soares and Warwick (1987).

Hart, Rutherford and Wickham (1986) used a FORM method to assess the reliability of stiffened plates in ship structures which were subjected to combinations of time varying loads representative of ship structures. The stiffened plates were idealised as a single-stiffener together with an effective width of plating. The strength of the plate element was predicted by the method of Chatterjee and Dowling (1977). The strength of the stiffening elements was determined from a Perry-Robertson type of expression:

$$\sigma_s^2 - \sigma_s \left(\sigma_o + \sigma_E + \delta \sigma_E \right) + \sigma_o \sigma_E = 0$$

in which σ_E is the Euler critical strength of the stiffener, σ_s is the strength of the stiffener and δ is a parameter that represents the imperfections. Two modes of failure were examined: plate induced, and stiffener induced failure. The reliability formulation was used to quantify the effect of corrosion on the reliability of the stiffened plates. The formulation adopted was similar to the approaches already described except that now the limit state equation was that for a stiffened plate, instead of for a plate element.

Recently Pu, Das and Faulkner (1996) have proposed a new approach to predict the strength of stiffened plates. They incorporated the plate strength model of Guedes Soares (1988b) in the stiffened plate formulation of Faulkner et al (1973). They compared the results of these two methods with experimental data and they concluded that the proposed method shows, on the average, better predictions than the original method of Faulkner et al if only experimental data are included in the calibration. If both experimental and numerical data are included, the proposed method has more or less the same accuracy as Faulkner's original method, although with a smaller variability.

They have conducted a reliability analysis using FORM and SORM, and for both cases they obtained differences in the relative error smaller than 3%.

5. Concluding Remarks

A discussion has been provided of the main probabilistic methods that have been used in the study of the compressive strength of flat structural components such as columns and plates. The approaches ranged from the sophisticated stochastic formulation of the initial imperfections, to the simpler second order methods. In general, the sophisticated probabilistic formulations have only been applied to idealized cases of little practical significance, like the elastic buckling of a column. The second order methods were shown to have a much greater usefulness, in that they are able to cope with realistic cases involving elasto-plastic collapse, and the influence of initial distortions and residual weld induced stresses.

6. References

Akita, Y (1988): "Reliability and Damage of Ship Structures", *Marine Structures*, Vol. 1, N° 2, pp. 89-114.

Ang, AH-S; Cornell, C.A. (1974): "Reliability Bases of Structural Safety and Design", *J. Struct. Div.*, ASCE, Vol.100, pp. 1955-1969.

Antoniou, A.C. (1980): "On the Maximum Deflection of Plating in Newly Built Ships", *J. Ship Research*, Vol. 24, pp. 31-39.

Antoniou, A.C., Lavidas, M; Karvounis, G (1984): "On the Shape of Post-Welding Deformations of Plate Panels in Newly Built Ships", *J. of Ship Research*, Vol. 28, pp. 1-10.

Augusti, G; Baratta, A (1971): "Probabilistic Theory of Slender Column Strength" (in French), *Construction Metallique*, Vol. 8, No. 2, pp. 5-20.

Augusti, G; Baratta, A (1975): "Reliability of Slender Columns: Comparison of Different Approximations", *Proceedings of the IUTAM Symposium*, pp. 183-198.

Beer, H; Schulz, G (1970): "Bases Theoriques de Courbes Europeennes de Flambement", *Construction Metallique*, Vol. 22, No. 6, Nov-Dec., pp. 436-452.

Bernard, MC; Bogdanof, JL (1971): "Buckling of Columns with Random Initial Displacements", *J. Eng. Mech., Div.*, ASCE, Vol. 97, pp. 755-771.

Bjorhovde, R (1978): "The Safety of Steel Columns", *J. Struct. Div.*, ASCE, Vol. 109, pp. 463-478.

Bjorhovde, R; Galambos, TV; Ravindra, MK (1978): "LRFD Criteria for Steel Beam-Columns", *J. Struct. Div.*, ASCE, Vol. 104, pp. 1371.

Bolotin, VV (1965): "Statitiscal Aspects in the Theory of Structural Stability", *Proc. Conf. on Dynamic Stability of Structures*, Inst. of Civil Engineers, London, pp. 67-81.

Bonelo, M.A., Chryssanthopoulos, M.K. and Dowling, P.J. (1991): "Probabilistic Strength Modelling of Unstiffened Plates under Axial Compression" *Proceedings of the 10ᵗʰ International Conference on Offshore Mechanics and Arctic Engineering* (OMAE), ASME, Vol. II, pp. 255-264

Bonelo, M.A. and Chryssanthopoulos, M.K. (1993a): "Buckling Analysis of Plated Structures Using System Reliability Concepts", *Proceedings of the 12ᵗʰ International Conference on Offshore Mechanics and Arctic Engineering (OMAE)*, ASME, Vol. II, pp. 313-321.

Bonelo, M.A., Chryssanthopoulos, M.K. and Dowling, P.J. (1993b): "Ultimate Strength Design of Stiffened Plates under Axial Compression and Bending", *Marine Structures*, Vol. 6, N° 5-6, pp. 533-552.

Boyce, WE (1961): "Buckling of a Column with Random Initial Displacements", *J. of the Aerospace Sciences*, pp. 308-320.

Brown, CB; Evans, R.J. (1972): "Safety of an Elastic Beam-Column", *J. of the Struct. Div.*, ASCE, Vol. 98, pp. 805-811.

Carlsen, CA (1977): "Simplified Collapse Analysis of Stiffened Plates" *Norwegian Maritime Research,* vol. 7, No. 4, pp. 2-0-36.

Chaterjee, S. and Dowling, P.J.: "The Design of Box Girder Compression Flanges", *Steel Plated Structures*, P.J. Dowling et. al. (Eds), Crosby Lockwood Staples, London, 1977 pp.

Chung, BT; Lee, GC (1971): "Buckling Strength of Columns Based on Random Parameters", *J. Struct. Div.*, ASCE, Vol. 97, No. ST7, July, pp. 1927-1945.

Ditlevsen, O (1979): " Generalized Second Moment Reliability Index", *J. Struct. Div. Mech.*, Vol. 7, pp. 435-451.

Djalaly, H (1977): "Reliability Strength of Compression Members", *Stability of Steel Structures,* Prelim. Rep. IABSE, pp. 107-112.

Dwight, JB; Little, G.H. (1976): "Stiffened Steel Compression Flanges-a Simpler Approach", *The Structural Engineer*, Vol. 54A, pp. 501-509.

Ellingwood, B; Reinhold, T.A. (1989): "Reliability Analysis of Steel Beam-Columns", *J. Struct. Div.*, ASCE, Vol. 106, pp. 2560-2566.

Faulkner, D. (1975): " A Review of Effective Plating for Use in the Analysis of Stiffened Plating in Bending and Compression", *J. of Ship Research,* Vol. 19, pp. 1-17.

Faulkner, D.; Adamchak, J.C.; Snyder, GJ; Vetter, MF (1973): "Synthesis of Welded Grillages to Withstand Compression and Normal Loads", *Computers and Structures*, Vol. 3, pp. 212-246.

Faulkner, D.; Guedes Soares, C.; Warwick, DM (1987): "Modelling Requirements for Structural Design and Strength Assessment", *Integrity of Offshores Structure-3*, Faulkner, D., Cowling, M.J. and Incecik, A. (Eds) Elsevier Applied Science, pp. 25-54.

Frangopol, D; Hung, N.D. (1977): "A Probabilistic Approach for Checking Safety of Centrally Loaded Steel Columns", *Stability of Steel Structures*, Prelim. Rep. IABSE, pp. 112-117.

Fraser, W.B.; Budiansky, B.(1969): "The Buckling of a Column with Random initial Deflections", *J. Appl. Mech.*, Vol. 36, pp. 233-240.

Fukumoto, Y.; Itoh, Y. (1983): "Evaluation of Multiple Column Curves Using the Experimental Data-Base Approach", *J. of Constructional Steel Research*, Vol. 3, No. 3, pp. 1-19.

Galambos, T.V. (1983): "Reliability of Axially Loaded Columns", *Eng. Struct.*, Vol. 5, pp. 73-78.

Gordo, J. M. and Guedes Soares, C. (1993) "Approximate Load Shortening Curves for Stiffened Plates Under Uniaxial Compression", *Integrity of Offshore Structures-5*, D. Faulkner, M.J. Cowling, A. Incecik, P.K. Das (Eds.) EMAS, pp. 189-211.

Guedes Soares, C. (1981): "Survey of Methods of Prediction of the Compressive Strength of Stiffened Plates", *Report MK/R 57*, Division of Marine Structures, Norwegian Institute of Technology.

Guedes Soares, C. (1988a): "Uncertainty Modelling in Plate Buckling", *Structural Safety*, Vol. 5, pp. 17-34.

Guedes Soares, C. (1988b): "Design Equation for the Compressive Strength of Unstiffened Plate Elements with Initial Imperfections", *J. Constructional Steel Research*, Vol. 9, pp. 287-310.

Guedes Soares, C., (1988c)"A Code Requirement for the Compressive Strength of Plate Elements", *Marine Structures*, Vol. 1, pp. 71-80.

Guedes Soares, C., (1992) "Design Equation for Ship Plate Elements Under Uniaxial Compression", *J. Constructional Steel Research*, Vol. 22, pp. 99-114.

Guedes Soares, C and Faulkner, D. (1987): "Probabilistic Modelling of the Effect of Initial Imperfections on the Compressive Strength of Rectangular Plates", *Proc. 3rd Int. Symp. on Practical Design of Ships and Mobile Units*, (PRADS' 87), Trondheim, June, Vol. II, pp. 783-795.

Guedes Soares, C. and Gordo, J. M. (1996a) "Compressive Strength of Rectangular Plates Under Transverse Loading.", *Journal of Constructional Steel Research*, Vol. 36, N° 3, pp. 215-234.

Guedes Soares, C. and Gordo, J. M. (1996b) "Compressive Strength of Rectangular Plates Under Biaxial Load and the Lateral Pressure.", *Thin-Walled Structures*, Vol. 24, pp. 231-259.

Guedes Soares, C and Kmiecik, M. (1993) "Simulation of the Ultimate Compressive Strength of Unstiffened Rectangular Plates", *Marine Structures*, Vol. 6, pp. 553-569.

Guedes Soares, C. and Silva, A. G. (1991) "Reliability of Unstiffened Plate Elements Under In-Plane Combined Loading", *Proceedings of the 10th Offshore Mechanics and Artic Engineering Conference (OMAE) ASME*, Vol.II, pp. 265-276.

Guedes Soares, C; Soreide, T.H. (1983): "Behaviour and Design of Stiffened Plates under Predominantly Compressive Loads", *International Shipbuilding Progress*, vol. 30, pp. 13-27.

Harding, J.E (1985): "The Interaction of Direct and Shear Stresses on Plate Panels", *Plated Structures*, R. Narayanan (Ed.), Applied Science Pub., pp. 221-255.

Hart, D.K; Rutherford, S.E.; Wickham, A.H.S. (1986): "Structural Reliability Analysis of Stiffened Panels", *Trans. Royal Institution of Naval Architects*, Vol. 128, pp. 293-310.

Hasofer, A.M; Lind, N.C (1974): "An Exact and Invariant First-Order Reliability Format", *J. Engng. Mech. Div.*, ASCE, Vol. 100, pp. 111-121.

Hawranek, R.; Rackwitz, P. (1976): "Reliability Calibration for Steel Columns", *Bulletin d'Information No. 112*, Comite Europeen du Beton, pp. 125-157.

Horne, M.R; Narayanan, R (1977): "Design of Axially Loaded Stiffened Plates", *J. Struct. Div.*, ASCE, Vol. 103, pp. 2243-2257.

Itoh, Y.; Fukumoto, Y, (1987): "Stochastic Evaluation of Compressive Strength of Unstiffened Plate Components", *4th Int. Colloquium on Stability of Plate and Shell Structures*, Belgium, April.

Ivanov, L.D; Rousev, SH (1979): "Statistical Estimation of Reduction Coefficient of Ship's Hull Plates with Initial Deflections", *The Naval Architect*, No. 98, pp. 1182.

Jacquot, R.G. (1972) "Nonstationary Random Column Buckling Problem", *J. Eng. Mech. Div.*, ASCE, Vol. 98, 1972, pp. 1182.

Jastrzebski, T.; Kmiecik, M.: (1986) "Statistical Investigations of the Deformations of Ship Plates", (in French), *Bulletin Association Technique Maritime et Aeronautique*, Vol. 86, pp. 325-345.

Kmiecik, M. (1971): "Behaviour of Axially Loaded Simply Supported Long Rectangular Plates Having Initial Deformations", *Report No. R84*, Ship Research Institute, Trondheim, 1971.

Kmiecik, M., Guedes Soares, C. (1994): "Response Surface Approach to the Probability Distribution of the Collapse Strength of Plates", *Report 2.2R-07(1), SHIPREL Project BRITE/EURAM 4554*, IST, Lisbon.

Kmiecik, M., Jastrzebski, T. and Kuzniar, J. (1995): "Statistics of Ship Plating Distortions", *Marine Strutures*, Vol.8, N° 2, pp. 119-132

Kotoguchi, H., Leonard, JW; Shiomi, H (1985): "Statistical Evaluation of Steel Beam-Column Resistance", *Engng. Struct.*, Vol. 3, pp. 573-588.

Lenz, J., Ravindra, M.K., Galambos, T.V. (1973): "Reliability Based Design Rules for Column Buckling". *Computers & Structures*, Vol. 3, 1973, pp. 573-588.

Perry, S.H; Chilver, AH (1976): "The Statistical Variation of The Buckling Strength of Columns", *Proc. Instn. Civ. Engrs.*, Part 2, Vol. 61, pp. 109-125.

Pu, Y., Das, P.K., Faulkner, D., (1996) "Ultimate Compression Strength and Probabilistic Analysis of Stiffened Plates", *Proceedings 15th Offshore Mechanics and Artic Engineering Conference*, Vol. II, ASME, pp. 151-157.

Rackwtiz, R., Fiessler, B. (1978): "Structural Reliability under Combined Random Load Sequences", *Computers Structures*, Vol. 9, pp. 489-949.

Rojiani, K.B.; Woeste, F.E. (1982): "A Probabilistic Analysis of Steel Beam-Columns", *Eng. Struct.*, Vol. 4, pp. 233-241.

Roorda, J. (1969): "Some Statistical Aspects of the Buckling of Imperfect Structures", *J. Mech. Phys. Solids*, Vol. 17, pp. 111-123.

Sfintesco, D. (1970): "Fondement Experimental des Coubres Europeenes de Flambement", *Construction Metallique*, Vol. 22, No. 6, Nov-Dec., pp. 409-415.

Smith, C.S. (1981): "Imperfection Effects and Design Tolerances in Ships and Offshore Structures", *Trans. Inst. Engineers and Shipbuilders in Scotland*, vol. 124, pp. 39-46.

Soreide, T.H. and Czujko, J. (1983): "Load-carrying Capacity of Plates Under Combined Lateral Load and Axial/Biaxial Compression", *Proceedings of the 2nd Int. Symposium on Practical Design in Shipbuilding (PRADS'93)*, Tokyo, 1983.

Strating, J., Vos, HJ (1973): "Computer Simulation of the ECCS Buckling Curve using a Monte-Carlo Method", *Report* Stevin Laboratory, Delft University of Technology.

Thompsom, J.M.T. (1967): "Toward a General Statistical Theory of Imperfection-Sensitivity in Elastic Post-Buckling", *J. Mech. Phys. Solids*, Vol. 15, pp. 413-417.

Ueda, Y., Yao, T. (1985): "The Influence of Complex Initial Deflection Modes on the Behaviour and Ultimate Strength of Rectangular Plates in Compression", *J. Constructional Steel Research*, Vol. 5, pp. 265-302.

RELIABILITY ANALYSIS WITH IMPLICIT FORMULATIONS

J.P. MUZEAU
LERMES, CUST
Blaise Pascal University, Clermont-Ferrand, France

M. LEMAIRE
LaRAMA, Blaise Pascal University
Institut Français de Mécanique Avancée, Clermont-Ferrand, France

1. Introduction

Methods of evaluating the reliability index are now well known, and more and more software is available for its calculation in the classical cases. Nevertheless, most require an explicit limit state function. As failures generally appear during severe and extreme loading conditions, they are often associated with strongly non-linear mechanical behaviour. Then, it is unrealistic to calculate a large number of realizations, because the existing mechanical models are generally too time consuming.

The aim of this chapter is to describe a method designed to decrease the number of calculations when the limit state function is implicit. It is called the SRQ method. In the first part, the bases of the method are described. In the second part, some test examples, solved to illustrate the process of calculations, are shown. Then, the method is applied to the evaluation of the reliability index in non-linear mechanical behaviour.

2. Statement of the problem

The formulation of algorithms necessary to calculate the Hasofer and Lind reliability index β supposes that the limit state function of the structural elements, $G(X) = 0$, and its gradients, $\dfrac{\partial G}{\partial X}$, are available in an explicit analytical form in the space of the random variables X_i.

Generally, this condition is satisfied only for some linear mechanical problems or if approximations are used. The first case includes the study of tension, compression without buckling or elastic bending without lateral buckling. The second corresponds to the use of simplified mechanical models such as plastic analysis by linear sequences

C. Guedes Soares (ed.), Probabilistic Methods for Structural Design, 141–160.
© 1997 *Kluwer Academic Publishers.*

(collapse mechanisms by plastic hinges [1]) which can be extended to the case of geometrically non-linear behaviour [2].

In the other cases, such as buckling, elastoplastic non-linear problems or structures with semi-rigid connections, the analytical relationships describing the limit state function are not available as functions of the basic random variables X_i. Then, it is possible to obtain the safety margin only in an implicit numerical form.

When a structure is subjected to buckling, for example, the effect of the geometrical non-linearity does not allow the explicit calculation of the value of the bending moment M due to the axial load N. The use of a non-linear mechanical model is the only way to obtain the exact relationship describing the behaviour of the structure, and this relationship is necessarily implicit.

There are various ways to solve this kind of problem: simulation techniques or importance sampling for instance. Direct simulation technique [3] allows the generation of a large number of random events. Then, under some hypothesis, it is possible to calculate the Cornell reliability index (but this index is not fully satisfactory). This method can also be used to count the frequency of failure events. Its main defect that it requires heavy computational costs and that, generally, it does not lead to a good estimation of the probability of failure P_f.

Importance sampling is another possibility. It provides an efficient way of significantly reducing the computational time, but it requires starting from the most probable failure point (which is unknown at the beginning of the calculation) and its use remains too expensive if the limit state function is implicit. Nevertheless, it can be used to compare a result provided by another method (or to increase its accuracy).

Thus, it is necessary to develop an appropriate method of calculating the structural reliability in order to estimate the Hasofer and Lind reliability index β_{HL} associated with many kinds of limit state functions. This method must be applicable for all kinds of mechanical models describing the real behaviour of a structure, and it must lead to a reasonable computation time [4], [5].

The most simple idea is to use the software FORM or SORM [6] in an implicit way (called *Implicit FORM*) by linking it to a mechanical model of calculation and by using numerical derivations. However, on the one hand, because of the rather large number of calls of the mechanical model, the computation time necessary for this iterative method is significant (it is function of the number of random variables) and, on the other hand, the numerical derivations can lead to an appreciable error.

Here, another method, called the *explicitation method* [1] is presented. It is based on the idea that, to limit the number of calls of the mechanical model, the best solution is to use Rackwitz's algorithm with an explicit function representative of the implicit limit state. It is then necessary to transform the implicit limit state function to an analytically explicit one by using an interpolation method. This *explicitation method* leads to a

---

[1]   *Method to approximate the implicit limit state function into an explicit one.*

number of calls of the mechanical model generally smaller than the one required when using *Implicit FORM*. Nevertheless, *Implicit FORM* will be used as a validation test.

3. Presentation of the *explicitation method*

3.1 PRINCIPLE OF THE METHOD

The method is based on replacing the implicit limit state function by an explicit one constructed from numerical values obtained from some efficient mechanical model. This allows the use of the classical algorithm for calculating the reliability index β_{HL} which requires an explicit limit state function.

The *explicitation method* relies on a local approximation of the limit state function near a critical point by a well-known mathematical function. This function must satisfy the following criteria:

- it has, obviously, an explicit form;
- it is simple;
- it takes into account the different kinds of limit states which can be found in the field of structural mechanics, and depends on the distribution of the basic random variables of the problem.

There are several ways of constructing an approximation $Q(X)$ of the limit state function $G(X)$ but, generally, they require a large number of calculations. For instance, if p is the number of Lagrange interpolations and n is the number of random variables, then Lagrange interpolation method requires p^n calls of the mechanical model.

To limit the computation time, a complete quadratic function $Q(X_1)$ is adopted as a base of approximation. So, in the space of n random variables X_i, an approximation *(k)* of the limit state function $G(X_1)$ possesses the following general form:

$$G(X_1) \approx Q^{(k)}(X_1) = c + \sum_{i=1}^{n} a_i X_i + \sum_{i=1}^{n} b_{ii} X_i^2 + \sum_{i=1}^{n-1} \sum_{j=i+1}^{n} b_{ij} X_i X_j \qquad (1)$$

where c, a_i and b_{ij} are unknown constant coefficients. Sometimes, the terms $b_{ij} X_i X_j$ do not exist.

To build the limit state function, the two following techniques can be used:
- construction by an intersection technique called the **SRQ method**,
- construction by directional interpolations.

In [4], it was shown that the SRQ method is more efficient than directional interpolations. So, only the SRQ method will be described here.

3.2 THE SRQ METHOD

3.2.1 General form

The approximate function $Q^{(k)}(X_l)$, defined in equation (1), is taken as an interpolation function on the hypersurface of the limit state $G(X_l)$ in a space of n random variables. It is defined by a minimum number R_{min} of independent realizations:

$$R_{min} = L = \frac{(n+1)(n+2)}{2} \tag{2}$$

A number of points R larger than L, associated with the least squares method, stabilise the solution of the calculation of the coefficients of the expression of $Q^{(k)}(X_l)$. So, this method consists in the minimization of the quantity:

$$\sum_{l=1}^{R} \left| Q^{(k)}(X_l^r) - G(X_l^r) \right|^2 \tag{3}$$

In this expression, $Q^{(k)}(X_l^r)$ is the value of the realization r of the approximate function $Q(X_i)$ at the point X_l^r, and $G(X_l^r)$ is the value of the realization r of the implicit function $G(X)$ at the same point X_l^r.

The minimization of the equation (3) leads to the solution of the following system of L equations with L unknowns:

$$[P]\{C\} = \{H\} \tag{4}$$

in which $[P]$ is a matrix function of the variables X_l^r, $\{C\}$ is the vector of the L unknown coefficients a_i, b_{ij} and c and $[H]$ is a vector function of the realizations $G(X_l^r)$ of the limit state.

The solution of the equation (4) leads to the coefficients of the function $Q(X)$. The approximate Cartesian equation of the limit state function is obtained by the intersection of the hypersurface $Q(X_l)$ with the hyperplane of the variables X_i, or, in other words: $Q(X_l) = 0$ is the sought approximate explicit function (fig. 1).

The explicit form of $Q(X)$ can be used in the classical algorithm to calculate β_{HL}, and then $X_l^{*(k)}$, an approximation (k) of the most probable failure point.

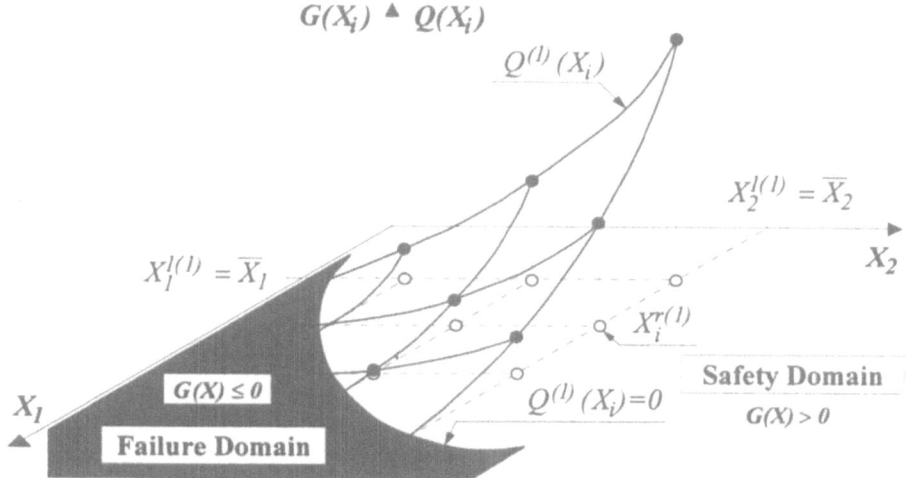

Figure 1. Second degree surface defined by several points (n=2, L=6, R=7)

To improve the approximation in the neighbourhood of the point $X_i^{*(k)}$, iterations are necessary. The iteration $(k+1)$ starts with a new set of points $X_i^{*(k+1)}$ chosen carefully closer to the failure surface calculated during the previous iteration (k), but trying to stay in the safety domain $G(X_l) > 0$ [2]. Repetition of this procedure allows the solution of the reliability problem to be approached.

3.2.2 Construction of the linear system

Let $\{C\}$ be the transposed vector of the unknown coefficients:

$$\langle C \rangle = \langle c, a_i(i = 1, ..., n), b_{ii}(i = 1, ..., n), ..., b_{ij}(i = 1, ..., n-1; j = i+1, ..., n) \rangle \quad (5)$$

and $\langle X^r \rangle$ the transposed vector of combinations of degree $d \le 2$ of the realizations of random variables:

$$\langle X^r \rangle = \langle 1, x_i(i = 1, ..., n), x_i^2(1 = 1, ..., n), ..., x_i x_j(i = 1, ..., n-1; j = i+1, ..., n) \rangle \quad (6)$$

---

[2] *Because the mechanical model may have problems to calculate a state of equilibrium of the structure in the failure domain. Depending on this mechanical model, it is generally possible to calculate a valuable solution in case of a geometrically non-linear behaviour but it can be much more difficult in the case of an elasto-plastic behaviour.*

Equation (1) becomes:

$$Q(X_i^r) = Q^r = \langle X^r \rangle \{C\} \tag{7}$$

The relationship to be minimised relatively to $\{C\}$ (equation (3)), is:

$$\sum_{r=1}^{R} (Q^r - G^r)^2$$

The condition for minimisation is:

$$\sum_{r=1}^{R} (Q^r - G^r) \left\langle \frac{\partial Q^r}{\partial \{C\}} \right\rangle d\{C\} = 0 \tag{8}$$

This can be rewritten as the equation:

$$\sum_{r=1}^{R} \left(\{X^r\} \langle X^r \rangle \right) \{C\} = \sum_{r=1}^{R} \left(G^r \{X^r\} \right) \tag{9}$$

The identification of each term leads to:

$$[P] = \sum_{r=1}^{R} \left(\{X^r\} \langle X^r \rangle \right) \tag{10}$$

and to:

$$\{H\} = \sum_{r=1}^{R} \left(G^r \{X^r\} \right) \tag{11}$$

3.2.3 Selection of a set of calculation points

The domain of localization of the points x_i^r must be chosen carefully in order to lead to a fast convergence.

First of all, a starting point $x_i^{I(k)}$ is to be chosen as the origin ($r = 1$) of the mesh at the iteration (k). It can be taken as the one corresponding to $\bar{x}_i$, the mean value of

x_i, which is the only point known *a priori* and located not far from the failure surface, but in the safe region.

Then, the set of points $x_i^{r(k)}$ is calculated from increments carried out at the point $x_i^{l(k)}$ according to several directions. The signs of these increments are chosen in such a way that the points get closer to the failure surface. They are chosen such that the failure region is approached when the values of the resistance variables R are decreasing and those of load effect variables S are increasing.

The different points are obtained by the following relationship (with $r = 2, ..., R$):

$$x_i^{r(k)} = \lambda \, d_{ij}^{(k)} \, \Delta x_j^{(k)} + x_i^{l(k)} \qquad (12)$$

where λ is an integer and $d_{ij}^{(k)}$ is a diagonal matrix defining the direction of the steps $\Delta x_j^{(k)}$. Its values are such as:

- if $i \neq j$: $d_{ii} = 0$
- if $\dfrac{\partial g(x_i)}{\partial x_i} > 0$: $d_{ii} = -1$ or 0 (3)
- if $\dfrac{\partial g(x_i)}{\partial x_i} < 0$: $d_{ii} = +1$ or 0 (3)

Physical and realistic considerations about the nature of the random variable x_i often allow the sign of the gradient to be known *a priori*.

- λ is an integer linked to the number R of points: $1 \leq \lambda \leq 10$.
- at one iteration $k > 1$, the starting point is chosen close to the most probable failure point calculated at the previous iteration:

$$x_i^{l(k+1)} = (1-\mu) \, x_i^{l(k)} + \mu \, x_i^{*(k)} \qquad (0 \leq \mu \leq 1) \qquad (13)$$

- Δx_i is the increment, chosen equal to $v \, x_i^{l(1)}$ when $k = 1$, with $0 < v < 1$ and, generally, $v = 0.1$.

At the following iterations, it becomes:

$$\Delta x_i^{(k+1)} = v\left(x_i^{*(k)} - x_i^{l(k+1)}\right) \qquad (0 < v < 1) \qquad (14)$$

Figure 2 shows the region where the points x_i^r are located in the simple case $n=2$.

---

(3) *The choice between -1 or 0 and +1 or 0 corresponds to the fact that the R points must represent a mesh of the safety domain (fig. 2).*

3.2.4 Flowchart of the SRQ method

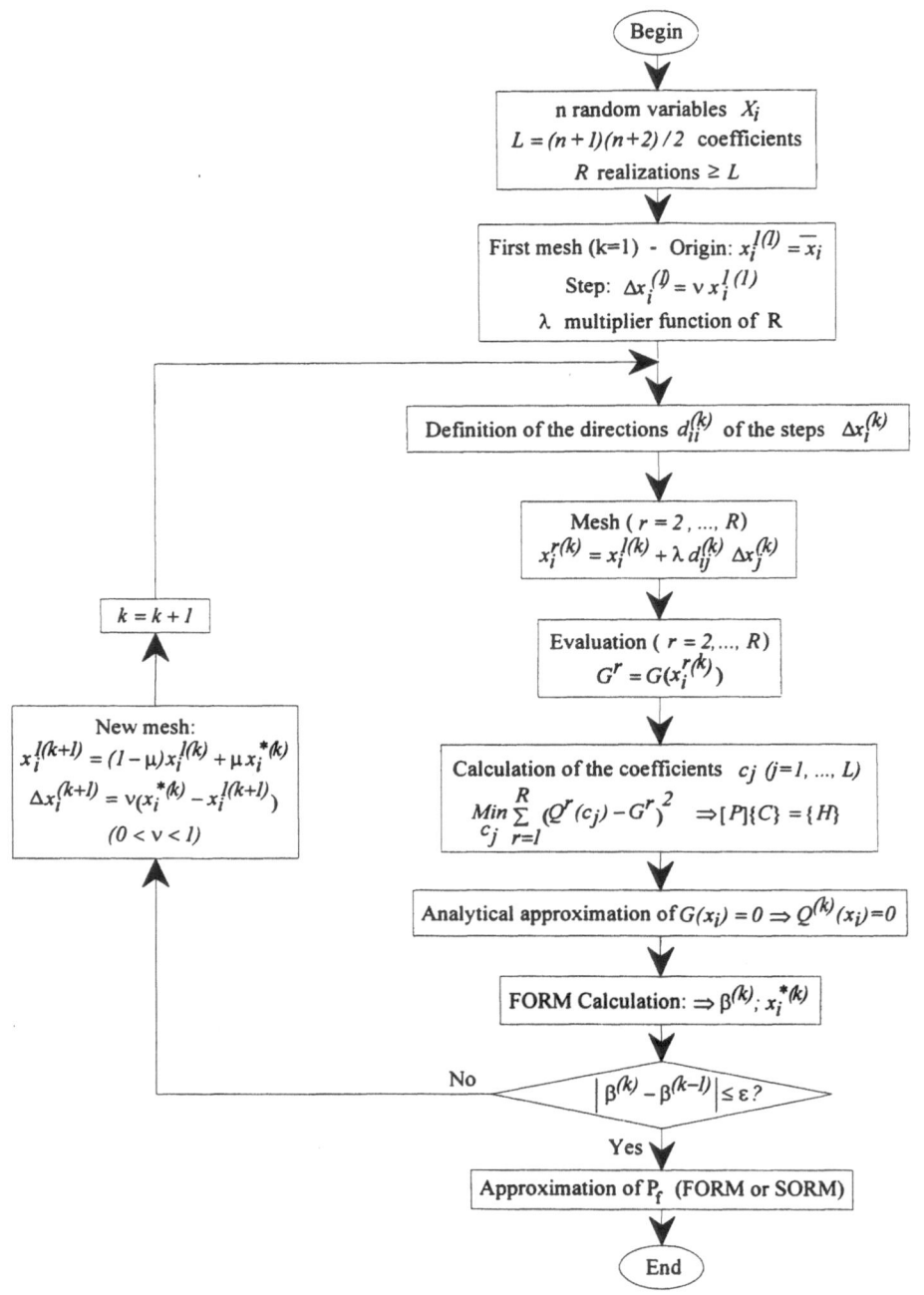

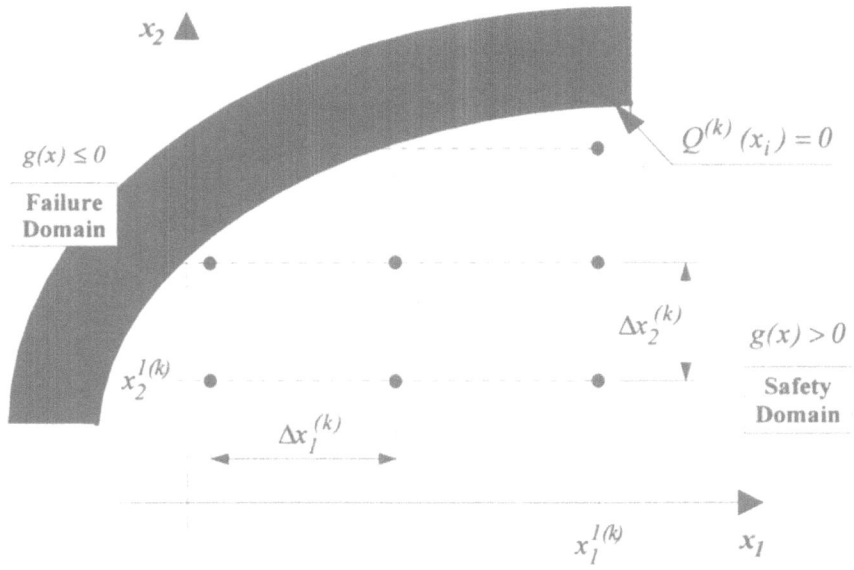

Figure 2. Position of calculation points (n=2, L=6, R=7)

4. Test example

In order to show the results of the method, a test example is given for which the exact theoretical result has been calculated by Shinozuka [7].

This example concerns a member loaded with an axial tension load F. The member has a circular cross-section with a random diameter X_1 and its yield strength, X_2, is random too. These two random variables are assumed to be Gaussian and statistically independent.

The elastic limit state of the member is defined by the following relationship:

$$G(X_1,X_2) = \frac{\pi(X_1 - 29)^2(X_2 - 170)}{875} - F = 0 \qquad (15)$$

The axial load is assumed to be deterministic and equal to 50 kN. The random variables are defined in table 1.

TABLE 1. Characteristics of the random variables

Variable		Mean value	Standard deviation
Diameter	$\varnothing$	29 mm	3 mm
Yield strength	f_y	170 MPa	25 MPa

With these data, the reliability index is equal to 2.879. It is denoted by β_{ref} , and will be compared with the values calculated with the SRQ method.

TABLE 2. Number of calls of the mechanical model

μ	$N_c = R \times$ realisations
0.10	R × 15
0.20	R × 9
0.30	R × 6
0.40	R × 5
0.50	R × 4
0.60	R × 4
0.70	R × 4
0.80	R × 3
0.90	R × 3

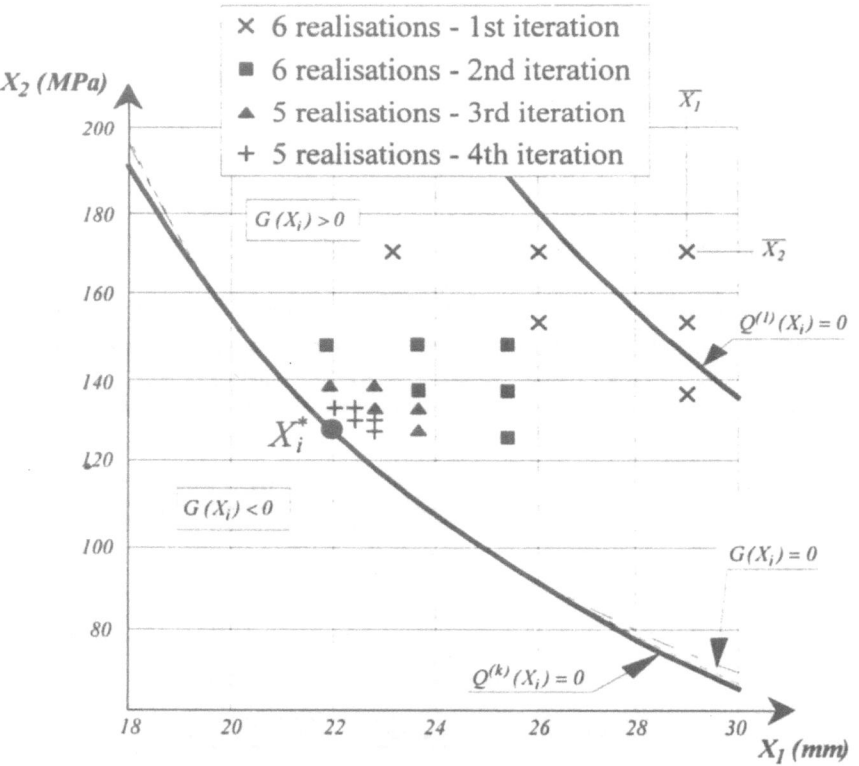

Figure 3. Example of approximation ($R = 6$, $\mu = 0.5$, $N_c = 22$)

A first calculation is carried out with $v = 0.1$ at the first iteration. Then, v is chosen to equal 0.5 but, in this example, its value is not significant. First, the number is taken to be the minimum points, $R = 6$; then, other values are considered. Table 2 shows the minimum number of calls of the mechanical model, as a function of μ, to obtain the value of β with an accuracy of 10^{-4}.

Final Figure 3 shows different approximations of the surface $G(X_1) = 0$ and the corresponding calculation points related to $\mu = 0.5$. Note that after the second iteration, the approximation of the exact surface is reasonably precise and much closer to the most probable failure point X_i^*. The total number of calculation points is equal to $6 \times 4 - 2 = 22$, because two points are the same in the two last iterations.

ly, to show the differences between various methods, table 3 shows the results of calculations with a Monte-Carlo numerical integration (MCI), the first order method (FORM), the second order method (SORM) with numerical calculation of gradients, the importance sampling methods (IS) and the SRQ method (SRQ) related to $\mu = 0.8$ ($N_c = R \times 3 = 18$).

TABLE 3. Comparison of the number of calls of the mechanical model

Method	Probability of failure	β	N_c
MCI	$2.32.10^{-3}$	2.831	1000
FORM	$1.9967.10^{-3}$	2.879	24
SORM	$2.4900.10^{-3}$	2.808	29
IS	$2.4910.10^{-3}$	2.808	74
SRQ	$1.9967.10^{-3}$	2.879	18

This table shows that the SRQ method is the one requiring the smaller number of calls of the mechanical model (18) in comparison with FORM (24), SORM (29) and the IS method (74). It is difficult to compare the results related to the MCI method, its precision depending on the number of simulations.

5. Examples of application

5.1 APPLICATION TO A SIMPLE STRUCTURE

We take a structure which members that may buckle, and calculate the failure probabilities using various methods.

The mechanical model which is used is very precise [5]. It takes into account the effect of large displacements in elastic behaviour.

The structure is a truss composed of three hinged members subjected to a vertical load P. All the members are identical and composed of circular pipes whose outer diameter $\varnothing$ is equal to *660.6 mm* and where thickness t is *10.31 mm* (fig. 4). They are made with **X52** steel grade with yield stress $f_y = 358\ MPa$, and Young's modulus $E = 21.10^4\ MPa$. The length l (fig. 4) is *27.59 m* and the angles θ_1, θ_2 and θ_3 are respectively $\pi/2$, $\pi/3$ and $\pi/4$.

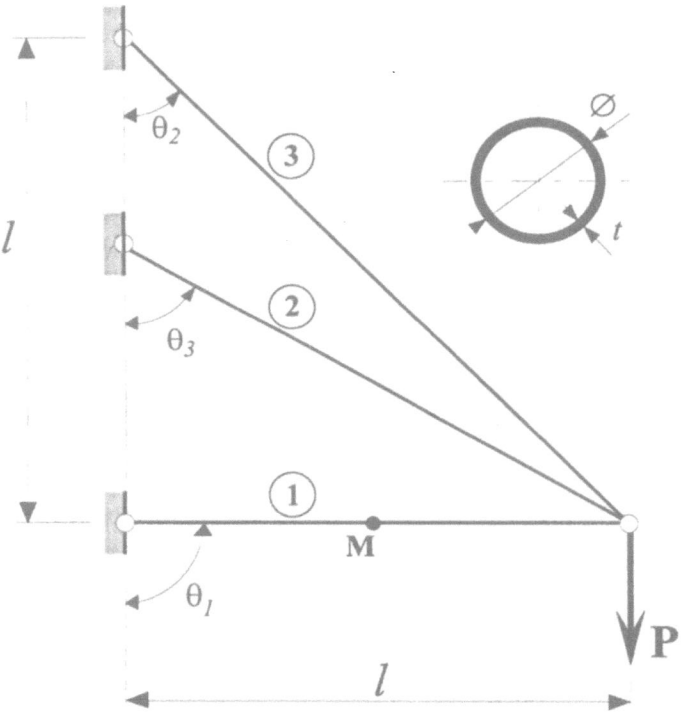

Figure 4. Three members truss

5.1.1 Deterministic study

To predict the physical behaviour of this structure, a study in a deterministic context when large displacements can occur is carried out. The slenderness of the member ① is equal to *120*. An initial defect of straightness is introduced in this member to start the phenomenon of buckling. It is a sinusoidal defect whose maximum amplitude u_0 is equal to *0.2 %* of its length *l*.

Figure 5 shows two displacement stages of the structure while figure 6 shows the vertical displacements of the mid-point *M* of the member ① as a function of the load *P*.

It should be noted that the member ② is in tension at the beginning of loading (**A**) and changes to compression after buckling of the member ① (**B**). Member ③ remains in tension throughout the loading.

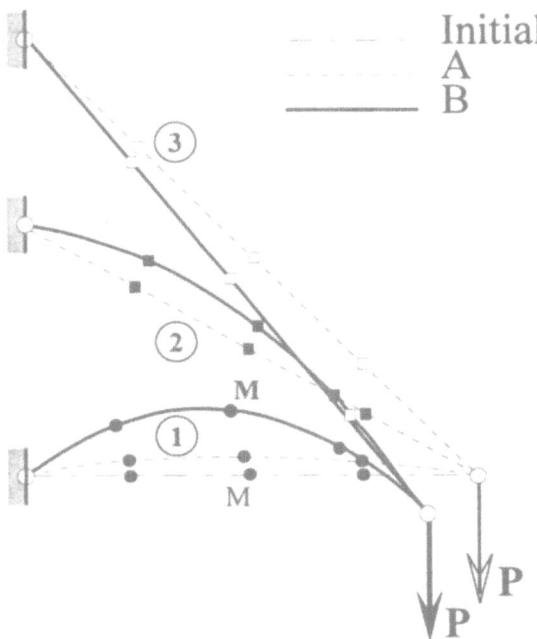

Figure 5. Displacements stages

5.1.2 Study in random context.

The considered limit state is the attainment of yield under the combined action of bending *M* and axial load *N*. For tubular cross-sections, it is possible to write the limit state function in the form:

$$E = cos\left(\frac{\pi}{2}\frac{N}{N_p}\right) \pm \frac{M}{M_p} \leq 0 \tag{16}$$

where N_p and M_p are respectively the plastic axial force and the fully plastic bending moment.

The three random variables are considered to be independent; they are shown in table 4.

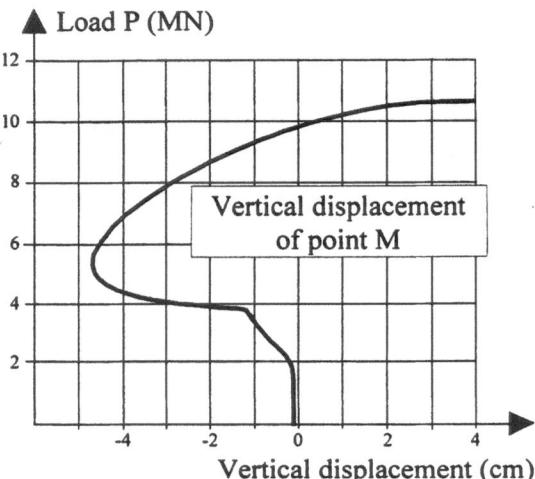

Figure 6. Deterministic behaviour of the structure

TABLE 4. Random variables

Variable		Law	Mean value	C.O.V.
Yield stress	f_y	Gaussian	308 MPa	13 %
Load	P	Gaussian	$\overline{P}(\lambda)$	
Diameter	$\varnothing$	Gaussian	660.6 mm	0.5 %

The mean value of the load P, $\overline{P}$, is obtained from the maximum design value of the member ① with respect to *Eurocode 3*:

$$\overline{P} = \frac{A\,f_y\,\chi}{\gamma\,1.192} \tag{17}$$

where A is the mean value of the cross sectional area, f_y the yield stress, χ the reduction factor due to buckling and is a function of λ (here $\lambda = 120$, so: $\chi = 0.342$)

and γ is the safety factor chosen to be *1.283*. N is the axial load in the member ①; it is equal to *1.192 P* in a linear calculation.

5.1.3 Results

The comparisons are conducted with a method of simulations [5] which gives the Cornell reliability index β_C, with *Implicit FORM,* and with the *explicitation method* which gives the Hasofer and Lind reliability index β_{HL}. To show the difference arising from the choice of mechanical behaviour, the reliability indices obtained relative to each member ⊗ are given with the precise non-linear elastoplastic model (table 5) but also with a linear one (table 6). The number of calls of the mechanical model (N_C) are included in these results. The value $N = 1.192\ P$ is used in the non-linear calculation as data for the example.

Table 5 shows that the results are very similar in the three methods, but with the number of calls N_C slightly smaller with the SRQ method than with Implicit FORM (24 compared with 34). With the classical simulation technique, this number is obviously larger.

TABLE 5. β_i obtained with a linear mechanical model

		Member ①	Member ②	Member ③
Simulations	β_C	5.391	6.560	5.727
	N_C	1000	1000	1000
Implicit Form	β_{HL}	5.488	6.791	5.787
	N_C	34	34	34
SRQ Method	β_{HL}	5.489	6.792	5.786
	N_C	24	24	24

Table 6 confirms the lower level of the reliability index when a geometrically non-linear model is taken into account, so that the effects of buckling are really considered. For member ①, the value of β_{HL} is now in the region of 2.4 compared to the previous value of 5.4. This shows that the reliability index is clearly affected by the phenomenon of instability. The values of N_C are smaller with the *explicitation method* than with Implicit FORM (for example, 135 compared to 36, for member ②). The *explicitation method* attains the goal of decreasing the computational costs. Note that when buckling is considered, the number of calls of the mechanical model is larger than in the linear case because the domain of failure becomes more complex.

TABLE 6. β_i obtained with a non-linear mechanical model

		Member ①	Member ②	Member ③
Simulations	β_C	3.712	7.228	5.531
	N_C	10000	1000	1000
Implicit Form	β_{HL}	2.393	6.893	5.235
	N_C	35	135	84
SRQ Method	β_{HL}	2.357	6.607	5.328
	N_C	36	36	36

The reliability indices for members ② and ③ are practically unaffected by the nature of the mechanical model, because they are subjected to tension before buckling of member ①.

Finally, a computation of the number of events in the failure domain with the simulation technique leads to a probability of failure equal to 0.92×10^{-2} for the member ①; this is very close to the values of $\Phi(-\beta)$ obtained with Implicit FORM: 0.84×10^{-2} and with the SRQ method: 0.92×10^{-2}.

5.2 COMBINATION OF ULTIMATE AND SERVICEABILITY LIMIT STATES

For the combination of two limit states (ultimate and serviceability), it is possible to obtain curves describing the evolution of the reliability of a simple compressed member, while taking account of the effect of non-linear displacements (very important for large values of slenderness) [8]. The SRQ method allows the values of the reliability index β to be obtained when a member may buckle, and the results can be compared together.

For example, we consider the calculation for a pin-ended IPE column axially loaded (fig. 7). Eccentricity of the axial load (e_{cc}) and defect of straightness (u_0) are taken into account.

The member is assumed to be made from S235 steel grade (yield stress $f_y = 235 \, MPa$) with an effective length equal to l. Table 7 shows the random variables taken into account.

If $\overline{\lambda}$ is the reduced slenderness parameter and A_g the gross area of the member, the LRFD Specifications (AISC 1986) give the maximum factored compressive load P_n as:

$$P_n = A_g \, F_{cr} \qquad (18)$$

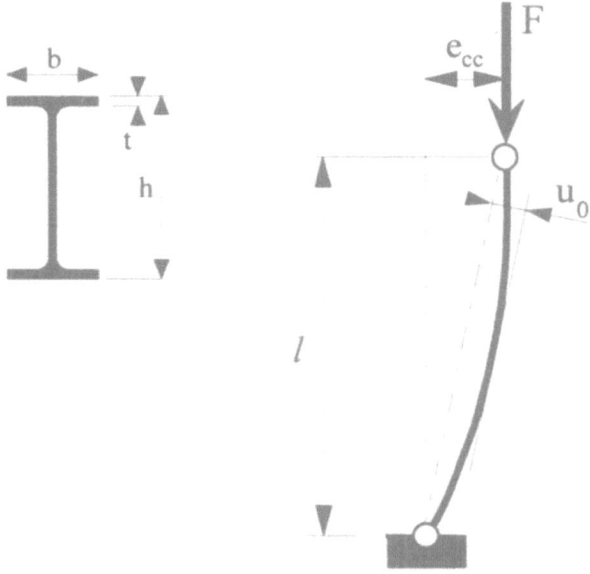

Figure 7. Pin-ended compression member

TABLE 7. Random variables

Random variables	Distribution	Mean value	COV	Tolerance
Cross-sectional dimensions				
Δh, Δb	Uniform	0 mm		± 0.2 mm
Δt	Uniform	0 mm		± 0.1 mm
Out-of-straightness u_0	Gaussian	0 mm	0.1 % l	± 0.3 % l
Eccentricity e_{cc}	Gaussian	0 mm	0.175 %	± 0.35 % l
Yield stress f_y	Log-normal	263.2 MPa	6 %	
Young Modulus E	Log-normal	21.10^4 MPa	6 %	
Axial load F	Log-normal	$\overline{F}$	10 %	

F_{cr} takes buckling into account; it is obtained from the following relationships:

$$F_{cr} = 0.658^{\lambda_c^2} f_y \qquad \text{if } \lambda_c \leq 1.5 \tag{19.a}$$

$$F_{cr} = \frac{0.877}{\lambda_c^2} f_y \qquad \text{if } \lambda_c > 1.5 \tag{19.b}$$

Then, the mean value $\overline{F}$ is obtained from:

$$\overline{F} = P_n / \gamma \tag{20}$$

where γ is the safety factor chosen here equal to 1.5 ($\overline{F}$ is the unfactored axial load).
With IPE shapes, the ultimate limit state (ULS) function has the form:

$$U_i = 1 - \frac{N}{N_p} \pm \frac{M}{\alpha M_p} = 0 \tag{21}$$

if N and M are respectively the axial load and the bending moment calculated with the non-linear mechanical model (and including implicitly the buckling behaviour), and N_p and M_p are the plastic resistances ($\alpha = 1.22$ with this shape).
    The serviceability limit state (SLS) function is:

$$S_i = \delta_{300} \pm \delta \tag{22}$$

if δ is the computed deflection due to buckling and δ_{300} is the allowable deflection chosen here equal to $\ell / 300$.
    The evolution of the reliability indexes as a function of the slenderness is shown in figure 8.
    Before interpreting these results, the difference between the two limit states must be emphasised. Reaching an ultimate limit state has much more severe consequences than reaching a serviceability limit state, and the probability of failure attached to each of them must be different. In the ECCS recommendations [9] for example, the considered values are $P_f(ULS) \leq 10^{-5}$ and $P_f(SLS) \leq 5 \times 10^{-2}$.
    Coming back to figure 8, it can be checked that, in the range of the small values of slenderness, the ultimate limit state is the most severe. Buckling does not really affect the behaviour of the column.

In the range of large slenderness ($\bar{\lambda} > 1$) or, in other words, when buckling creates a significant effect, if β_{ULS} is relatively stable ($2.6 \leq \beta_{ULS} \leq 2.9$), β_{SLS} decreases very rapidly ($\beta_{SLS} \approx 0.8$ when $\bar{\lambda} > 1.5$) and a serviceability limit state may have to be considered.

Generally, standard codes do not require to check such a limit state under axial loading but, considering that steel makers are trying to produce columns with higher yield stress, it may become necessary to be careful in the range of large slenderness. An increasing of resistance will decrease $P_f(ULS)$ but will not affect $P_f(SLS)$ and the serviceability limit state may become dominant.

The SRQ method allowing to take into account the second order effect due to buckling, it is possible to determine β_{SLS} as well as β_{ULS} and to define new limits of acceptability for compressed columns, from the point of view of strength but also from the one of displacement.

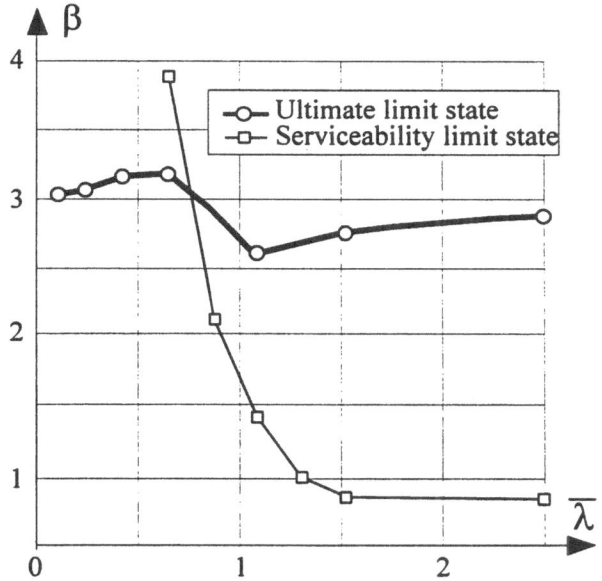

Figure 8. AISC-LRFD: Curves $\beta = f(\bar{\lambda})$ obtained with two kinds of limit states

6. Conclusion

This paper presents a reliability method of calculation which allows the solution of the problem of implicit limit states met in the case of non-linear behaviour.

It is applied to an example where a comparison is carried out with *Implicit FORM* and simulation methods. The SRQ method provides a better approach to real structural reliability: it can account for possible defects in a non-linear context (buckling for example) and uses less computation time than other methods: simulation, *Implicit FORM*...

Being not carried out in the space of the standard Gaussian variables but in the one of the original random variables, the limit state functions are not much distorted and can be easily mathematically treated.

The problem which can be found with such a method is that there are complex limit state functions which cannot be approximated by a quadratic form without introducing too many errors but it is very rare in the field of structural mechanics.

Having a strong effect on the convergence of the method, the choice of the calculation points is particularly important and they must be carefully chosen. Nevertheless, further works must be carried out to improve their choice.

7. References

1. Thoft-Christensen, P. and Murotsu, Y.: *Application of structural systems reliability.* Springer-Verlag, Berlin, Heidelberg, 1986.
2. Lemaire M., Chung Fook Mun J.F., Mohamed Mohamed A., Muzeau J.P.: *Model of negative strain-hardening to evaluate the reliability of beam-columns.* 6th International Conference on Applications of Statistics and Probability in Civil Engineering, CERRA-ICASP'6, Mexico, June 1991.
3. Muzeau J.P.: *Reliability of steel columns subjected to buckling. Comparison of international standard codes of design.* 5th International Conference on Structural Reliability, ICOSSAR'89, San-Francisco.
4. El-Tawil K., Lemaire M., Muzeau J.P.: *Reliability method to solve mechanical problems with implicit limit states.* 4th WG 7.5 Working Conference on Reliability and Optimization of Structural Systems, IFIP, Munich, September 1991, Springler-Verlag, Lecture notes in Engineering.
5. Muzeau J.P., Lemaire M. and El-Tawil K.: *Modèle fiabiliste des surfaces de réponse quadratiques (SRQ) et évaluation des règlements.* Construction Métallique, n°3, 1992.
6. Rackwitz R. and Fissler B.: *Structural reliability under combined random load sequence.* Computer and Structures, Vol. 9, 1978.
7. Shinozuka M.: *Basic analysis of structural safety.* Journal of Structural Engineering, Vol. 109, n°3, March 1983.
8. Muzeau J.P., Lemaire M. and Besse P.: *Association of serviceability and strength limit states in the evaluation of reliability in non-linear behaviour.* 6th International Conference on Structural Safety and Reliability, ICOSSAR'93, Vol. III, pp. 1957-1960, Innsbruck, Austria, August 1993.
9. E.C.C.S.: *European recommendations for steel construction,* ECCS-EG 77-1E, 1977.

METHODS OF SYSTEM RELIABILITY

IN MULTIDIMENSIONAL SPACES

R. RACKWITZ
Technical University Munich
Arcisstr. 21, 80290 Munich, Germany

1. Introduction – Notions and Definitions

In general, a **system** is understood as a technical arrangement of clearly identifiable (system–) **components** whose functioning depends on the proper functioning of all or a subset of its components. For a reliability analysis, a number of idealizations are convenient if not necessary. It is assumed that the components can attain only two states, i.e. one **functioning** (safe, working, active,....) and one **failure** (unsafe, defect, inactive,...) state. This is a simplification which is not always appropriate but we will maintain it throughout the text. If there is a natural multi–state description of a component or a system we shall assume that this is reduced to a two–state description in a suitable manner. In practice, this step of modeling might be not an easy task. It is, nevertheless, mandatory in practical system reliability analyses. Several attempts have been made to establish concepts for analyzing systems with **multi–state components** (see, for example, Caldarola, 1980; Fardis and Cornell, 1981). It should be clear that systems then have also multiple states and the definition of safe or failure states requires great care. Such relatively recent extensions of the classical concepts cannot be dealt with herein.

A representation of component performance by only two states is called a **Boolean** representation, but we shall avoid the explicit use of Boolean algebra as far as possible. As a consequence of the Boolean component representation, systems can be only in either the **functioning** or in the **failure** state. We shall only deal with so–called **coherent** systems, i.e. systems which remain intact if an additional functioning component is added.

One can distinguish two basic types of systems, the **series** and the **parallel** system. Later, we shall add other related types whose separate definition is useful for classification and calculation purposes. A series system consisting of n components is said to fail if **any** of its components fail. Classical examples are the chain whose failure is a consequence of the failure of any of its links, or a four–wheel car where any flat tire prohibits further use of the car (usually). A parallel system of n components is said to fail if **all** components fail. As an example, assume that a town is supplied by several electrical lines and each one is capable of delivering the required power. Or, in aircraft control, two computers

C. Guedes Soares (ed.), Probabilistic Methods for Structural Design, 161–212.
© 1997 *Kluwer Academic Publishers.*

are installed in ideal stand–by redundancy. If the first fails the second takes over and can fulfill all demands. The control system fails if the second computer fails, too. Remember, however, that in most technical systems such as structures, failure of some components causes higher loads on the remaining components. This problem will require some thought. A system is called **redundant** if it can remain intact in spite of the failures of some of its components.

In general, systems are built up by many components in a complex logical arrangement of series and parallel subsystems. Let $F_i = \{X \in V_i\}$ be the failure event of the i–th system component, and V_i the failure domain. Denote by F the system failure event. Clearly, for a series system ("or"–connection) we have F as the union of the individual failure events (see figure 1.1)

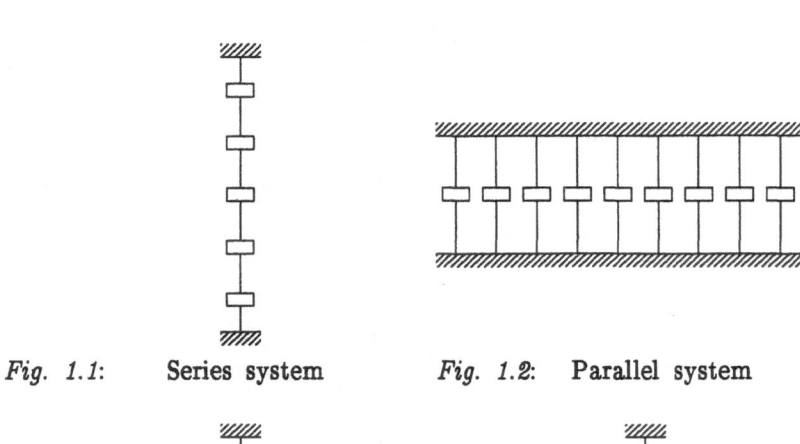

Fig. 1.1:   Series system   Fig. 1.2:   Parallel system

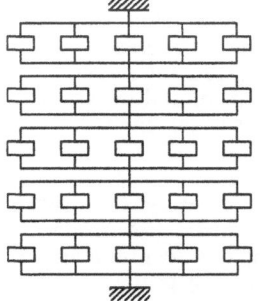
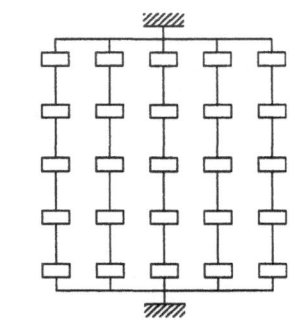

Fig. 1.3:   Parallel systems   Fig. 1.4:   Series systems
             in series   in parallel

$$F_s = \cup F_i \tag{1.1}$$

while for the parallel system ("and"–connection) F is the intersection of the F_i's (figure 1.2)

$$F_p = \cap F_i \tag{1.2}$$

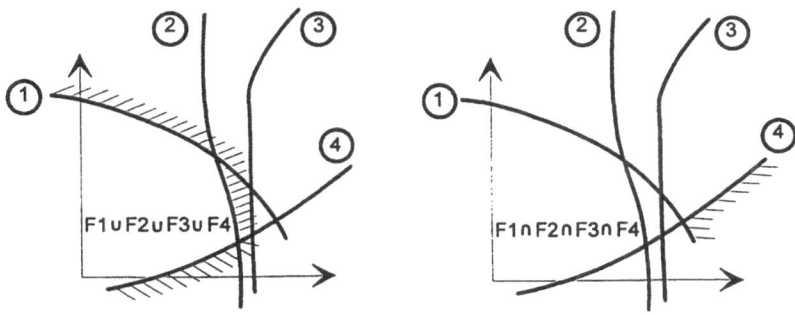

Fig. 1.5:   Series system   *Fig. 1.6:*   Parallel system

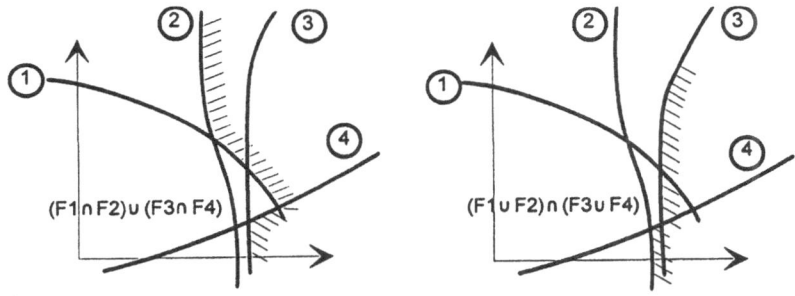

Fig. 1.7:   Parallel systems   *Fig. 1.8:*   Series systems
              in series   in parallel

Correspondingly, for parallel systems in series (unions of intersections) we have (figure 1.3)

$$F = \cup \cap F_{ij} \qquad (1.3)$$

whereas series systems in parallel (intersections of unions) are described by (figure 1.4)

$$F = \cap \cup F_{ij} \qquad (1.4)$$

For convenience, the same systems are also presented in figures 1.5 to 1.8 when the F_i's are given by certain domains in the space of uncertain variables $X = (X_1, X_2)$.

Of utmost importance in reliability theory is the fact that any system can be reduced to either of the last two forms in making extensive use of the distributive laws of set algebra

$$F_i \cap (F_j \cup F_k) = (F_i \cap F_j) \cup (F_i \cap F_k) \qquad (1.5)$$

$$F_i \cup (F_j \cap F_k) = (F_i \cup F_j) \cap (F_i \cup F_k) \qquad (1.6)$$

Furthermore, essential reductions are usually possible by applying the so–called absorption rules, i.e. for $F_i \subseteq F_j$ there is

$$F_i \cup F_j = F_j \quad \text{and} \quad F_i \cap F_j = F_i \qquad (1.7a)$$

or for $F_i \subset F_j$ and $F_k \subset F_j$ there is

$$(F_i \cup F_k) \subset F_j \text{ and } (F_i \cap F_k) \subset F_j \qquad (1.7b)$$

It follows that the union or intersection of an event with itself is the event. The absorption rules are important when making certain sets **minimal**. If, in particular, eq. (1.3) is a minimal set it is denoted by a **minimal cut set**. Cut sets are minimal if they contain no other cut set as a genuine subset. Analogously, representation (1.4) is called a **tie set**. Such sets are minimal if no tie set contains another tie set as a genuine subset.

The analysis has four steps:
1) investigate the logical structure of the interaction of the components of the system
2) reduce the system to a minimal one
3) evaluate probabilities
4) determine sensitivity and importance measures of parameters, components and subsystems.

The first step involves classical engineering evaluations and, probably, is the most difficult task. It requires much care and experience to model components and the system realistically and, in a reliability sense, completely. This modeling phase must be done with due consideration of the various consecutive steps. The second step will be highly formalized; a few hints will be given later, and the reader is referred to the vast literature in this area for additional information. The third step will be the main subject of this review. The fourth step, although important and informative, will not be considered.

2. Formal Logical Analysis of Systems

The logical structure of simple system can usually be assessed directly and we shall illustrate this by a simple example. Complex systems require more formal tools when assessing and reducing the logical structure because a direct analysis can be prone to error, and lengthy.

Illustration 2.1: Water supply system

We consider a simple supply system as shown in figure 1. Two sources S1 and S2 supply two consumers (town areas) A1 and A2. The arrows indicate the possible direction of flow. The system is said to fail if one of the consumers is no longer supplied. Failure could be brought on by some extraordinary event such as a flood, earthquake or fire. Here, it is easy to write down all possible connections leading to system failure.

$$F = \{[F_1 \cap (F_2 \cup F_3) \cap (F_3 \cup F_4 \cup F_5)] \cup$$
$$\cup [(F_1 \cup F_5) \cap (F_3 \cup F_4) \cap (F_2 \cup F_3 \cup F_5)]\} \quad (1)$$

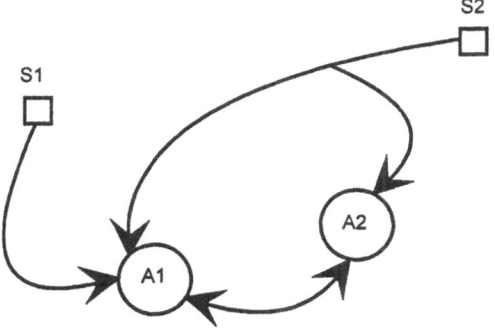

System failure is when A1 or A2 are not supplied

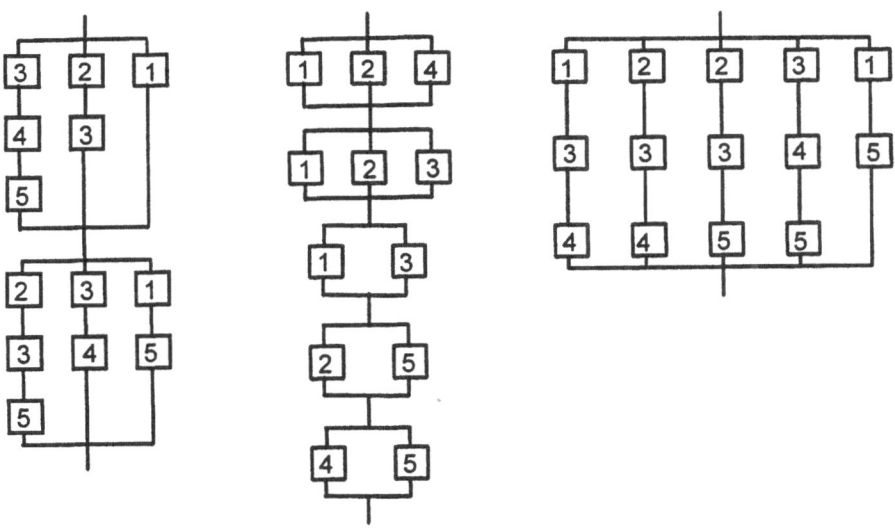

Figure 2.1:   System representation by block diagrams

The system can be represented in terms of a block diagram in which one can easily recognize the logical structure. As in eq. (1) one considers the supply of A1 and A2 separately. For example, A1 is not supplied if line 1 **and** (line 2 or 3) **and** (line 3 or line 4 or line 5) are broken.

This system of events in eq. (1) is not yet a cut set system and not yet minimal. One may now apply the laws (1.5) and (1.6) and obtain for the supply of A1

$$F_{A1} = \{(F_1 \cap F_2 \cap F_3) \cup (F_1 \cap F_2 \cap F_4) \cup (F_1 \cap F_2 \cap F_5) \cup$$

$$\cup (F_1 \cap F_3 \cap F_3) \cap (F_1 \cap F_3 \cap F_4) \cup (F_1 \cap F_3 \cap F_5)\} \qquad (2)$$

(Carry out all **cuts** in the upper half of the block diagram which make the system fail). Next the absorption laws are applied. First all multiple events in a cut set are deleted, except one. Next, multiple cut sets are deleted except one. Finally, all cut sets which are subsets of other cut sets are deleted. In doing so one arrives at

$$F = \{(F_1 \cap F_3) \cup (F_3 \cap F_5) \cup (F_4 \cap F_5) \cup$$

$$\cup (F_1 \cap F_3 \cap F_4) \cup (F_1 \cap F_3 \cap F_5)\} \qquad (3)$$

Quite analogously, one can produce tie sets. We remember that according to de Morgan's law, it is $\overline{A \cap B} = \overline{A} \cup \overline{B}$ and $\overline{A \cup B} = \overline{A} \cap \overline{B}$. Therefore, for the representations (1.3) and (1.4) we have $\cup \cap F_{ij} = \Omega \backslash (\cap \cup \overline{F}_{ij})$ and $\cap \cup F_{ij} = \Omega \backslash (\cup \cap \overline{F}_{ij})$. Hence, having found the minimal cut set for the failure events yields by passing over to the complementary events (by reversing the set operators) we find the minimal tie set of safe events and vice versa. We recommend this as an exercise. The result is

$$F = \{(F_1 \cup F_5) \cap (F_1 \cup F_3 \cup F_4) \cap (F_2 \cup F_3 \cup F_4) \cap$$

$$\cap (F_2 \cup F_3 \cup F_5) \cap (F_3 \cup F_4 \cup F_5)\} \qquad (4)$$

#

The general case requires more systematic tools. A first possibility is a complete analysis of all possible sequences of events. This type of analysis is called **event tree analysis**. For larger systems this can become quite cumbersome but it will be seen later that in a number of applications this is the only and natural way to arrive at a suitable representation of the situation. Another famous approach is called **fault tree analysis**. The two types of analysis are complementary. The first starts the analysis from the intact system, the second from the failed system. They will be illustrated at examples.

Illustration 2.2 (cont.): Water supply system (failure and fault trees)

In our artificial water supply system the time—sequence of failures of components is irrelevant for the final system states (but not necessarily for the corresponding probabilities). Here, we develop sequences of events starting from component No.1

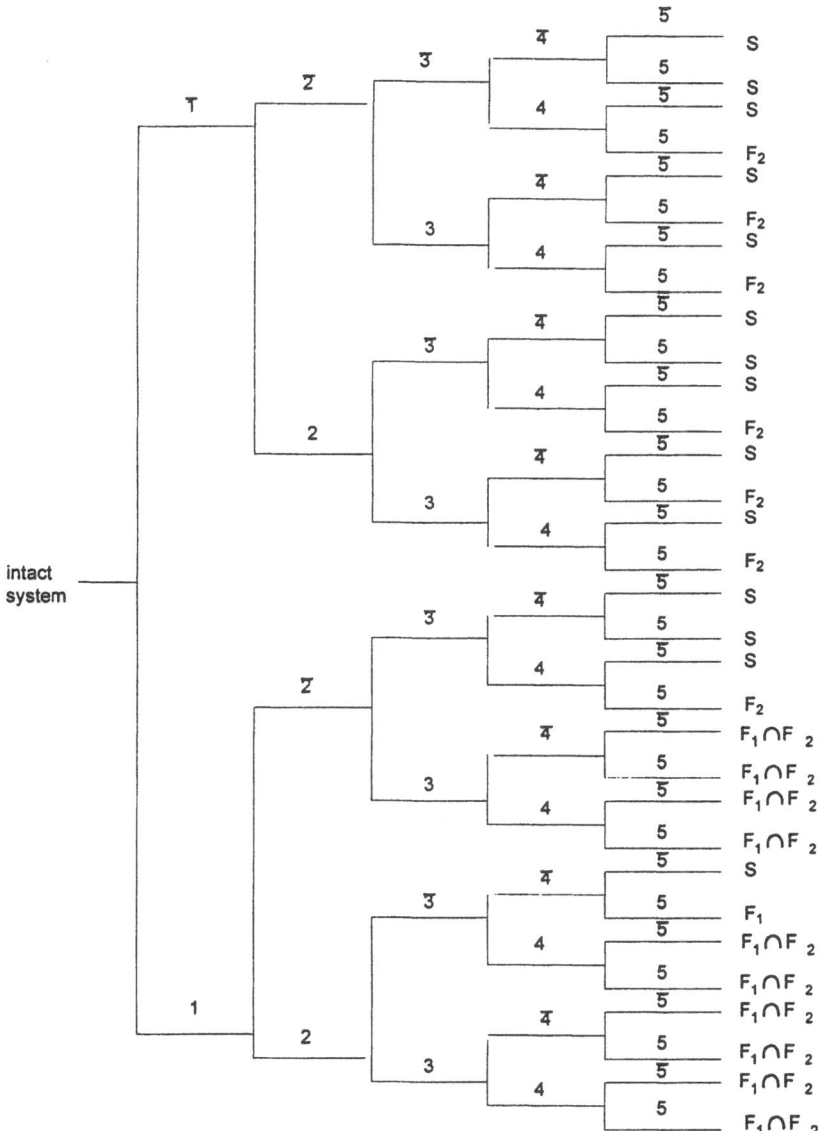

Figure 2.2:   Event tree for water supply system

(compare figure 1). The reader may verify that the same result is obtained by starting at another component. One observes that at the end of each branch one arrives at four types of events: Survival (S), Failure of supply of A1 (F_1), Failure of supply of A2 (F_2), Failure of supply of A1 and A2 ($F_1 \cap F_2$).

The complete assembly of survival and failure events constitutes an exhaustive, disjoint system of events. In applications one might wish to differentiate between the different failure types because they are associated with different consequences. Usually only the failure branches of the **event tree** are of interest. In this case the event tree may be called the **failure tree**, and it is sufficient to investigate only those branches which lead to failure (failure **branches**, failure **paths**).

Another possibility of system analysis is by so–called **fault** trees which is a backwards analysis technique of the failure branches of an event tree. System failure is the top event. Then, a next level of subsystem and its logical connection is defined by **or** or **and** gates. In this manner one pursues all possibilities until one arrives at the componental basic events. The designation of **and** or **or** gates is not standardized. Here, we use a + for the **or** gate and a • for the **and** gate.

Neither the result of the event tree nor of the fault tree has been reduced to a minimal form. Many formal algorithms exist for these reductions but they resemble each other to a large degree. Their differences can frequently only be recognized for very large systems. An account of several methods and some special tasks is given in Yen (1975) (see also Barlow and Proschan (1975)).

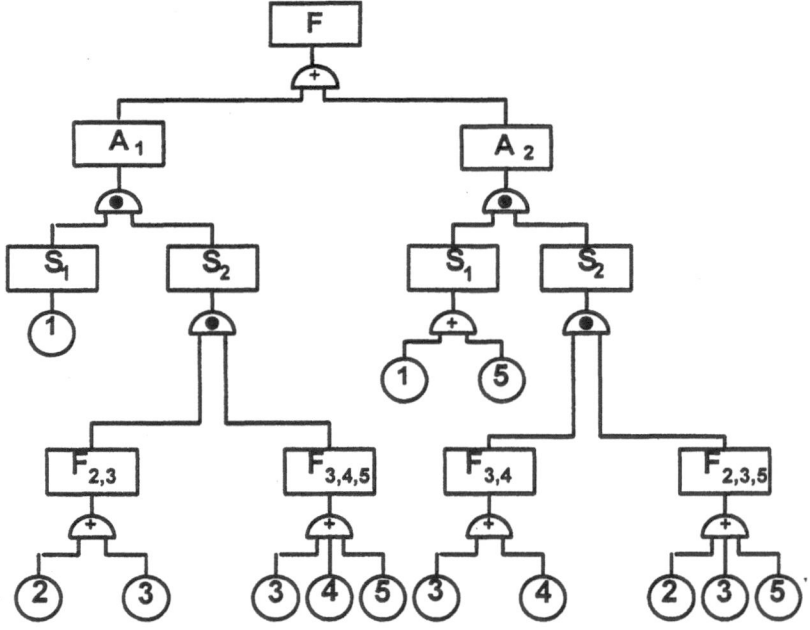

Figure 2.3:   Fault tree for water supply system

#

3. Elementary Probabilistic Evaluation of Systems

In this section we compute the failure probability of systems as analyzed before, under the more or less restrictive assumption that the componental failure events are either independent or fully dependent.

For a series system (or–connection) made up of independent events, we may use the complementary events F_i to give

$$P_{f,S} = P(\underset{i}{\cup} F_i) = 1 - P(\underset{i}{\cap} \bar{F}_i)$$

$$= 1 - \underset{i}{\Pi} P(\bar{F}_i) = 1 - \underset{i}{\Pi}(1 - P(F_i)) \tag{3.1}$$

and, analogously, for the parallel system

$$P_{f,P} = P(\cap F_i) = \underset{i}{\Pi} P(F_i) \tag{3.2}$$

In the fully dependent case we have

$$P_{f,S} = P(\underset{i}{\cup} F_i) = \underset{i}{\max} \{P(F_i)\} \tag{3.3}$$

$$P_{f,P} = P(\underset{i}{\cap} F_i) = \underset{i}{\min} \{P(F_i)\} \tag{3.4}$$

Systems with more complex structure are more difficult to handle. If we assume that the cut sets are disjoint, then, the third Kolmogorovian theorem of probability theory applies directly. Remember, that disjoint cut sets usually consist of failure and survival events which here are denoted by F_i^*.

$$P_f = P(\underset{}{\dot\cup} \cap F_{ij}^*) = \underset{i}{\Sigma} P(\underset{j}{\cap} F_{ij}^*) \tag{3.5}$$

$\dot\cup$ stands for a disjoint union. In general, disjoint sets are larger than minimal cut sets. Also, the presence of failure and survival events in the same cut set can make their evaluation difficult. Therefore, they are seldom used.

If the cut sets are not disjoint, one may use the well–known expansion formula for the probability of unions of events, i.e.:

$$P(\cup F_i) = \underset{i}{\Sigma} P(F_i) - \underset{i<j}{\Sigma \Sigma} P(F_i \cap F_j)$$

$$+ \underset{i<j<k}{\Sigma \Sigma \Sigma} P(F_i \cap F_j \cap F_k) - ... (-1)^{n+1} P(F_1 \cap ... F_n) \tag{3.6}$$

Several cut sets in a cut system representation can share the same components. Therefore, cut set failures are no longer mutually exclusive, and a more detailed analysis is required as before. The use of the sum of componental failure probabilities (the first term m (3.6)) always provides an upper bound P_U, as is easily verified. It may be shown that the first two terms in (3.6) always give a lower bound, P_L, to the probability. Thus $0 \le P_L \le P(\cup F_i) \le P_U \le 1$. Unfortunately, these bounds are unsatisfactorily wide for large systems. Consideration of higher order terms in eq. (3.6) is rather laborious, and it is not obvious where to truncate the expansion. The terms first tend to increase, so that the bounds established by considering partial sums increase also; only later do the terms decrease so that the sum converges to the exact result.

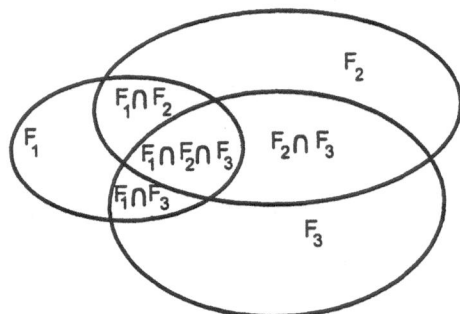

Figure 3.1:   Derivation of Ditlevsen's bounds

It is possible, however, to derive simple bounds of increasing order and increasing narrowness. The idea can be deduced from figure 3.1. For the first two events we have

$$P(F_1 \cup F_2) = P(F_1) + P(F_2) - P(F_1 \cap F_2)$$

For the third event in a union, an upper bound is obtained if the intersection with the larger probability is subtracted, i. e. $P(F_1 \cap F_3)$ or $P(F_2 \cap F_3)$, from the additional term $P(F_3)$. A lower bound is to subtract the sum of these joint probabilities provided that they are not larger than $P(F_3)$. Repeated application of this scheme for more than three events yields:

$$P(\bigcup_{i=1}^{n} F_i) = \begin{cases} \le P(F_1) + \sum_{i=2}^{n} \{P(F_i) - \max_{j<i} \{P(F_i \cap F_j)\} \\ \ge P(F_1) + \sum_{i=2}^{n} \max \{0, P(F_i) - \sum_{i=2}^{n} P(F_i \cap F_j)\} \end{cases} \quad (3.7)$$

This elementary result has been derived repeatedly, in one form or another, e.g. by Konias (1966), Hunter (1976,1977), Ditlevsen (1979). The narrowness of these bounds depends on the ordering of the events. A different ordering may be necessary for the upper and the lower bound. An algorithm for a best ordering is given by Dawson and Sankoff (1967). Hohenbichler (1980) (see also Hohenbichler and Rackwitz (1983)) generalized these bounds to include more than two—dimensional intersections (but not all higher dimensional intersections). It was found by numerical studies that little is gained by those extensions except for small (!) systems.

The (two–dimensional) bounds are exact for fully dependent events, are very close to the exact result for independent events, become less satisfactory with increasing number of events, but are generally of high quality for small probability events.

Irrespective of the dependence structure of the events, we have, as mentioned shortly after eq. (3.6), the so–called trivial bounds

$$\max_{i=1}^{n} \{P(F_i)\} \leq P(\bigcup_{i=1}^{n} F_i) \leq \sum_{i=1}^{n} P(F_i) \leq 1 \qquad (3.8)$$

for the series system and

$$0 \leq P(\bigcap_{i=1}^{n} F_i) \leq \min_{i=1}^{n} \{P(F_i)\} \qquad (3.9)$$

for the parallel system; these are direct consequences of the elementary probability theorem. The bounds (3.8) become obsolete for larger systems and/or larger individual event probabilities. The bound (3.9) generally is of little use. Slightly better is the following obvious relationship

$$0 \leq P(\bigcap_{i=1}^{n} F_i) \leq \min_{j<i=1}^{n} \{P(F_i \cap F_j)\} \qquad (3.10)$$

which is of the same nature as eq. (3.7) since it involves the probability of the intersection of any two events. As for eq. (3.7), higher order intersection terms could be included in bounds of the type of eq. (3.10) but with moderate improvement of the upper bound and, without further information about the dependence structure of the componental failure events, no possibility of sharpening the lower limit. This is the reason why we shall not further investigate tie sets of failure events. In addition, if the events F_i and F_j in eq. (3.10) correspond to unions we have, for example, $F_i \cap F_j = (\cup_r F_{ir}) \cap (\cup_s F_{js})$ whose probability in turn is only easily computed if first a (minimal) cut set representation is found and one of the formula (3.5) to (3.7) are applicable.

4. First–Order Reliability Methods for System Analysis (FORM)

4.1 COMPONENTS

Let $U = (U_1,...,U_n)^T$ be an independent standard normal vector and the failure domain be given as $V = \{g(U) \leq 0\} = \{a^T U + \beta \leq 0\}$ with $g(0) > 0$ and $\|a\| = 1$. Then, the exact failure probability is

$$P_f = P(F) = P(U \in V) = \Phi(-\beta) \qquad (4.1.1)$$

$u^* = -\beta a$ the β–point, a its vector of direction cosines and β the geometrical safety index $\beta = + \|u^*\|$.

The standard normal integral $\Phi(c)$ may be determined by one of the expansions given in the literature. Note that in eq. (4.1.1) we distinguished the failure event F and the failure domain V. This is formally correct, but at the moment not really necessary. Therefore, we shall use $P(V)$ instead of $P(U \in V)$ for $P(F)$ in the sequel.

Next, we generalize this result for non–linear, differentiable failure surfaces $g(u) = 0$ by expanding it to first–order in the so–called β–point which, for the moment, is defined as the minimal distance of points u on $g(u) = 0$ to the coordinate origin. For β we use the convention

$$\beta = \begin{cases} + \|u^*\| & \text{for } g(0) > 0 \\ \\ - \|u^*\| & \text{for } g(0) \leq 0 \end{cases} \tag{4.1.2}$$

while

$$\|u^*\| = \min \{\|u\|\} \quad \text{for } \{u: g(u) \leq 0\} \tag{4.1.3}$$

Finding u^* is a problem of optimization (minimizing $\|u\|$) under an inequality constraint. The inequality condition is written here only as a reminder that the failure surface $g(u) = 0$ resp. the failure domain $V = \{g(U) \leq 0\}$ is assumed to be non–degenerate, i.e. that the failure set in a sufficiently small neighborhood of u^* is non–empty and has non–zero probability. Practically, the equality condition is sufficient. The search for u^* will be discussed later. Here, it is further assumed that $g(u^*)$ possesses all first–order derivatives so that a tangent linear approximation of the failure surface is uniquely defined; this expansion is possible because the multi–normal density drops off with $\exp[- 1/2 \|u\|^2]$. At the β–point $\|u\|$ is a minimum and $u^* = \max \{\varphi(u)\}$. As a consequence, u^* is also denoted as the **most likely failure point**. In the non–linear case,

$$P_f \approx \Phi(- \beta) \tag{4.1.4}$$

This approximation is widely accepted, even though there is an unquantified error. Again, $u^* = -\beta \, \alpha$, where $\alpha = g(u^*)/\|g(u^*)\|$ is the normalized gradient of $g(u^*) = 0$ at u^* and the linear approximation to $g(u) = 0$ is $h(u) = \alpha^T(u - u^*) = \alpha^T u + \beta = 0$.

4.2 UNIONS AND INTERSECTIONS

The failure event (domain) can also be given as a union of intersections of individual (componental) failure domains. Let

$$V = \bigcap_{i=1}^{m} V_i$$

with $V_i = H_i = \{\alpha_i^T U + \beta_i \leq 0\} = \{Z_i \leq -\beta_i\}$. If the individual failure domains V_i are originally bounded by non–linear failure surfaces, we understand that these failure domains have been replaced by linearly bounded half–spaces H_i as described in section 4.1. The covariance matrix for the vector Z is given by $\Sigma_Z = \{\alpha_i^T \alpha_j ; i,j = 1,...,m\}$. This is equal to the correlation coefficient matrix R, because Z is a zero mean, unit variance vector. The failure probability becomes (Hohenbichler and Rackwitz, 1983):

$$P_f = P(\bigcap_{i=1}^{m} \{Z_i \le -\beta_i\}) = \Phi_m(-\beta ; R) \qquad (4.2.1)$$

Φ_m is the multinormal integral. Similarly, for a union of events

$$V = \bigcup_{i=1}^{m} V_i$$

we have

$$P_f = P(\bigcup_{i=1}^{m} \{Z_i \le -\beta_i\}) = 1 - P(\bigcap_{i=1}^{m} \{Z_i > -\beta_i\})$$

$$= 1 - P(\bigcap_{i=1}^{m} \{Z_i \le \beta_i\}) = 1 - \Phi_m(\beta ; R) \qquad 4.2.2)$$

The numerical evaluation of the standard multinormal integral $\Phi_m(c ; R)$ is essential for eqs. (4.2.1) and (4.2.2) to be of any practical use. Unfortunately, there are no simple general and exact results.

4.3 CUT SET SYSTEMS

As outlined in section 2, more general failure domains (systems) must be given either in terms of a disjoint or a (minimal) cut set representation. In the first case the cuts can also contain safe events, and there will be no additional difficulties. The system failure probability is simply the sum of all cut probabilities. If, for example, a cut set is given by $\{V_i, V_j\}$ where $V_i = \{g_i(U) \le 0\}$ and $V_j = \{g_j(U > 0\}$ it is clear that by multiplying $g_j(U)$ by (-1) one obtains $V_j = \{-g_j(U \le 0\}$ as required for formulae (4.1) and (4.2). In contrast to our general assumption, V_j now is a large probability event but its intersection with small probability events V_i may still yield small joint probabilities. In a first–order context the multiplications of $g_j(U)$ by (-1) leads to a sign–change of the original correlation coefficient.

If the system is represented by minimal cut sets, one straightforward calculation method is to use the Ditlevsen bounds as derived in section 3 in eq. (3.7). They require the evaluation of the intersection of any two intersections of failure domains, i.e. the probabilities

$$P(V_i \cap V_j) = P(\overset{m_i}{\underset{r=1}{\cap}} V_r \cap \overset{m_j}{\underset{s=1}{\cap}} V_s) = \Phi_{m_i + m_j}(\beta_{i+j}; R_{i+j})$$

$$(4.3.1)$$

with $\beta_{i+j} = (\beta_1, ..., \beta_{m_i}, \beta_{m_i+1}, ..., \beta_{m_i+m_j})^T$ and

$$R_{I \setminus J} = \{\alpha_p^T \alpha_q \; ; \; p,q = 1, ..., m_i, m_i + 1, ..., m_i + m_j\}$$

Alternatively, a formally exact result is also obtained by applying the expansion theorem (3.6), but the numerical effort may become great for large systems. The same is true for the calculation of disjoint cut set probabilities, so that, in practice, one prefers the bounds (3.7). These, in turn, may be weakened in that, after arranging the cut sets according to their (descending) probabilities, the intersection probabilities of any two cut sets are only computed for the k first few dominating sets, while the rest of the cut sets are taken into account by either their upper bound or their lower bound.

Illustration 4.3.1 (cont.): Water supply system

We are ready to apply the above results to the water supply system discussed previously. Assume that the componental failure events are now given by $V_i = \{X_i - Y \le 0\}$ where the X_i represent some **resistance** variables which are assumed to be independent and normally distributed with mean $m_i = m$ and standard deviation $\sigma_i = \sigma$. Y is a normal **loading** variable with mean μ and standard deviation τ. Therefore, the componental failure probabilities are

$$P(F_i) = P(V_i) = P(X_i - Y \le 0) = P(Z_i \le -\beta_i) = \Phi(-\beta_i)$$

$$(1)$$

It follows that $\beta = \beta_i = (m - \mu)(\sigma^2 + \tau^2)^{-1/2}$ and $\rho = \rho_{ij} = Cov[Z_i, Z_j] = \tau^2(\sigma^2 + \tau^2)^{-1}$. Let the parameters be chosen such that $\beta = 3$ and $\rho = 0.5$. The state variables are represented by

$$Z_i = \sqrt{\rho} \, U + \sqrt{1 - \rho} \, U_i$$

$$(2)$$

By conditioning first on the variable $U = u$, one recognizes that the variables Z_i are conditionally independent and the results of section 3 apply. In particular, it can easily be verified that

$$P(F) = \int_{-\infty}^{+\infty} [3 \, p^2(u) - 4 \, p^4(u) + 2 \, p^5(u)] \, \varphi(u) \, du$$

$$(3)$$

with $p(u) = \Phi((-\beta - \sqrt{\rho}u)/(1-\rho)^{1/2})$. The term in square brackets would be the exact system failure probability if the components were independent. $\varphi(c)$ is the standard normal density. Numerical integration yields $P(F) = 2.31 \cdot 10^{-4}$ with component probability $P(F_i) = 1.35 \cdot 10^{-3}$. Note that $P(F) = 5.47 \cdot 10^{-6}$ would have been obtained if the F_i's had been independent. This emphasizes the significance of stochastic dependencies among componental failures.

The same system is used to illustrate the material in this section. One determines $P_{f,ik} = P(F_i \cap F_k) = 8.19 \cdot 10^{-5}$ and $P_{f,ijk} = P(F_i \cap F_j \cap F_k) = 1.51 \cdot 10{-5}$. The trivial bounds for the system failure probability eq. (3.8) become

$$\max\{P_{f,k}\} = 8.19 \ 10^{-5} \leq P_f \leq 2.76 \ 10^{-4} = \sum_{k=1}^{5} P_{f,k} \tag{4}$$

Ditlevsen's bounds require the probabilities of the intersections of any two cut sets in the system. These probabilities are collected in the following matrix $P = \{p_{ij}: i,j = 1,...,5\}$:

$$P = \begin{bmatrix} 1.51 \cdot 10^{-5} & 4.65 \cdot 10^{-6} & 4.65 \cdot 10^{-6} & 1.90 \cdot 10^{-6} & 4.65 \cdot 10^{-6} \\ & 1.51 \cdot 10^{-5} & 4.65 \cdot 10^{-6} & 4.65 \cdot 10^{-6} & 4.65 \cdot 10^{-6} \\ & & 8.19 \cdot 10^{-5} & 1.51 \cdot 10^{-5} & 4.65 \cdot 10^{-6} \\ & \text{symm.} & & 8.19 \cdot 10^{-5} & 1.51 \cdot 10^{-5} \\ & & & & 8.19 \cdot 10^{-5} \end{bmatrix}$$

For example, the element p_{12} is computed from
$p_{12} = P((F_1 \cap F_2 \cap F_4) \cap (F_1 \cap F_2 \cap F_5)) = P(F_1 \cap F_2 \cap F_4 \cap F_5)$
$= \Phi_4(\{-3\};\{0.5\}) = 4.65 \cdot 10^{-6}$. The sharper bounds eq. (3.7) give

$$2.26 \ 10^{-4} \leq P_f \leq 2.42 \ 10^{-4}$$

with essentially the same numerical values, whatever sequence of the five cut sets is considered. These bounds are appreciably narrower than the trivial bounds and, of course, contain the exact result.

#

This chapter describes the first–order reliability method. Beginning with the fundamental work of Breitung (1984) and Hohenbichler et al. (1987), asymptotic concepts have been applied, leading to an improvement and justification of the first–order theory. It is outlined in appendix A.

4.4 PROBABILITY DISTRIBUTION TRANSFORMATIONS

The foregoing results are very special as they apply only to independent standard normal variates. However, if the **distribution** of the original basic variables is continuous, it is always possible to find a probability distribution transformation

$$X = T(U) \tag{4.4.1}$$

such that

$$P_f = P(h(X) \leq 0) = P(h(T(U)) \leq 0) = P(g(U) \leq 0) \tag{4.4.2}$$

where we used the abbreviation $h(T(U)) = g(U)$. Such transformations are well known from simulation. Remember that if a random number generator is available for producing uniformly distributed variables G_i in $[0,1]$, then, we use the identity $P(G_i \leq g) = F_G(g) = F_X(x) = P(X_i \leq x)$ to produce random numbers for the variable X. By solving for X_i we obtain $X_i = F_X^{-1}[G_i]$ as random numbers distributed according to F_X. A similar concept is applied to eq. (4.4.1). Let X be an independent vector with marginal distribution functions $X_i \sim F_i(x)$. It follows that the identity

$$F_i(x_i) = \Phi(u_i) \tag{4.4.3}$$

holds and, therefore (Rackwitz/Fiessler,1978),:

$$X_i = F_i^{-1}[\Phi(U_i)] \tag{4.4.4a}$$

or

$$U_i = \Phi^{-1}[F_i(X_i)] \tag{4.4.4b}$$

The multidimensional dependent case is more involved. If X has distribution function $F_X(x) = P(X \leq x) = P\left(\cap_{i=1}^n \{X_i \leq x_i\}\right)$, then it is always possible to represent this distribution function as a product of conditional distribution functions, i.e.

$$F_X(x) = F_1(x_1) \, F_2(x_2|x_1) \dots F_n(x_n|x_1,\dots,x_{n-1})$$

where

$$F_i(x_i|x_1,\dots,x_{i-1}) = \int_{-\infty}^{x_i} \frac{f_i(x_1,\dots,x_{i-1},s)}{f_{i-1}(x_1,\dots,x_{i-1})} \, ds$$

and

$$f_j(x_1,\dots,x_j) = \int_{-\infty}^{+\infty} \dots \int f_X(x_1,\dots,x_j,s_{j+1},\dots,s_n) \, ds_{j+1}\dots ds_n$$

This elementary result is used to construct a transformation which has been proposed by Hohenbichler and Rackwitz (1981) following an idea by Rosenblatt (1952). It will be denoted by the Rosenblatt−transformation in the sequel. We transform sequentially using the identities

$$\Phi(u_1) = F_1(x_1) \tag{4.4.5a}$$
$$\Phi(u_2) = F_2(x_1|x_2) \tag{4.4.5b}$$
$$\Phi(u_n) = F_n(x_n|x_1,...,x_n) \tag{4.4.5c}$$

Hence,

or

$$X = T(U) = (T_1(U_1), T_2(U_1,U_2),...,T_n(U_1,...,U_n))^T \tag{4.4.6}$$

$$X_1 = F_1^{-1}[\Phi(U_1)] \tag{4.4.6a}$$
$$X_2 = F_2^{-1}[\Phi_2)|F_1^{-1}[\Phi(U_1)]] \tag{4.4.6b}$$
$$X_3 = F_3^{-1}[\Phi(U_3)|F_2^{-1}[\Phi(U_2)|F_1^{-1}[\Phi(U_1)]],F_1^{-1}[\Phi(U_1)]] \tag{4.4.6c}$$

and the inverse transformation

$$U = T^{-1}(X) = (T_1^{-1}(X_1), T_2^{-1}(X_1,X_2),...,T_{n-1}(X_1,...,X_n))^T \tag{4.4.7}$$

$$U_1 = \Phi^{-1}[F_1(X_1)] \tag{4.4.7a}$$
$$U_2 = \Phi^{-1}[F_2(X_2|X_1)] \tag{4.4.7b}$$
$$U_3 = \Phi^{-1}[F_3(X_3|X_1,X_2)] \tag{4.4.7c}$$

In words: In the first step, the first variable is transformed. In the second step, all variables conditioned on the first are transformed, and so forth.

Illustration 4.4.1: Correlated normal variables

Let $X \sim N_n(m;\Sigma)$. We first standardize X by applying $Y_i = (X_i - m_i)/\sigma_{ii}$ so that $Y \sim N_n(0;R)$ with $\rho_{ij} = \sigma_{ij}/(\sigma_{ii}\sigma_{jj})^{1/2}$. We now transform according to

$$Y = A\,U \tag{1}$$

where $A = \{a_{ij}\ ;\ 1 \le i,j \le n\}$ and $a_{ji} = 0$ for $j > i$. The a_{ij}'s are determined from $\text{Var}[Y_i] = \sum_{k=i}^{i} a_{ik}^2 = 1$ and $\text{Cov}[Y_i,Y_j] = \sum_{k=i}^{i} a_{ik}a_{jk} = \rho_{ij}$. Clearly, $a_{11} = \rho_{11} = 1$. One finds with $\rho_{ii} = 1$.

$$a_{i1} = \rho_{i1}\ ;\ 2 \le i \le n \tag{2a}$$

$$a_{ii} = (\rho_{ii} - \sum_{k=1}^{i-1} a_{ik}^2)^{1/2}\ ;\ 2 \le i \le n \tag{2b}$$

$$a_{ij} = (\rho_{ij} - \sum_{k=1}^{j-1} a_{jk})/a_{jj}\ ;\ 1 < j < i \le n \tag{2c}$$

The Rosenblatt–transformation precisely corresponds to Cholesky's triangularization procedure for symmetric, positive definite matrices. If $\mathbf{Y}$ is **only centralized** beforehand, the ρ_{ij}'s must be replaced by the σ_{ij}'s, in which case we denote the transformation matrix by $\mathbf{C}$. Hence, the complete transformation is

$$\mathbf{X} = \mathbf{CU} + \mathbf{m} \tag{3}$$

If lognormal variables are correlated, an (original) matrix of (numerical) covariances has to be transferred into a matrix of covariances of the logarithms of the variables. For X_i and X_j being both lognormal it is:

$$\text{Cov}[X_i, X_j] := \ln\left(1 + \text{Cov}[X_i, X_j]/(E[X_i]\,E[X_j])\right) \tag{4}$$

For X_i being normal and X_j being lognormal it is:

$$\text{Cov}[X_i, X_j] := \text{Cov}[X_i, X_j]/E[X_j] \tag{5}$$

Note that if the matrix of covariances in the original space is positive definite, these formulae do not ensure that the covariance matrix in the log space is positive definite. However, if the covariance matrix is directly evaluated from the logarithms of the variables it is always positive definite.

\#

More recently, a number of alternative and somewhat less general probability distribution models and the corresponding transformations have been proposed (Der Kiureghian and Liu, 1986). Following Nataf (1962), a joint distribution is assigned to any two variables X_1 and X_2 such that two transformed variables $Z_1 = \Phi^{-1}[F_{X_1}(X_1)]$ and $Z_2 = \Phi^{-1}[F_{X_2}(X_2)]$ are jointly normal, i.e. according to the rules of probability calculus

$$f_{X_1, X_2}(x_1, x_2) = \varphi_2(z_1, z_2; \rho_{0,12})\, \frac{f_{X_1}(x_1)\, f_{X_2}(x_2)}{\Phi(z_1)\, \Phi(z_2)} \tag{4.4.8}$$

where $\varphi_2(z_1, z_2; \rho_{0,12})$ is the bivariate normal density, and the correlation coefficient is obtained from the integral equation

$$
\begin{aligned}
\rho_{12} &= \int\!\!\int_{-\infty}^{+\infty} \left[\frac{x_1 - m_1}{\sigma_1}\right]\left[\frac{x_2 - m_2}{\sigma_2}\right] \varphi_2(z_1, z_2; \rho_{0,12})\, \frac{f_{X_1}(x_1)\, f_{X_2}(x_2)}{\Phi(z_1)\, \Phi(z_2)}\, dx_1\, dx_2 \\[2mm]
&= \int\!\!\int_{-\infty}^{+\infty} \left[\frac{x_1 - m_1}{\sigma_1}\right]\left[\frac{x_2 - m_2}{\sigma_2}\right] \varphi_2(z_1, z_2; \rho_{0,12})\, dz_1\, dz_2 \tag{4.4.9}
\end{aligned}
$$

and where ρ_{12} is the correlation coefficient between X_1 and X_2. This model is valid for strictly increasing and continuous marginal distribution functions $F_{X_i}(x_i)$, and for $-1 \leq \rho_{0,12} \leq +1$. In the multivariate case the matrix of correlation coefficients R_0 must be positive definite. In general, $\rho_{0,12}$ differs only slightly from ρ_{12}. The final transformation is

$$X = T(Z(U)) = T(A \, U) \qquad (4.4.10)$$

or

$$X_1 = F_{X_1}^{-1}(a_{11}U_1)$$
$$X_2 = F_{X_2}^{-1}(a_{21}U_1 + a_{22}U_2)$$
$$X_3 = F_{X_3}^{-1}(a_{31}U_1 + a_{32}U_2 + a_{33}U_3)$$
$$\vdots$$
$$X_n = F_{X_n}^{-1}(\sum_{i=1}^{n} a_{ni}U_i)$$

with triangular matrix A determined from the equivalent correlation matrix R_0.

4.5 COMPUTATION OF THE MULTINORMAL INTEGRAL

In the general case, one has to evaluate the multi–normal integral. Unfortunately there are only a few analytical solutions. It is, however, possible to derive very good approximations and an asymptotic formula. In view of its many applications the properties of a multi–normal vector are discussed first. The density of the multi–normal vector $Y = (Y_1,...,Y_n)^T$ is

$$\varphi_n(y) = (2\pi \det(\Sigma))^{-1/2} \exp[-\tfrac{1}{2}[(y-m)^T \Sigma^{-1}(y-m)]] \qquad (4.5.1)$$

and after standardization by $X_i = (Y_i - \mu_i)/\sigma_i$ such that $\Sigma = D^T R D$ with $D = \text{diag}\{\sigma_i\}$ the diagonal matrix of the standard deviations and R the matrix of the correlation coefficients:

$$\varphi_n(x) = (2\pi)^{-n/2}(\det(R))^{-1/2} \exp[-\tfrac{1}{2}[(x^T R^{-1} x)] \qquad (4.5.2)$$

If $R = I$ (I = unit matrix), the vector Z is uncorrelated:

$$\varphi_n(z) = (2\pi)^{-n/2} \exp[-\tfrac{1}{2} z^T z] \qquad (4.5.3)$$

This also implies independence of the components of the vector. Let now $Y = AZ + m$. The covariances σ_{ij} are

$$\sigma_{ij} = \text{Cov}[Y_i, Y_j] = \sum_{i=1}^{n} a_{ij} \, a_{ik} \, E[X_i^2] - m_j \, m_k$$

Consequently $A = R$. The multi–normal distribution function can now be written as:

$$\Phi_n(x;R) = \int_{-\infty}^{x} \varphi_n(t;R) \, dt \qquad (4.5.4)$$

The normal density is symmetric in the sense that:

$$\varphi_n(x;R) = \varphi_n(-x;R) \qquad (4.5.5)$$

No such simple symmetry relation can be established for the distribution function. An important property is that for $\{\rho_{ij}\} \leq \{\kappa_{ij}\}$ (Slepian, 1962; Sidak, 1964)

$$\Phi_n(x;R) \leq \Phi_n(x;K) \qquad (4.5.6)$$

Unfortunately only the two and three dimensional cases have simple solutions (Owen, 1956). If the variables can be represented by

$$X_i = \lambda_i \, Y_0 + \sqrt{(1 - \lambda_i^2)} \, Y_i \qquad (4.5.7)$$

where $Y_0, Y_1, ..., Y_n$ are independent standard normal variables and, therefore, $\rho_{ij} = \kappa_i \kappa_j$, it is (Dunnet and Sobel, 1955):

$$\Phi_n(x;R) = \int_{-\infty}^{+\infty} \varphi(y_0) \prod_{i=1}^{n} \Phi\left(\frac{x_i - \lambda_i y_0}{\sqrt{(1 - \lambda_i^2)}}\right) dy_0 \qquad (4.5.8)$$

For the special case of equicorrelation we have $\sqrt{\rho} = \lambda_i = \lambda_j \geq 0$. On the basis of eq. (4.5.8) bounds can be constructed which, however, are not always sufficiently narrow.

The two dimensional case is needed more frequently. It can be computed as a special case of eq. (4.5.8) or by numerical integration according to

$$\Phi_2(x,y;\rho) = \Phi(x) \, \Phi(y) + \int_{0}^{\rho} \varphi_2(x,y;t) \, dt \qquad (4.5.9)$$

with

$$\varphi_2(x,y;\rho) = \frac{1}{2\pi \, (1 - \rho^2)^{1/2}} \exp\left[-\frac{1}{2} \frac{x^2 - 2\rho xy + y^2}{1 - \rho^2}\right]$$

For the general case the following scheme has been proposed (Hohenbichler and Rackwitz, 1985):

$$\Phi_m(\mathbf{c};\mathbf{R}) = P(\bigcap_{i=1}^{n} \{Z_i \leq c_i\}) = P(Z_1 \leq c_1) \, P(\bigcap_{i=2}^{n} \{Z_i \leq c_i\} | \{Z_1 \leq c_1\})$$

$$(4.5.10)$$

The Z_i's have the Rosenblatt–transformation

$$Z_i = \sum_{i=1}^{n} a_{ij} U_j \qquad (4.5.11)$$

with $a_{11} = 1$ and $a_{i1} = \rho_{i1}$. The condition in the second term of eq. (4.5.11) can be removed by observing that it affects only the first variable. The distribution function of a new conditional (truncated) variable $\bar{U}_1$ is, for $\bar{U}_1 \leq c_1$

$$F_{\bar{U}_1|c_1}(\bar{u}_1) = P(U_1 \leq \bar{u}_1 | U_1 \leq c_1) = \frac{P(\{U_1 \leq \bar{u}_1\} \cap \{U_1 \leq c_1\})}{P(U_1 \leq c_1)} = \frac{\Phi(\bar{u}_1)}{\Phi(c_1)}$$

$$(4.5.12)$$

Using $F_{\bar{U}_1|c_1}(\bar{u}_1) = \Phi(u_1)$ with U_1 a new auxiliary standard normal variable

$$\bar{U}_1 = \Phi^{-1}[\Phi(U_1)\Phi(c_1)] \qquad (4.5.13)$$

in eq. (4.5.11), one obtains:

$$
\begin{aligned}
P(\bigcap_{i=2}^{m} \{Z_i \leq c_i | Z_1 \leq c_1\} &= P(\bigcap_{i=2}^{m} \{\alpha_{i1} \bar{U}_1 + \sum_{j=2}^{i} \alpha_{ij} U_j \leq c_i\}) \\
&= P(\bigcap_{i=2}^{m} \{\alpha_{i1} \Phi^{-1}[\Phi(U_1) \Phi(c_1)] + \sum_{j=2}^{i} \alpha_{ij} U_j \leq c_i\}) \\
&= P(\bigcap_{i=2}^{m} \{g_i(\mathbf{U}) \leq 0\} = P(\bigcap_{i=2}^{m} \{\alpha_i^{(2)T} \mathbf{U} \leq c_i^{(2)}\}) \\
&= \Phi_{m-1}(\mathbf{c}^{(2)};\mathbf{R}^{(2)})
\end{aligned}
$$

$$(4.5.14)$$

so that eq. (4.5.10) can be written as

$$\Phi_m(\mathbf{c};\mathbf{R}) = \Phi(c_1) \, \Phi_{m-1}(\mathbf{c}^{(2)};\mathbf{R}^{(2)}) \qquad (4.5.15)$$

Hence the dimension of the multinormal integral has been diminished by one. In line 2 of eq. (4.5.14) one recognizes that only the first variable enters non–linearly. The functions $g_i(\mathbf{u})$ in the third line can be linearized around their respective β–points. Repeated application of this scheme leads to the approximation

$$\Phi_m(\mathbf{c};\mathbf{R}) \approx \Phi(c_1) \, \Phi(c_2^{(2)})...\Phi(c_m^{(m)}) \qquad (4.5.16)$$

Several improvements are possible which cannot be discussed here; their effect in general is only small. The probability content of $\{g_i(u) \leq 0\}$ can be computed exactly in terms of the bivariate normal integral, from which

$$c_i{}^{(i)} = \Phi^{-1}[\Phi_2(c_1, c_i; \rho_{i1})/\Phi(c_1)] \qquad (4.5.17)$$

is the distance to the origin of an equivalent hyperplane having the same probability content. For Φ_2 formula (3.2.6.9) can be used. More accurate results can be obtained (Gollwitzer and Rackwitz, 1988), with the asymptotic theory outlined in appendix A. The accuracy of the computation scheme (4.5.15) using asymptotic concepts has been tested up to dimension m = 200 with excellent results under asymptotic conditions. For non—asymptotic conditions, the equivalent hyperplane concept (Gollwitzer and Rackwitz, 1983) with eq. (4.5.19) yields results that are only slightly less satisfactory.

One special asymptotic result due to Ruben (1964) is given for its simplicity. If all c_i's are negative and the solution of

$$\gamma = R^{-1}c \qquad (4.5.18)$$

leads to a vector γ with positive elements then, for $\|c\| \to \infty$,

$$\Phi(-c; R) \sim \varphi(c; R)\,(\det(R))^{1/2}\left(\prod_{i=1}^{n} \gamma_i\right)^{-1} \qquad (4.5.19)$$

The condition of negative c_i's but positive γ_i's restricts the domain of application to a certain extent.

4.6 SEARCH ALGORITHMS

The mathematical basis for the search for the β—point given only one failure domain (restriction), or a cut set of failure domains (several restrictions), is the existence of an optimum point in the admissible domain. In the first order context, discussed before, the optimum point corresponds to the maximum density of the standard normal vector in the failure domain. Due to the rotational symmetry of the standard space, the point is also the point inn or on the boundary of V that is closest to the origin. The existence of such a point is defined by Lagrange's theorem. The Lagrangian function is defined as:

$$L(u) = f(u) + \sum_{j=1}^{m} \lambda_j\, g_j(u) \qquad (4.6.1)$$

and the necessary conditions for an optimal point u^* are

$$\nabla L(u^*) = \nabla f(u^*) + \sum_{j=1}^{m} \lambda_j^* \nabla g_j(u^*) = 0 \qquad (4.6.2)$$

$$g_j(u^*) = 0 \text{ for } j = 1,2....m \qquad (4.6.3)$$

m is the number of restrictions and the λ_j are the so called Lagrangian multipliers. If n is the dimension of the vector $\mathbf{u}$, then eqs. (4.6.2) and (4.6.3) form a non–linear system of equations with $n + m$ unknowns (the point $\mathbf{u}^*$ and m λ^*–values). For inequality restrictions in the form $g_j(\mathbf{u}) \leq 0$ the first–order conditions are called the **Kuhn–Tucker** conditions:

$$\nabla L(\mathbf{u}^*) = \nabla f(\mathbf{u}^*) + \sum_{j=1}^{t} \lambda_j^* \, \nabla g_j(\mathbf{u}^*) = 0 \tag{4.6.4a}$$

$$g_j(\mathbf{u}^*) = 0 \quad j = 1,2....t \tag{4.6.4b}$$

$$\lambda_j^* \geq 0 \quad j = 1,2....t \tag{4.6.4c}$$

$$g_k(\mathbf{u}^*) \leq 0 \quad k = t+1,...m \tag{4.6.4d}$$

t is the number of active restrictions at the point $\mathbf{u}^*$. The index k runs over all inactive restrictions. If the Hessian of the Lagrangian function is positive definite and the point $\mathbf{u}^*$ fulfills the Kuhn–Tucker conditions then $\mathbf{u}^*$ is a local optimum point.

This analysis forms the basis for fast algorithms which are globally convergent when started from an arbitrary initial point (see Gill et al., 1981; Hock and Schittkowski, 1983 and Arora, 1989). In the following we can only outline the main features of a suitable algorithm.

The following function has to be minimized

$$\mathbf{u}^* = \min\{f(\mathbf{u})\} = \min\{\| \mathbf{u} \|^2\} \tag{4.6.5}$$

given the constraints:

$$g_j(\mathbf{u}) \leq 0 \quad \text{für } j=1,2........m \tag{4.6.6}$$

The Lagrangian function with linearized constraints becomes:

$$L(\mathbf{u},\lambda) = \|\mathbf{u}^o\|^2 + 2\,\mathbf{u}^{oT}\,\Delta\mathbf{u} + \Delta\mathbf{u}^T\,\Delta\mathbf{u} + \sum_{j=1}^{m} \lambda_j \,\{g_j(\mathbf{u}^o) + \nabla g_j^{oT}\,\Delta\mathbf{u}\} \tag{4.6.7}$$

and the Kuhn–Tucker conditions are:

$$\nabla L(\mathbf{u},\lambda) = 2\,\mathbf{u}^o + 2\,\Delta\mathbf{u} + \sum_{j=1}^{t} \lambda_j \,\nabla g_j^o = 0 \tag{4.6.8a}$$

$$g_j(\mathbf{u}) = g_j(\mathbf{u}^\circ) + \nabla g_j^{oT}\,\Delta\mathbf{u} = 0 \quad j=1,2,..t \tag{4.6.8b}$$

Let G be the matrix of the gradients of the (active) constraints and Γ a vector with the values of the constraint functions

$$G = [\nabla g_1^o,...\nabla g_j^o.......\nabla g_t^o]_{n x t} \qquad (4.6.9)$$

The system of equations can be written in matrix form as

$$\begin{bmatrix} 2 \, I & G \\ G^T & F \end{bmatrix} \begin{bmatrix} \Delta u \\ \lambda \end{bmatrix} = \begin{bmatrix} -2 \, u^o \\ -\Gamma \end{bmatrix} \qquad (4.6.10)$$

Solution of this system yields the following iteration scheme:

$$u^{k+1} = G_k \, (G_k^T \, G_k)^{-1} \, (G_k^T \, u^k - \Gamma_k) \qquad (4.6.11)$$

The matrices G are given by:

$$G_k = A_k \, N_k \qquad (4.6.12)$$

with

$$A_k = [a_1^k,...a_j^k,......a_t^k] \qquad (4.6.13)$$

$$a_j^k = \frac{1}{\|\nabla g_j^k\|} \, \nabla g_j^k \qquad (4.6.14)$$

and N_k = diagonal matrix with the norms of $\|\nabla g_j^k\|$. With this notation, the algorithm can be written as

$$u^{k+1} = A_k \, \Sigma_k^{-1} \, (A_k^T \, u^k - N_k^{-1} \, \Gamma_k) \qquad (4.6.15)$$

with $\Sigma_k = A_k^T \, A_k$ the covariance matrix of the linearized (active) constraint functions in the point u^k.

Specialization of this scheme to only one constraint yields the algorithm already given by Hasofer and Lind (1974) and Rackwitz and Fießler (1978). Those algorithms are not yet surely convergent. They can be made convergent by introducing either a deceleration scheme or a suitable step length procedure (see, for example, Abdo and Rackwitz, 1990). The convergence rate can be made especially high if at least approximate information about the curvature properties of the Lagrangian function are used for the step length procedure (Schittkowski, 1983). This, however, is suitable only if the problem dimension is not too high.

5. Applications

We present several representative applications to demonstrate the theory. Emphasis is on methodological aspects. The mechanical models are somewhat simplified. The first few examples concern structural systems. The last few examples deal with various cases in which conditional probabilities have to be computed.

Illustration 5.1: Chain with n links (Ditlevsen, 1982)

Consider a chain with n links whose resistances X_i are independently normally distributed with mean $m = m_i$ and standard deviation $\sigma = \sigma_i$ and which is loaded by a normally distributed load Y with mean μ and standard deviation τ. Failure is for

$$F = \bigcup_{i=1}^{n} \{Z_i \leq 0\} = \bigcup_{i=1}^{n} \{X_i - Y \leq 0\} \tag{1}$$

One determines $\beta = (m - \mu)/(\sigma^2 + \tau^2)^{1/2}$ and $\rho = \rho_{ij} = \text{Corr}[Z_i, Z_j] = \sigma^2/(\sigma^2 + \tau^2)$. Therefore, using the simple correlation structure of the Z_i's and the exact formula for the multinormal integral with equicorrelation one determines

$$P_f = 1 - \Phi_n(\beta ; \rho) = 1 - \int_{-\infty}^{\infty} \varphi(u) \, \Phi^n(\frac{\beta - u\lambda}{\sqrt{1 - \lambda^2}}) \, du \tag{2}$$

with $\lambda = +\sqrt{\rho}$ and $U_i = (Z_i - (m - \mu))/(\sigma^2 + \tau^2)^{1/2}$.

Alternatively, eq.(3.7) can be used. It receives the form

$$\Phi(-\beta_1 + \sum_{i=2}^{n} \max\{0 , \Phi(-\beta_i) - \sum_{j=1}^{i-1} \Phi(-\beta_i , -\beta_j ; \rho_{ij}\} \leq P_f$$

$$\leq \Phi(-\beta_1) + \sum_{i=2}^{n} \Phi(-\beta_i) - \max_{j=1}^{i-1}\{\Phi(-\beta_i , -\beta_j ; \rho_{ij})\} \tag{3}$$

where Φ_2 is the two–dimensional normal integral. Ditlevsen (1982) gave the following bounds to Φ_2. For $\rho_{ij} \geq 0$

$$\max\{\Phi(-\beta_i) \, \Phi(-\frac{\beta_i - \rho_{ij}\beta_i}{\sqrt{1 - \rho_{ij}^2}}) , \Phi(-\beta_j) \, \Phi(-\frac{\beta_i - \rho_{ij}\beta_i}{\sqrt{1 - \rho_{ij}^2}})\} \tag{4}$$

is a lower bound, and the sum of the two terms in brackets an upper bound. For $\rho_{ij} < 0$ the **max** operation has to be changed into **min** to produce an upper bound while the lower bound is zero in this case. For $\rho = \rho_{ij} \geq 0$ and $\beta = \beta_i$ we have

$$\Phi(-\beta)\ \Phi(-\beta\frac{\sqrt{1-\rho}}{\sqrt{1+\rho}})\leq \Phi(-\beta,-\beta;\rho)\leq 2\ \Phi(-\beta)\ \Phi(-\beta\frac{\sqrt{1-\rho}}{\sqrt{1+\rho}})\qquad(5)$$

Application to eq. (3) produces

$$\Phi(-\beta)\ \max_{i=2}^{n}\ \{i[1-(i-1)\Phi(-\beta\frac{\sqrt{1-\rho}}{\sqrt{1+\rho}})]\}\leq P_f\leq \Phi(-\beta)$$

$$[(n-(n-1)\Phi(-\beta\frac{\sqrt{1-\rho}}{\sqrt{1+\rho}})]\qquad(6)$$

The lower bound becomes largest for:

$$i=\mathrm{int}\{(1/\Phi(-\beta\frac{\sqrt{1-\rho}}{\sqrt{1+\rho}})+1)/2\}$$

#

Illustration 5.2: Ideal series and parallel systems

Figure 5.2.1 shows the influence of the number of components and their correlation in series and parallel systems. The componental safety index is $\beta = 4.26$. The components are assumed to be equicorrelated; this is a special case, but it can illustrate the general behavior. The figure was calculated by formula (4.5.8) and by (4.5.16); there were only minor numerical differences. The upper part of the figure presents the results for parallel systems and the lower part for series systems. One recognizes that for series systems neither the number of components in the system nor their correlation is very important unless the system becomes very large. This justifies the use of Ditlevsen's bounds or even the first order bounds. The contrary is the case for parallel systems. Their failure probability decreases dramatically with the number of components, especially for small correlation coefficients. We conclude that the study of parallel systems requires a rigorous evaluation of the multinormal integral; correlations between components must be properly taken into account. Note that if the system components have non–linear limit state functions, equivalent linearization techniques are sufficient for series systems, but exact location of the β–point and hence a correct determination of their dependence structure is important. This aspect will be demonstrated later in more detail in illustration 5.4.

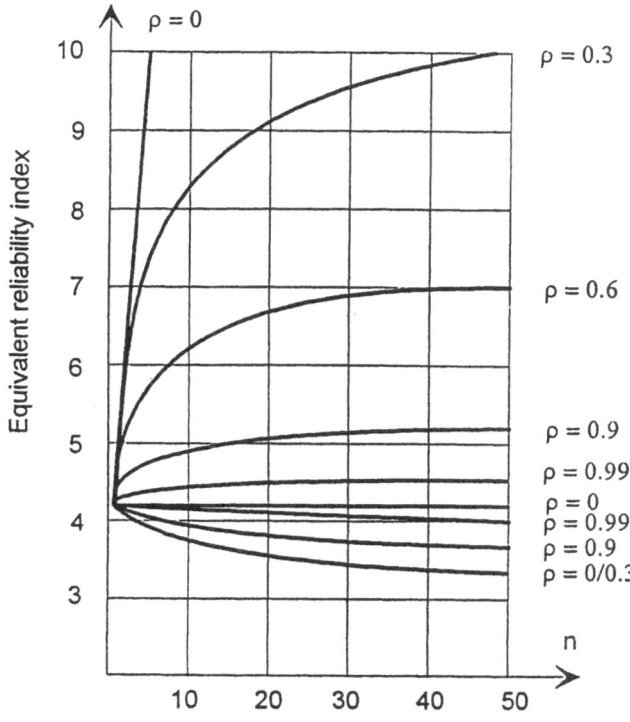

Figure 5.2.1: Failure probability (equivalent reliability index) versus number of components and componental correlation coefficients in series and parallel systems

\#

Illustration 5.3: Rigid–plastic portal frame (Madsen et al. 1987)

Figure 3.1 shows that a portal frame can fail in three modes.

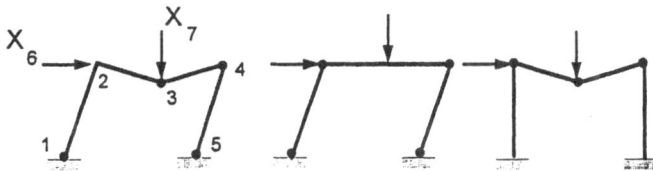

Figure 5.3.1: Failure mechanisms of a portal frame

By the principle of virtual work one obtains the three state functions

$$M_1 = X_1 + X_2 + X_4 + X_5 - X_6 h$$
$$M_2 = X_1 + 2 X_3 + 2 X_4 + X_5 - X_6 h - X_7 h$$
$$M_3 = X_2 + 2 X_3 + X_4 - X_7 h$$

The uncertain variables are assumed to be independent and log–normally distributed

Variable	Distr.	$E[X_i]$	$D[X_i]$
$X_1,...X_5$	LN	134.9 [kNm]	13.49 [kNm]
X_6	LN	50 [kN]	15 [kN]
X_7	LN	40 [kN]	12 [kN]

The height is taken to be 5 [m]. Transformation, and iterative determination of the β–point yields

Mode	β	α_1	α_2	α_3	α_4	α_5	α_6	α_7
1	2.71	.084	.084	–	.084	.084	−.986	–
2	2.88	.077	–	.150	.150	.077	−.827	−.509
3	3.44	–	.084	.164	−.084	–	–	−.979

Due to the probability transformation, the limit state surfaces are slightly curved. The correlation coefficient matrix of the safety margins M_i is computed to be

$$\mathbf{R} = \begin{bmatrix} 1.00 & .841 & .014 \\ .841 & 1.00 & .536 \\ .014 & .536 & 1.00 \end{bmatrix}$$

The matrix of cut set probabilities is

$$\mathbf{P} = \begin{bmatrix} 3.4 \cdot 10^{-3} & & \text{symm.} \\ 9.2 \cdot 10^{-4} & 1.99 \cdot 10^{-3} & \\ 1.14 \cdot 10^{-6} & 4.25 \cdot 10^{-5} & 2.9\,1 \cdot 10^{-4} \end{bmatrix}$$

The first–order bounds are

$$3.36 \cdot 10^{-3} \leq P_f \leq 5.64 \cdot 10^{-3}$$

and the bounds of second order

$$4.67 \cdot 10^{-3} \leq P_f \leq 4.67 \cdot 10^{-3}$$

Of course the bounds are not strict bounds because the mode probabilities themselves are approximated to first order. A slight improvement could be achieved if the second order corrections were considered for each mode. The cut set probabilities are also calculated to first order. Investigations by Madsen (1985), however, have shown that a second order refinement for the joint probabilities in Ditlevsen's formula is unnecessary.

#

Illustration 5.4:   Small Daniels–system (Hohenbichler and Rackwitz, 1983, and Hohenbichler et al., 1987))

The figure below shows a so–called Daniels–system. At loading, all fibers experience the same strain. The weakest fiber fails first, then the second weakest, etc. At failure of a fiber, the load is equally distributed to the remaining intact fibers. The failure load of the system is reached when the remaining fibers can no longer sustain the load carried by the fiber which just failed

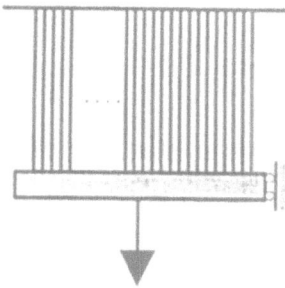

Figure 5.4.1:   Daniels system

The fibers have independent, identically distributed strengths X_i. If ordered according to $(\hat{X}_1 \leq \hat{X}_2 \leq \dots \leq \hat{X}_n)$, the system strength can be given as

$$R = \max_{i=1}^{n}\left\{(n - k + 1)\,\hat{X}_k\right\} \tag{1}$$

Hence, there is

$$P_f(s) = P\left[\bigcap_{i=1}^{n}\left\{(n - k + 1)\,\hat{X}_k - s \leq 0\right\}\right] \tag{2}$$

which is a parallel system. The distribution function of $\hat{X}_1$ is

$$F_1(x_1) = 1 - [1 - F_X(x_1)]^n \tag{3}$$

The distribution functions of the X_i are the distribution functions of the original variables truncated at $\hat{X}_i = x_1$.

$$P(X_i \leq x | \hat{X}_i = x_1) = \frac{F_X(x) - F_X(x_1)}{1 - F_X(x_1)} \quad \text{for } x > x_1,\ i = 2,...,n$$

Accordingly, the distribution function of $\hat{X}_2 | \hat{X}_1 = x_1$ is

$$F_2(x_2 | x_1) = 1 - \left[1 - \frac{F_X(x_2) - F_X(x_1)}{1 - F_X(x_1)}\right]^{n-1}; \ x_2 > x_1 \tag{4}$$

and generally

$$F_i(x_2 | x_1,...,x_{i-1}) = 1 - \left[1 - \frac{F_X(x_i) - F_X(x_{i-1})}{1 - F_X(x_{i-1})}\right]^{n-i-1} \quad \text{for } x_i > x_{i-1} > ... > x_1 \tag{5}$$

This allows us to perform the Rosenblatt–transformation and to compute the reliability indices $\beta = -\Phi^{-1}(p_f)$ for systems of different size.

Reliability indices for Daniels–system (elastic–brittle components)

n (1)	I (2)	$\|u^*\|$ (3)	IIa (4)	IIb (5)	IIc (6)	Exact (7)
1	2.00	2.00	2.00	2.00	2.00	2.00
3	1.50	1.50	1.82	1.82	1.82	1.82
5	1.58	1.34	1.76	1.76	1.83	1.87
10	1.59	1.19	1.72	1.72	1.84	2.03
15	–	1.16	1.64	1.64	2.05	2.19
20	–	1.16	1.60	1.61	2.20	2.32

For the numerical calculations, the load is assumed to be $s = n\,(m_X + a\,\sigma_X)$ with $a = 2$. The fiber strength is normally distributed according to $N(m_X; \sigma_X)$ with coefficient of variation $V_X = \sigma_X / m_X = 0.2$. Column (2) lists the results obtained by an individual linearization (without second order correction) and subsequent evaluation of the multinormal integral. Comparison with the exact result in the last column obtained from the following recursion formula due to Daniels (1945)

$$P_f(s) = F^{(n)}(s) = \sum_{k=1}^{n} (-1)^{k+1} \binom{n}{k} F_X^k(s)\, F_X^{(n-k)}\left(\frac{s}{n-k+1}\right) \tag{6}$$

shows that in this case a simple first–order theory with individual linearization is insufficient. Column (3) contains the distances of the joint β–points from the origin (geometrical reliability indices). Note that they decrease with n down to a value of 1.16 for which asymptotic conditions are hardly valid. Column (4) contains the reliability indices if the active constraints are linearized at the joint β–point, and no second order correction is performed. Interestingly, those values differ only slightly from the values in column (5) where the second order correction is included. Apparently the limit state surfaces are only slightly curved. Only column (6), where additionally all inactive constraints have been considered in their linearized form, yields acceptable results for larger n. If the geometric reliability indices are increased beyond 2, say, for all n the strategies followed for column (4) or (5) already yield very accurate results. The contribution of inactive constraints becomes insignificant. For geometrical reliability indices larger than 3, the exact results are reproduced.

Although this example is used to discuss methodological aspects, it should be observed that the overall reliability of a Daniels system with brittle components strongly depends on the number of components. The indices first decrease, and increase only for larger numbers of components. Daniels systems with a small number of components are therefore less safe than an appropriately designed single component. Only systems with a large number of components show a pronounced effect of redundancy.

Similar computations can be performed for perfectly elastic–plastic material

Reliability indices for Daniels–system (elastic–plastic components)

n (1)	I (2)	$\|u^*\|$ (3)	IIa (4)	IIb (5)	IIc (6)	Exact (7)
1	2.00	2.00	2.00	2.00	2.00	2.00
3	3.21	3.21	3.21	3.45	3.50	3.46
5	4.01	4.01	4.01	4.45	4.48	4.47
10	5.47	5.47	5.47	6.28	6.33	6.33
15	6.59	6.59	6.59	7.69	7.73	7.75
20	7.54	7.54	7.54	8.87	8.90	8.94

In this case the geometric reliability index (column (3)) is already relatively close to the exact results whose probabilities are computed from

$$P_f(s) = \Phi\left[-\frac{s - n\ m_X}{\sqrt{n}\ \sigma_X}\right] \tag{7}$$

The reliability indices increase monotonically with n, indicating the significant difference of ductile and brittle componental behavior. Asymptotic conditions are met for all n. Individual linearization is moderately successful (column 2) as is linearization at the joint β–point (column 4), which for linear constraints must yield the same results. However, the strategy followed in column (5) or (6) produces very accurate results. In fact, it can be shown that the curvatures of the limit state surfaces are relatively large in this case. In column (6) all formally inactive components are considered, too. For the plastic case one has to include always the full set of components. Therefore, the differences between column (5) and (6) also indicate the achievable level of accuracy. The remaining differences must be attributed to numerical noise. Note that a reliability index of 8.9 corresponds to a failure probability of about $3 \cdot 10^{-19}$.

It should be mentioned that the small Daniels system is one of the few mechanical redundant systems which have an exact solution. For very large Daniels systems there are a number of asymptotic results, the first of which is also due to Daniels (1945).

#

Illustration 5.5: Failure trees of structures (Moses, 1990)

The vast majority of researchers use the so called failure tree approach for the analysis of general redundant structural systems starting with the work of Murotsu (1981). This approach has been derived from well known approaches in classical reliability. It rests on the identification and analysis of the sequence of component-al state changes (failures) from the initial intact state of the structure to at least the dominating failure modes of the system (see figure 5.5.1 for an illustration of the method). The method requires discretization of the structure into a finite number of components (members, cross sections, joints). Those components can change their state either from elastic behavior to plastic behavior or can fail in a brittle manner, possibly after having experienced some plasticity. Componental failures are viewed as state changes. In particular, if in a certain failure branch i failures (state changes) have already taken place, the mechanics for the i+1-th failure have to take account of the i previous componental state changes. The i+1-th failure event can be written as

$$F^{i+1} = F^1 \cap F^{2|1} \cap F^{3|1 \cap 2} \cap ... \cap F^{i|1 \cap 2 \cap ... \cap i-1} \cap F^{i+1|1 \cap 2 \cap ... \cap i} \tag{1}$$

Note that in failure events on the right hand side of this equation the changes in the mechanical behavior of the system introduced by the failures left to the considered have to be taken into account appropriately. This is indicated by notations of the kind $F^{i|1 \cap 2 \cap ... \cap i-1}$.

The probabilities in eq. (1) can be computed by FORM/SORM. Mechanically, provisions must be taken that the sequence of state changes is a valid sequence which could occur along the possible load paths. Murotsu (1981) introduced the concept of **branch and bound**. The basic idea is to concentrate on the analysis of failure branches with largest failure probability. The book by Thoft–Christensen and Murotsu (1986) contains many details about the approach. Guenard (1984) and Gollwitzer (1986) improved the scheme, in that they not only computed the full intersection along a failure path by FORM/SORM, but continued the algorithm at a branching point in any of the already investigated failure branches with momentarily largest intersection probability. System failure is defined by singularity of the stiffness matrix, or equivalently, by formation of a mechanism. For example, in figure 5.5.1 the branches starting with (6−), (2+) and (5−) are not continued beyond failure of the first component. The path $(1+) \cap (6-) \cap (7+)$

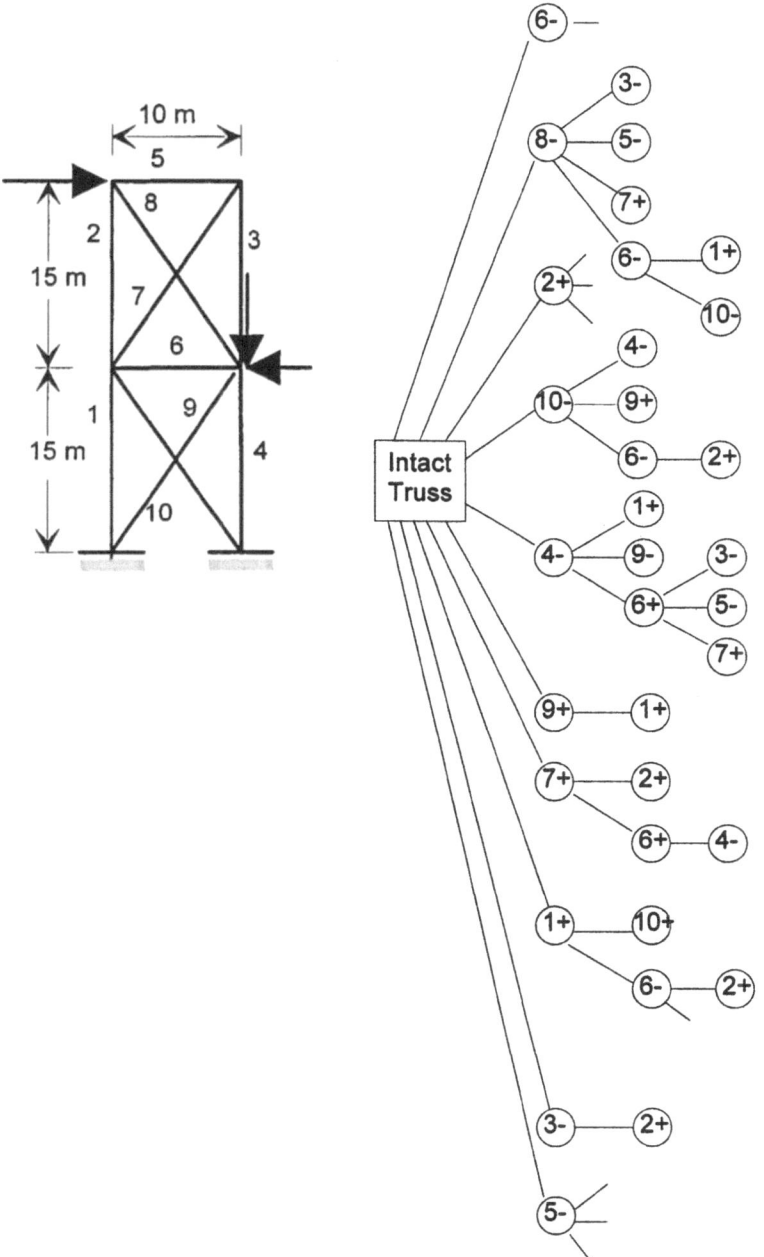

Figure 5.5.1:   Failure Tree for Truss Structure (after Moses, 1990)

is also not completed, while all other paths are completed. For a given monotonic-ally increasing load path it is important to observe the condition that along a failure path the load must increase. This algorithm leads to a number of completed failure paths which should include the most dangerous failure path and a generally much larger number of uncompleted failure paths (see fig. 5.5.1). The probability of the union of all completed failure paths is a lower bound to the system failure probability, and the union of all completed and uncompleted failure paths provides an upper bound to the failure probability. If the analysis is terminated before a complete failure path is found, the probability of the union of all investigated uncompleted failure paths is an upper bound to the system failure probability which in general is much smaller than the first—order series system bound. Quite frequently it suffices to investigate only rather short incomplete failure branches in practical applications.

Illustration 5.6:   **Updating** by observations of a structural component
                             damaged by cracks (after Schall et al. 1991)

Cracks frequently develop in the hatch corners of container ships. Crack growth can be described by the Paris/Erdogan relationship

$$\frac{da}{dn} = C \ Y(a) \ (\Delta s(n) \ \sqrt{(\pi \ a(n))})^m \qquad (1)$$

where $a(n)$ is the crack length, C and m material parameters, and $\Delta s(n)$ the far field stress range. $Y(a)$ is a geometry factor, which for edge cracks in steel plates can be set equal to $Y(a) = 1.12$. Integration of eq. (1) for $m > 2$ leads to

$$a(n) = \left\{ a_0^{\frac{2-m}{2}} + \frac{2-m}{2} C \ \pi^{m/2} \ n \ E[(1.12 \ \Delta S)^m] \right\}^{\frac{2}{2-m}} \qquad (2)$$

where $s(n)$ is introduced as a random load process, and the time integral on the right hand side of eq. (1) is replaced by the ensemble mean according to the ergodic theorem for random processes. a_0 is the initial crack length. Failure is the exceedance of a critical crack length a_{cr}.

$$V(n) = \{a_{cr} - a(n) \leq 0\} \qquad (3)$$

a_0 and C are considered as random variables. At time t_1 the hatch corner is inspected. There is a certain probability that a crack remains undetected. The probability of detection is usually given in the form $P(D) = P(a \leq a_D) = 1 - \exp[- \lambda_D (a_D - a_{D,0})]$ and the detection event is

$$D(n) = \{a(n) \geq a_D\} \qquad (4)$$

If the crack is detected, its length can be measured. This event is described by

$$B(n) = \{a(n) - \hat{a} \ \delta = 0\} \qquad (5)$$

where $\hat{a}$ is the observation and δ a measurement error. Let the inspection time be n_1. Then, the **a priori** failure probability (before inspection) is

$$P_f(n_1) = P(V(n_1)) \tag{6}$$

Formally, the **a posteriori** failure probability (after inspection) is

$$P_f(n_1 | B(n_1) \cap R(n_1)) = \frac{P(V(n_1) \cap B(n_1) \cap R(n_1))}{P(B(n_1) \cap R(n_1))} \tag{7}$$

It is now of interest to determine the failure probability until the next inspection at time n_2 in order to decide whether to do repairs already at time n_1 or possibly at time n_2. Now we also include the event, that there has been no failure until n_1.

$$P_f(n_1 + n_2 | B(n_1) \cap R(n_1)) = \frac{P(V(n_1+n_2) \cap B(n_1) \cap R(n_1) \cap \bar{V}(n_1))}{P(B(n_1) \cap R(n_1) \cap \bar{V}(n_1))} \tag{8}$$

For numerical evaluation of eq. (8) or (9) a theory for computation of conditional probabilities is needed. Numerator and denominator are computed separately. Conditions occurring in any cut set in the denominator must occur in all cut sets in the numerator. If the denominator and thus also the numerator contains equality constraints, one may use the additional theory based on asymptotic concepts given in appendix B. Note that denominator and the numerator then must not be interpreted as probabilities, but the quotient is a conditional probability.

Variable	Distribution function	Mean	Coefficient of variation
a_0	Rayleigh	1 [mm]	0.5
C	Lognormal	10^{-14} [–]	0.4
m	Constant	3.0	–
δ	Lognormal	1 [–]	0.2
a_D	Exponential	4 [mm]	0.75
$E[\Delta s]$	Constant	150 [Mpa]	–

For a numerical example, the following assumptions are made. The critical crack length is taken as 50 [mm]. The service time in terms of number of cycles is $n_{service} = \nu \, t_{service} = 0.2 \cdot 5 \, 10^8 = 10^8$ with $\nu = 0.2$ the cycle frequency. At inspection, a crack length of 2.5 [mm] is observed. The upper part of figure 5.6.1 shows the **a priori** failure probability versus service time. It is seen that it is unacceptable for times larger than about 10^8. In the lower part of figure 5.6.1 the

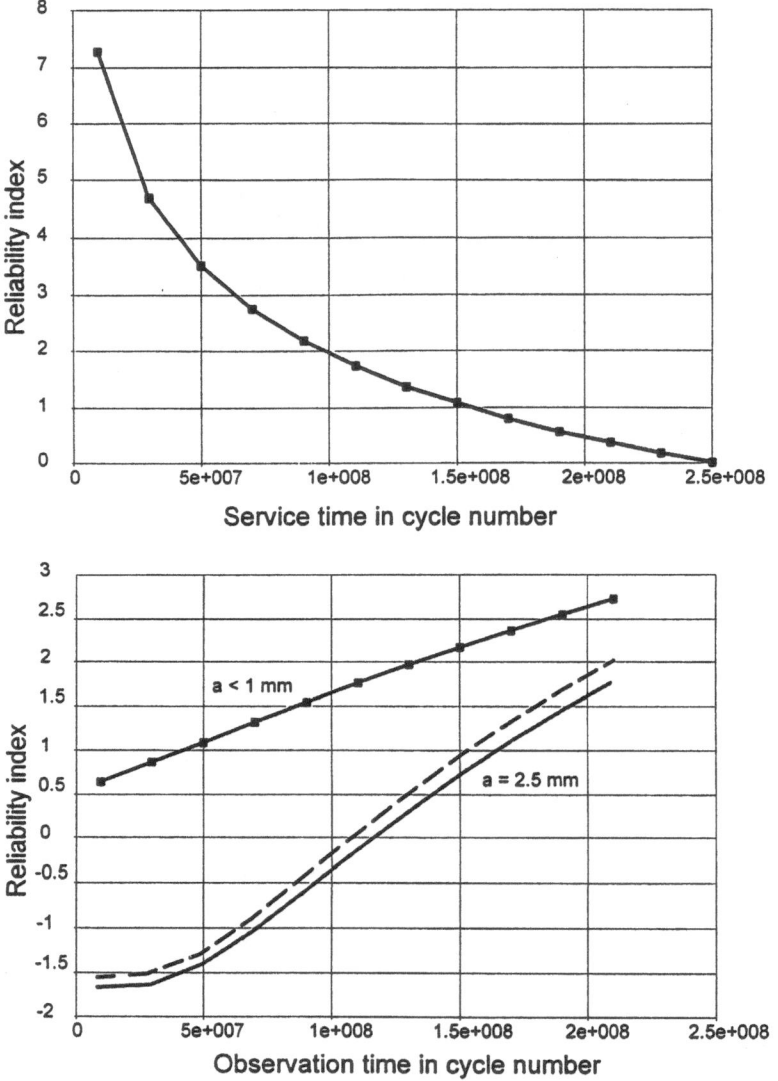

Figure 5.6.1:   Reliability indices versus service time/observation time

observation time is varied. It is seen that early observations of a crack of length
2.5 [mm] make the system less reliable. This is, of course, due to the fact that 2.5
[mm] is larger than the expected crack size at those times. It reaches this value at
about $1.2 \cdot 10^8$ cycles. Only at inspection times well above $1.2 \cdot 10^8$ cycles does the

a posteriori conditional failure probability become smaller than its **a priori** value, due to the fact that the observation then is smaller than expected.

If the crack were always detected the updating would become a little more efficient (dashed line). If, instead, a small crack is observed but cannot be measured, i.e. the observation event in eq. (6) is replaced by

$$B(n) = \{a(n) - \hat{a}_{th} \, \delta \le 0\} \tag{9}$$

with $\hat{a}_{th} = 1$ [mm], some reliability improvement is obtained for earlier observations.

#

Illustration 5.7:   Failure probability of existing structures with load and strength observations (Rackwitz and Schrupp, 1985)

The state function of an important component of an existing building, which is to be used for higher loads is

$$V(t) = \{R - (D + L(t)) \le 0\} \tag{1}$$

R is the componental resistance, D the dead load effect and $L(t)$ the time–variant live load effect. At time t_1 it is known that the component has not failed and that the highest live load was smaller or equal to ℓ. In addition, a non–destructive test for the resistance yielded the result ρ. The future load $L_2(t)$ can be modeled by a random sequence. Hence it is

$$\bar{V}(t_1) = \{R - (D + \max\{L_1(t_1)\}) > 0\} \tag{2}$$

$$B_\ell = \{\max\{L_1(t\} \le \ell\} \tag{3}$$

$$B_\rho = \{R = \rho + \epsilon\} \tag{4}$$

with ϵ a measurement error. For the future with future load we have $L_2(t)$

$$V(t) = \{R - (D + \max\{L_2(t)\}) \le 0\} \tag{5}$$

Then the failure probability in $[t_1,t]$ is

$$P_f(V(t)\,|\,B_\ell \cap B_\rho \cap \bar{V}(t_1)) = \frac{P(V(t) \cap B_\ell \cap B_\rho \cap \bar{V}(t_1))}{P(B_\ell \cap B_\rho \cap \bar{V}(t_1))} \tag{6}$$

Note that the various events are highly dependent. Also, the knowledge that the structure has survived under $\max\{L_1(t\} \le \ell$ is only important if this load level is rather high.

#

Illustration 5.8:   Optimization of proof–load level (Gollwitzer et al., 1990)

The boosters for a space craft can be modeled as cylindrical containers under high inner pressure. In order to save weight, one uses high–strength flow–turned steels. Even the most careful manufacturing and subsequent quality control cannot exclude the possibility that a small crack will occur. In particular, very small cracks (about 1 mm) can hardly be detected by non–destructive test methods. Therefore, proof–load tests must performed for each segment. However, this test is problematic in various respects. A test under low pressure will not give much information about the presence of a crack. High pressure close to and above the operating pressure will potentially indicate whether cracks are present. Then, a certain proportion of sections of the booster will fail already during proof loading. But even if potential cracks do not become unstable they may grow and will diminish the reliability under operating conditions. The level of proof pressure must therefore be such that cracks are detected, but do not grow too much, and the proportion of segments failing under proof pressure remains small.

For a inner circular crack with radius a the following instability criterion can be assumed

$$V = \left\{ \frac{L_r}{[8/\pi^2 \ln(\sec(\pi\,L_r/2))]^{1/2}} - K_r \leq 0 \right\} \tag{1}$$

where

$K_r = K_1/K_{1c} = \sigma \sqrt{(\pi a)}\ Y(a)$
K_{1c} = fracture toughness
a = crack length
$Y(a) \approx 1.0$ = geometrical factor
$L_r = p_i/p_0$
p_i = inner pressure
$p_0 \approx [t/r\ (R_y + R_m)/2]\ 1.07\ [1 - \pi a^2/(4t(a + t))]$
R_y = yield stress of steel
R_m = rupture strength of steel
t = wall thickness
r = radius of cylinder

During proof–load testing the crack grows according to

$$\Delta a = \frac{\pi}{6\ (R_y\ K_{1c})^2} \frac{K_{1,q}^4}{\left[1 - (\frac{K_{1,q}}{K_{1c}})^2\right]} \tag{2}$$

with $K_{1,q} = \sigma_q (\pi a)^{1/2}$ the stress intensity factor at pressure q and stress $\sigma_q = q\ r/t$. For failure during proof loading eq. (1) holds with p_i replaced by q and a by a + Δa. For survival of proof loading there is

$$D = \left\{ \frac{L_r}{[8/\pi^2 \ln(\sec(\pi\,L_r/2))]^{1/2}} - K_r > 0 \,|\, p_i = q, a = a + \Delta a \right\} \tag{3}$$

Consequently, the conditional failure probability is

$$P(V|D) = \frac{P(V \cap D)}{P(D)} \qquad (4)$$

For the numerical calculations the following assumptions are made.

Variable	Distribution function	Mean	Coefficient of variation
p_i	Normal	125 [bar]	0.08
R_y	Normal	1450 [MPa]	0.05
R_m	Normal	1700 [MPa]	0.05
K_{1c}	Normal	105 [MPa$\sqrt{m}$]	0.15
a	Rayleigh	1 [mm]	0.52
t	Normal	8.50 [mm]	0.02

Further, it is assumed that at least one crack is present in each test piece. A more rigorous treatment would possibly assume a Poissonian distribution of cracks, and concentrate on the largest of those cracks. Some typical results are given in the following figure. The required reliability level was $\beta = 4.26$ ($P_f = 10^{-5}$) (dashed line). It is recognized that this reliability cannot be reached for moderate proof pressures and for the conservative assumptions made.

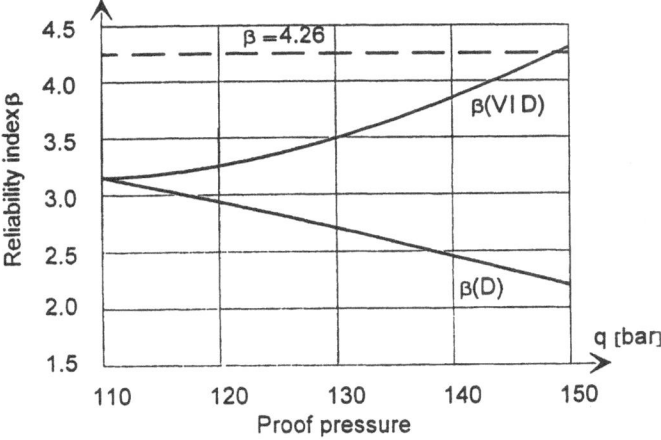

Figure 5.8.1:   Reliability indices for proof load test

Only if the proof pressure is about the maximum expected operating pressure (MEOP = 155 bar) can the required reliability be achieved. The probability of failure during proof loading is also shown. This means that, at a proof pressure of 150 bar, about 1 % of the test pieces will fail.

#

6. Summary and Discussion

The purpose of this review was to develop some of the basic notions and mathematics for the analysis of systems and elaborate on methodological aspects. The notion **system** is used in a general sense, i.e. a system problem is present as soon as the failure event is a union or an intersection or a combination of these. It was shown that complex systems can be reduced to a minimal cut set representation. Probability evaluations are straightforward only for either independent or fully dependent componental failure events. Otherwise serious computational difficulties arise. The easiest way to handle dependent failure events in complex systems is by use of certain concepts of first–order reliability (FORM). Those concepts require a probability distribution transformation, an algorithm to find the most likely failure point (β–point) in the so called standard space, and some non–trivial evaluations of the multinormal integral. For a first approximation, one can linearize all components individually. Better results are obtained if the joint β–point is found. These results have been improved in two ways. On the one hand a second order reliability method (SORM) has been developed which can be shown to be asymptotically ($P_f \to 0$ or $P_f \to 1$) exact (see Breitung, 1974; Hohenbichler et al., 1987; Breitung and Hohenbichler, 1989; Breitung, 1994). A summary of the relevant results is given in appendix A. Very recently it has been proposed to apply those asymptotic concepts directly in the so called original space (Breitung, 1994); this has some advantages but also disadvantages. Numerical studies have shown that the corrections to FORM by SORM are usually insignificant in practical applications, so that the additional effort for SORM is not always required. On the other hand, there has been considerable development to combine FORM/SORM concepts with importance sampling. Unfortunately, while good schemes are available for simple components and unions of components, very little has been done for intersections (Gollwitzer and Rackwitz, 1987). From the limited experience it is concluded that importance sampling is inefficient for parallel systems.

The computational tools developed so far rest on efficient search algorithms, and this is the key point for successful analyses. Those available in various programs require smooth differentiable failure surfaces and continuous distribution functions. Usually, the analyses are then successful. Yet, especially when conditional probabilities have to be determined and equality constraints are present, those algorithms can fail under extreme parameter combinations. However, individual linearization will work almost always, but only crude probability estimates can be expected.

Several examples illustrate typical applications of the theory.

7. References

Abdo, T., Rackwitz, R., A New β—Point Algorithm for Large Time—Invariant and Time—Variant Reliability Problems, A. Der Kiureghian, P. Thoft—Christensen (eds.): Reliability and Optimization of Structural Systems '90, Proc. 3rd WG 7.5 IFIP Working Conf., Berkeley, March 26—28, 1990, Springer, Berlin, 1991

Arora, J. S., Introduction to Optimum Design, McGraw—Hill, New York, 1989

Barlow, R.E., Proschan, F., Statistical Theory of Reliability and Life Testing, Holt, Rinehard and Winston, New York, 1975

Bleistein, N., Handelsman, R.A., Asymptotic Expansions of Integrals, Holt, Rinehard and Winston, New York, 1975

Breitung, K., Asymptotic Approximations for Multinormal Integrals, Journ. of the Eng. Mech. Div., Vol. 110, No.3, 1984, pp. 357—366

Breitung, K., Hohenbichler, M., Asymptotic Approximations for Multivariate Integrals with an Application to Multinormal Probabilities, Journ. of Multivariate Analysis, Vol. 30, No.1, 1989, pp. 80—97

Breitung, K., Asymptotic Approximations for Probability Integrals, Springer—Verlag, Berlin, 1994

Caldarola, L., Coherent Systems with Multistate Components, Nuclear Eng. and Design, Vol. 58, 1980, pp. 127—139

Copson, E.I., Asymptotic Expansions, Cambridge University Press, Cambridge, 1965

Dawson, D.A., Sankoff, D., An Inequality for Probabilities, Proc. Am. Math. Soc., 18, 1967, pp. 504—507

Der Kiureghian, A., Liu, P.—L., Structural Reliability under Incomplete Probability Information, Journal of Engineering Mechanics, ASCE, 1986, Vol. 112, No. 1, 85—104.

Ditlevsen, O., Narrow Reliability Bounds for Structural Systems, Journ. of Struct. Mech., Vol.7, No.4, 1979, pp. 453—472

Dunnet, C.W., Sobel, M., Approximation to the Probability Integral and Certain Percentage Points of Multivariate Analogue of Student's Distribution, Biometrika, Vol. 42, 1955, pp. 258—260

Fardis, M.N., Cornell, C.A., Analysis of Coherent Multistate Systems, IEEE Trans. on Reliability, Vol. R—30, 2, 1981, pp. 117—122

Gill, P.E., Murray, W., Wright, M.H., Practical Optimization, Academic Press, London, 1981.

Gollwitzer, S., Rackwitz, R., Equivalent Components in First—Order System Reliability, Rel. Eng., Vol. 5, 1983, pp. 99—115

Gollwitzer, S., Rackwitz, R., First—Order Reliability of Structural Systems, Proc. ICOSSAR'85, Vol. 1, 1985, pp. 171—180

Gollwitzer, S., Rackwitz, R., Comparison of Numerical Schemes for the Multinormal Integral, IFIP—TC7 Conf. Aalborg, May 1987, in: P.Thoft—Christensen, (Ed.), Lecture Notes in Engineering, Springer Verlag, Berlin, 1987

Gollwitzer, S., Rackwitz, R., An Efficient Numerical Solution to the Multinormal Integral, Prob. Eng. Mech., 3, 2, 1988, pp. 98—101

Gollwitzer, S., Grimmelt, M., Rackwitz, R., Brittle Fracture and Proof Loading of Metallic Pressure Vessels, Proc.: 7th Int. Conf. on Reliability and Maintainability, Brest, 1990,

Guenard, Y.F., Application of System Reliability Analysis to Offshore Structures, John A. Blume Engineering Center, Thesis, Report No. 71, Stanford University, 1984

Hasofer, A.M., Lind, N.C., An Exact and Invariant First Order Reliability Format, Journ. of Eng. Mech. Div., ASCE, Vol. 100, No. EM1, 1974, p. 111—121

Hock, W., Schittkowski, K., A Comparative Performance Evaluation of 27 Nonlinear Programming Codes, Computing, Vol. 30, 1983, pp. 335—358.

Hohenbichler, M., Zur zuverlässigkeitstheoretischen Untersuchung von Seriensystemen, Berichte zur Zuverlässigkeitstheorie der Bauwerke, SFB 96, Technische Universität München, H. 48, 1980

Hohenbichler, M., Rackwitz, R., Non—Normal Dependent Vectors in Structural Safety, Journ. of the Eng. Mech. Div., ASCE, Vol.107, No.6, 1981, pp.1227—1240.

Hohenbichler, M., Rackwitz, R., First—Order Concepts in System Reliability, Struct. Safety, 1, 3, 1983, pp. 177—188

Hohenbichler, M., Rackwitz, R., A bound and an approximation to the multivariate normal distribution function, Math. Jap., Vol. 30, 5, 1985

Hohenbichler, M., Gollwitzer, S., Kruse, W., Rackwitz, R., New Light on First— and Second—Order Reliability Methods, Structural Safety, 4, pp. 267—284, 1987

Hunter, D., An Upper Bound for the Probability of a Union, J. Appl. Prob., 13, 1976, p. 597

Hunter, D., Approximating Percentage Points of Statistics Expressible as Maxima, TIMS Studies in the Management Sciences, Vol. 7, 1977, pp. 25—36

Kounias, E.G.: Bounds for the Probability of a Union, with Applications, Ann. Math. Statist., Vol. 39, 1968, pp. 2154—2158

Madsen, H.O., First Order vs. Second Order Reliability Analysis of Series Systems, Structural Safety, Vol. 2, 3, 1985, pp. 207—214

Moses, F., New Directions and Research Needs in System Reliability, Structural Safety, 7, 2—4, 1990, pp. 93—100

Murotsu, Y., Okada, H., Yonezawa, M., Taguchi, K., Reliability Assessment of Redundant Structure, Proc. ICOSSAR'81, Structural Safety and Reliability, Trondheim, Elsevier, Amsterdam, 1981, pp. 315—329

Nataf, A., Determination des Distribution dont les Marges sont donnees, Comptes Rendus de l'Academie des Sciences, 1962, Vol. 225, pp. 42—43

Owen, D.B., Tables for Computing Bivariate Normal Probabilities, Ann. Math. Statist., Vol. 27, 1956, pp. 1075—1090

Rackwitz, R., Fiessler, B., Structural Reliability under Combined Random Load Sequences, Comp. & Struct., Vol. 9, 1978, pp. 484—494

Rackwitz, R., Schrupp, K., Quality Control, Proof Testing and Structual Reliability, Structural Safety, 2, 1985, pp. 239 — 244

Rosenblatt, M., Remarks on a Multivariate Transformation, Ann. Math. Statistics, Vol. 23, 1952, pp. 470—472

Ruben, H., An Asymptotic Expansion for the Multivariate Normal Distribution and Mill's Ratio, Journ. of Research NBS, Vol.68B, 1, 1964, pp.3—11

Schall, G., Gollwitzer, S., Rackwitz, R., Integration of Multinormal Densities on Surfaces, 2nd IFIP WG—7.5 Working Conference on Reliability and Optimization of Structural Systems, London, September, 26.—28., 1988, pp. 235—248

Schall, G., Scharrer, M., Östergaard, C., Rackwitz, R., Fatigue Reliability Investigation for Marine Structures Using a Response Surface Method, Proc. 10th OMAE Conf., Stavanger, 1991, Vol.II, pp. 247—254

Schittkowski, K., Theory, Implementation and Test of a Nonlinear Programming Algorithm. In: Eschenauer, H., Olhoff, N. (eds.), Optimization Methods in Structural Design, Proceedings of the Euromech/Colloquium 164, University of Siegen, FRG, Oct. 12—14, 1982, Zuerich 1983

Sidak, Z., On Multivariate Normal Probabilities of Rectangles, their Dependence and Correlations, Ann. Math. Statist., Vol. 39, 1968, pp. 1425—1434

Slepian, D., The One—sided Barrier Problem for Gaussian Noise, Bell System Tech. Journ., Vol. 41, 1962, pp. 463—501

Thoft—Christensen, P., Murotsu, Y., Application of Structural Systems Reliability Theory, Springer, Berlin, 1986

Yen, J.Y., Shortest Path Network Problems, Anton Hain, Meisenheim am Glan, 1975

Appendix A:
Asymptotic Approximations to Probability Integrals — A Summary of Results

INTRODUCTORY SCALAR CASE

In 1820 the French mathematician P.S. de Laplace proposed to approximate the following integral

$$I(\lambda) = \int_a^b h(x) \exp[-\lambda f(x)] \, dx \qquad (A.1)$$

where $f(x)$ and $h(x) > 0$ are certain sufficiently smooth functions. λ is a parameter. Assume that $f(x)$ is a monotonically increasing function in the interval $[a,b]$ and has a minimum in a, at which $f'(a) > 0$. For increasing λ, the integral will be dominated by values of the integrand in the vicinity of $x = a$. According to the mean value theorem, the function $h(x)$ can be approximated by the term $h(a)$ which is put in front of the integral, and $f(x)$ is developed into a Taylor series truncated after the first non–vanishing term. Then

$$
\begin{aligned}
I(\lambda) \quad &= \int_a^b h(x) \exp[-\lambda f(x)] \, dx \\
&\approx h(a) \int_a^b \exp[-\lambda \, (f(a) + f'(a) \, (x-a) + \ldots)] \, dx \\
&\approx h(a) \exp[-\lambda \, (f(a) - f'(a) \, a)] \int_a^b \exp[-\lambda \, (f'(a) \, x + \ldots)] \, dx \\
&\approx h(a) \exp[-\lambda(f(a)-f'(a)a)] \, (-\lambda f'(a))^{-1} \, \{\exp[-\lambda f'(a)b] - \exp[-\lambda f'(a)a]\}
\end{aligned}
$$
$$(A.2)$$

For $a = 0$ and $b < \infty$ a λ it is always possible to choose such that

$$I(\lambda) \approx h(0) \, \exp[-\lambda f(0)] \, |\lambda \, f'(0)|^{-1} \qquad (A.3)$$

In the second case it is assumed that $f(x)$ has a minimum in $[a,b]$ at x^*. Without loss of generality we assume that $x^* = 0$ and $h(0) > 0$. Again $f(x)$ is developed into a Taylor series truncated after the first non–vanishing term. With $f'(0) = 0$ and $f''(0) > 0$ as well as $a = -\epsilon_1$ and $b = \epsilon_2$ we find

$$
\begin{aligned}
I(\lambda) \quad &= \int_a^b h(x) \exp[-\lambda f(x)] \, dx = h(0) \int_{-\epsilon_1}^{+\epsilon_2} \exp[-\lambda \, (f(0) + \tfrac{1}{2} f''(0) \, x^2 + \ldots)] \, dx \\
&\approx h(0) \exp[-\lambda f(0)] \int_{-\epsilon_1}^{+\epsilon_2} \exp[-\tfrac{1}{2} \lambda \, f''(0) \, x^2] \, dx
\end{aligned}
$$
$$(A.4)$$

With the substitution

$$\xi = \sqrt{(\tfrac{\lambda}{2} f''(0))}\ x$$

one obtains

$$I(\lambda) \approx h(0)\ \exp[-\lambda\ f(0)]\ \sqrt{(2/(\lambda\ f''(0)))}\ \int_{-\sqrt{(\lambda f''(0)\epsilon_1/2)}}^{+\sqrt{(\lambda f''(0)\epsilon_2/2)}} \exp[-\xi^2]\ d\xi$$

Even if ϵ_1 and ϵ_2 are small it is always possible to choose a λ such that the integration limits can be set at $\pm\infty$ without too much error

$$I(\lambda)\quad \approx h(0)\ \exp[-\lambda f(0)]\ \sqrt{(2/(\lambda f''(0)))} \int_{-\infty}^{+\infty} \exp[-\xi^2]\ d\xi$$

$$\approx h(0)\ \exp[-\lambda f(0)]\ \sqrt{((2\pi)/(\lambda\ f''(0)))} \tag{A.5}$$

These approximations can be shown to be asymptotically exact, i.e. for $\lambda \to \infty$ (Copson, 1965).

GENERAL CASE

Recently, these results have been generalized to multivariate integrals of various forms (Bleistein, 1975). A fairly general result for the integral

$$I(\lambda) = \int_D h(\mathbf{y})\ \exp[-\lambda\ f(\mathbf{y})]\ d\mathbf{y} \tag{A.6}$$

for $\lambda \to \infty$ where $\mathbf{y} = (y_1, y_2,...,y_n)^T$, and D a simply connected domain containing the origin has been given in (Breitung and Hohenbichler, 1989). Herein $f(\mathbf{y})$ is at least twice differentiable and has a minimum at $\mathbf{y} = \mathbf{y}^* \neq 0$. $h(\mathbf{y})$ is a slowly varying function and $h(0) \neq 0$. D is given by $D = \cap_{i=1}^{k} D_i$ with $D_i = \{\mathbf{y}\colon g_i(\mathbf{y}) \leq 0\}$ and $k \in \{1,2,...,n\}$. $f(\mathbf{y})$ as well as the functions $g_i(\mathbf{y})$ are at least twice differentiable at $\mathbf{y}^*$ and the function $h(\mathbf{y})$ is continuous at $\mathbf{y}^*$ and $h(\mathbf{y}^*) \neq 0$. At $\mathbf{y}^*$, $g_i(\mathbf{y}^*) = 0$ for i=1,2,...,k. The gradients $\mathbf{a}_i = \nabla g_i(\mathbf{y}^*)$ (i=1,2,...,k) are linearly independent. This implies that $a_{ij} = 0$ for i=1,2,...,k and j=k+1,...,n which always can be achieved by a suitable orthogonal transformation. It also means that $\partial f(\mathbf{y}^*)/\partial y_i = 0$ for i=k+1,...,n and the gradient can be represented as

$$\nabla f(\mathbf{y}^*) = \sum_{i=1}^{k} \gamma_i\ \mathbf{a}_i$$

with $\gamma_i < 0$ for i=1,2,...,k. Then, it has been proved that

$$I(\lambda) = \int_D h(\mathbf{y}) \exp[-\lambda f(\mathbf{y})]\, d\mathbf{y}$$

$$\sim h(\mathbf{y}^*) \exp[-\lambda f(\mathbf{y}^*)]\, (2\pi)^{\frac{n-k}{2}} \lambda^{-\frac{n+k}{2}} |\det(\mathbf{A})|^{-1} \left(\prod_{i=1}^{k} |\gamma|^{-1}\right) |\det(\mathbf{D})|^{-1/2}$$

$$\text{(A.7)}$$

for $\lambda \to \infty$ with $\mathbf{A} = \{a_{ij};\ i,j=1,2,\dots,k\}$ and $\mathbf{D} = \{d_{ij};\ i,j=k+1,\dots,n\}$ with elements

$$d_{ij} = \frac{\partial^2 f(\mathbf{y}^*)}{\partial y_i\, \partial y_j} - \sum_{s=1}^{k} \gamma_s \frac{\partial^2 g_s(\mathbf{y}^*)}{\partial y_i\, \partial y_j}$$

and $\det(\mathbf{D}) = 1$ for $k = n$.

APPLICATION TO PROBABILITY INTEGRALS

Probability integrals can always be written in the following form

$$P(V) = \int_V \psi_X(\mathbf{x})\, d\mathbf{x} = \int_V \exp[\ln\psi(\mathbf{x})]\, d\mathbf{x} = \int_V \exp[\ell(\mathbf{x})]\, d\mathbf{x} \qquad \text{(A.8)}$$

where $\ell(\mathbf{x}) = \ln\psi(\mathbf{x})$ is the likelihood function of the probability density $\psi(\mathbf{x})$. The integration domain is given by $V = \cap_{i=1}^{n} V_i$ and $g_i(0) > 0$ for at least one $i \in \{1,\dots,n\}$. The **critical point** is the point for which the log-likelihood function is maximal in V. The essential idea for applying the above results to probability integrations is a central scaling by a factor b, as shown in fig. A1 (Breitung, 1984). We define

$$\beta = (-\max\{\ell(\mathbf{x})\})^{\frac{1}{2}} = (-\ell(\mathbf{x}^*))^{\frac{1}{2}} \quad \text{for } \{\mathbf{x} \in V\} \qquad \text{(A.9)}$$

and

$$f(\mathbf{x}) = \beta^{-2}\, \ell(\mathbf{x}) \qquad \text{(A.10)}$$

At $\mathbf{x}^*$, $f(\mathbf{x}^*) = 1$. Also we assume $\ell(\mathbf{x}^*) < 0$. We consider the integral

$$P(b) = \int_V \exp[-b^2 f(\mathbf{x})]\, d\mathbf{x} \qquad \text{(A.11)}$$

and apply eq. (A.7) with $h(\mathbf{x}) = 1$ and $\lambda = b^2$

$$P(b) \sim (2\pi)^{\frac{n-k}{2}} b^{-n-k} \exp[-b^2]\, |\det(\mathbf{A})|^{-1} \left(\prod_{i=1}^{k} |\gamma_i|^{-1}\right) |\det(\mathbf{D})|^{-1/2}$$

$$\text{(A.12)}$$

By noting that

$$\nabla f(\mathbf{x}^*) = \beta^{-2} \nabla \ell(\mathbf{x}^*)$$

$$\nabla\nabla f(\mathbf{x}^*) = \beta^{-2} \nabla\nabla \ell(\mathbf{x}^*)$$

$$\nabla f(\mathbf{x}^*) = \sum_{i=1}^{k} \gamma_i \, a_i$$

and

$$\nabla \ell(\mathbf{x}^*) = \sum_{i=1}^{k} \delta_i \, a_i$$

with $\delta_i = \beta^{-2} \gamma_i$ we find

$$|\det(\mathbf{D})| = \beta^{-2(n-k)} |\det(\mathbf{L})|$$

where

$$\mathbf{L} = \left\{ \frac{\partial^2 \ell(\mathbf{x}^*)}{\partial x_i \, \partial x_j} - \sum_{s=1}^{k} \gamma_s \frac{\partial^2 g_s(\mathbf{x}^*)}{\partial x_i \, \partial x_j} \; ; \, i,j = k+1, \, ..., \, n \right\}$$

Hence

$$P(b) \sim (2\pi)^{\frac{n-k}{2}} \frac{\beta^{n-k}}{b^{n+k}} \beta^{2k} \exp[-b^2] \, |\det(\mathbf{A})|^{-1} \left(\prod_{i=1}^{k} |\gamma_i|^{-1} \right) |\det(\mathbf{L})|^{-1/2}$$

$$(A.13)$$

and with $b = \beta$ (β is already large)

$$P(V) \approx (2\pi)^{\frac{n-k}{2}} \exp[-\beta^2] \, |\det(\mathbf{A})|^{-1} \left(\prod_{i=1}^{k} |\gamma_i|^{-1} \right) |\det(\mathbf{L})|^{-1/2}$$

$$(A.14)$$

In particular, for $k = 1$ we have

$$P(V) \approx (2\pi)^{\frac{n-1}{2}} \exp[-\beta^2] \, |\gamma_1|^{-1} |\det(\mathbf{L})|^{-1/2}$$

$$(A.15)$$

with $\gamma_1 = \|\nabla\ell(\mathbf{x}^*)\| / \|\nabla g_1(\mathbf{x}^*)\|$. For $k = n$ we have

$$P(V) \approx \exp[-\beta^2] \, |\det(\mathbf{A})|^{-1} \, (\prod_{i=1}^{n} |\gamma_i|^{-1}) \qquad (A.16)$$

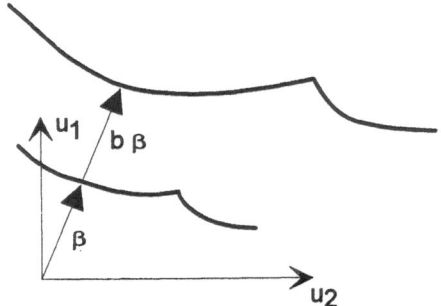

Figure A1: Scaling of the integration domain

The special case of multinormal integrals yields nicely compact results. Let $\mathbf{Y}$ be a standard normal vector with probability density $\varphi(\mathbf{y}) = (2\pi)^{-n/2} \exp[-\|\mathbf{u}\|^2/2]$. The scaling of the integral

$$P(V) = \int_V \varphi(\mathbf{y}) \, d\mathbf{y} \qquad (A.17)$$

yields upon substitution with $\mathbf{u} = \mathbf{y} \, b^{-1}$

$$P(bV) = \int_{bV} \varphi(\mathbf{y}) \, d\mathbf{y} = b^n \int_V \varphi(b\mathbf{u}) \, d\mathbf{u} = (2\pi)^{-n/2} b^n \int_V \exp[-b^2\|\mathbf{u}\|^2/2] \, d\mathbf{u}$$
$$\qquad (A.18)$$

where $h(\mathbf{u}) = 1$, $f(\mathbf{u}) = \|\mathbf{u}\|^2/2$ and $\lambda = b^2$. The critical point $\mathbf{u}^*$ has distance $\beta = \|\mathbf{u}^*\|$ from the origin.

Application of eq. (A.7) is again straightforward. Let the integration domain be given by $V = \{\cap_{i=1}^{n} V_i\}$ with $V_i = \{g_i(\mathbf{u}) \leq 0\}$. At the critical point $\mathbf{u}^*$, $g_i(\mathbf{u}) = \mathbf{a}_i^T (\mathbf{u} - \mathbf{u}^*) = 0$ for $i = 1,2,...,n$. $\mathbf{u}^*$ can be represented as $\mathbf{u}^* = \Sigma_{i=1}^{n} \gamma_i \mathbf{a}_i$ with $\mathbf{a}_i = \nabla g_i(\mathbf{u}^*)$. Then

$$P(V) \approx (\det(\mathbf{R}))^{-1/2} \prod_{i=1}^{n} \frac{\varphi_1(\mathbf{u}_i^*)}{(-\gamma_i)} \qquad (A.19)$$

This formula is a asymptotic approximation for the multinormal integral; note that $\beta_i = -\mathbf{a}_i^T \mathbf{u}^*$ and $\mathbf{R} = \{\mathbf{a}_i^T \mathbf{a}_j\}$ and therefore $P(V) \sim \Phi_n(-\boldsymbol{\beta}; \mathbf{R})$ (Ruben, 1964). For $k = 1$ with $g_1(0) \geq 0$ and $V_1 = \{g_1(\mathbf{u} \leq 0\}$ one obtains (Breitung, 1984)

$$P(V) \approx \|\mathbf{u}^*\|^{-1} \, |\det(\mathbf{D})|^{-1/2} \; \varphi_1(\mathbf{u}^*) \approx \Phi(-\beta) \, |\det(\mathbf{D})|^{-1/2}$$

$$\text{(A.20)}$$

with

$$\mathbf{D} = \{\delta_{ij} - \frac{\|\mathbf{u}^*\|}{\|\nabla g(\mathbf{u}^*)\|} \frac{\partial^2 g_1(\mathbf{u}^*)}{\partial u_i \, \partial u_j}; \; i,j = 2,...,n\}$$

In the second equation, use was made of $\Phi(-x) \sim \varphi(x)/x$ and $\beta = \|\mathbf{u}^*\|$. If suitable orthogonal transformations are performed, the correction term can also be given in terms of the main curvatures κ_i, i.e.

$$\mathbf{D} = \prod_{i=1}^{n-1} (1 - \beta \, \kappa_i)^{-1/2}$$

Further, if $g_i(0) > 0$ for at least one i $\in$ {1, 2,...,m}, $g_i(\mathbf{u}^*) = 0$ for i = 1, 2,...,k (k $\leq$ m), and $g_i(\mathbf{u}^*) > 0$ for i = k+1,...,m, then (Hohenbichler et al., 1987)

$$P(V) \approx |\det(\mathbf{D})|^{-1/2} \, P(\bigcap_{i=1}^{k} V_i) \qquad \text{(A.21)}$$

with

$$\mathbf{D} = \{\delta_{ij} - \sum_{s=1}^{k} \gamma_s \frac{\partial^2 g_s(\mathbf{u}^*)}{\partial u_i \, \partial u_j}; \; i,j=k+1,...,n\}$$

For the cut set probability the result for the multi–normal integral in eq. (A.19) with n = k or other suitable computation schemes for the normal integral (Gollwitzer and Rackwitz, 1988) can be used.

The terms with det($\mathbf{D}$) or det($\mathbf{L}$) are second order corrections which, in general, are small compared to the leading remaining first order term. The two types of results are mathematically equivalent. They reduce probability integration to a problem of nonlinear programming (the search for the critical point) and some simple algebra.

Appendix B: Computation of conditional probabilities

Conditional probabilities

$$P_f(F|A \cap B \cap C \,...) = \frac{P(F \cap A \cap B \cap C \cap ...)}{P(A \cap B \cap C \cap ...)} \qquad \text{(B.1)}$$

can be evaluated by separate computations of the numerator and the denominator. Conditioning events can, for example, be observations on some state of the system. It is possible that this must be formulated as an equality constraint. In this case surface integrals need to be computed. The typical integral is

$$I(D) = \int_D \varphi_n(x) \, ds(x) \tag{B.2}$$

with

$$D = E \cap F = \bigcap_{i=1}^{\ell} \{E_i\} \cap \bigcap_{j=\ell+1}^{m} \{F_j\} = \bigcap_{i=1}^{\ell} \{e_i(X) = 0\} \cap \bigcap_{j=\ell+1}^{m} \{g_j(X) \le 0\}$$

$X = (X_1,...,X_n)^T$ is the n–dimensional random vector with given distribution function. $ds(x)$ means surface integration. E_i denotes equality constraints. We perform all calculations in the standard space and assume $\ell < n$. We consider first the case

$$I(E) = \int_{\mathbb{R}^n \cap E} \varphi_n(u) \, du \tag{B.3a}$$

or

$$I(E) = \int_E \varphi_n(u) \, ds(u) \tag{B.3b}$$

Assume that there exist a maximum point (β–point) of $\varphi_n(u)$ $u^* = \min\{\|u\|\}$ on E for $E = \{u: e(u) = 0\}$ with $\beta = \|u^*\|$. By a orthogonal transformation, the β–point can be shifted to the negative u_n–axis, i.e. $u^* = (0,0,...,-\|u^*\|)^T$. Further we assume that $e(u) = 0$ is at least twice differentiable in u^*, and the mixed derivatives of $e(u^*)$ vanish; this can be achieved by an appropriate orthogonal transformation. The curvatures of $e(u)$ at u^* are then

$$\kappa_i = -\frac{1}{\frac{\partial e(u^*)}{\partial u_n}} \frac{\partial^2 e(u^*)}{\partial u_i^2} \quad \text{for } i = 1,2,...,n-1 \tag{B.4}$$

Then, asymptotically the following parameterization of u_n around u^* can be chosen

$$u_n = e^{-1}(u_1,...,u_{n-1}) = \psi(u_1,...,u_{n-1}) \approx -\beta + \frac{1}{2}\sum_{i=1}^{n-1} \kappa_i u_i^2 \tag{B.5}$$

The surface integral can be expressed by the following volume integral

$$I(E) = \int_{\mathbb{R}^{n-1}} \varphi_n(u_1,..,u_{n-1},\psi(u_1,..,u_{n-1})) \, (1 + \sum_{i=1}^{n-1}(\frac{\partial\psi(u^*)}{\partial u_i})^2)^{1/2} \prod_{i=1}^{n-1} du_i \tag{B.6}$$

Scaling with a factor $b > 1$ and transformation with $v_i = u_i/b$ yields

$$I(bE) = b^{n-1} \int_{\mathbb{R}^{n-1}} \varphi_n(b\mathbf{v},\psi(b\mathbf{v}))\, (1 + \sum_{i=1}^{n-1} (\frac{\partial \psi(b\mathbf{v}^*)}{\partial v_i})^2)^{1/2} \prod_{i=1}^{n-1} dv_i$$

(B.7)

In the expression

$$(2\pi)^{n/2}\, \varphi_n(b\mathbf{v},\psi(b\mathbf{v})) = \exp[-\tfrac{1}{2}\,(\sum_{i=1}^{n-1} (bv_i)^2 + (-\beta + \tfrac{b^2}{2} \sum_{i=1}^{n-1} (\kappa_i v_i^2))^2)]$$

the second term is developed as

$$(-\beta + \tfrac{b^2}{2} \sum_{i=1}^{n-1} (\kappa_i v_i^2))^2 \approx \beta^2 - b^2\beta \sum_{i=1}^{n-1} (\kappa_i v_i^2) + \ldots$$

Then, the last term equals unity.

$$I(bE) \sim b^{n-1}\, \varphi(\beta) \prod_{i=1}^{n-1} (1-\beta\kappa_i)^{-1/2} \prod_{i=1}^{n-1} \int_{-\infty}^{+\infty} (1-\beta\kappa_i)^{1/2}\, \varphi(bv_i\,(1-\beta\kappa_i)^{1/2})\, dv_i$$

(B.8)

Hence, there is asymptotically for large β ($b = 1$)

$$I(E) \approx \varphi(\beta) \prod_{i=1}^{n-1} (1 - \beta\,\kappa_i)^{-1/2}$$

(B.9)

We consider the more general case

$$I(E \cap F) = \int_{\mathbb{R}^n \cap E \cap F} \varphi_n(\mathbf{u})\, d\mathbf{u}$$

(B.10)

with $E = \{e_1(\mathbf{u}) = 0\}$ and $F = \cap_{i=2}^{m} F_i$ with $F_i = \{g_i(\mathbf{u}) \leq 0\}$ and $g_i(0) > 0$ for at least one i. With $u_i^* = \mathbf{a}_1^T \mathbf{u}^*$ one obtains on similar lines

$$I(E \cap F) \approx \varphi(\mathbf{a}_1^T \mathbf{u}^*)\, \Phi_{k-1}(\mathbf{c};R)\, \frac{1}{(\det(\mathbf{I} - \mathbf{D}))^{\frac{1}{2}}}$$

(B.11)

with

$$c = \{u^{*T}a_s - (u^{*T}a_1)(a_s^T a_1); s = 2,...,k\}$$

$$R = \{a_s^T a_t - (a_s^T a_1)(a_t^T a_1); s,t = 2,...,k\}$$

$$D = \{\sum_{s=1}^{k} \gamma_s \frac{\partial^2 g_s(u^{*})}{\partial u_i \partial u_j}; i,j = k+1,...,n\}$$

$$I = \{\delta_{ij}; i,j = k+1,...,n\}$$

a_r $(r = 1,...,k)$ is the normalized gradient of condition r in u^{*} and the $\gamma_s < 0$ are the solutions of the equations

$$u^{*} = \sum_{s=1}^{k} \gamma_s a_s \quad \text{for } s = 1,2,...,k$$

k is the number of in inequality condition. Generalization to multiple equality constraints is straightforward. The result is invariant under orthogonal transform-ations. Therefore, recursive application for the case of more than one equality constraint yields the general result. One can condition on all equality constraints simultaneously so that the result obtains the form

$$I(\bigcap_{i=1}^{\ell} (e_i = 0) \cap \bigcap_{j=\ell+1}^{m} (g_j \leq 0)) \approx \varphi_\ell(c_\ell; \Sigma_{\ell\ell}) \, \Phi_q(c_{q|\ell}; \Sigma_{q|\ell}) \frac{1}{(\det(I-D))^{\frac{1}{2}}}$$

$$\text{(B.12)}$$

with

$$q = m - \ell; \quad \ell \leq k \leq m$$

$$c = \begin{bmatrix} c_\ell \\ c_q \end{bmatrix}; \quad R = \begin{bmatrix} R_{\ell\ell} & R_{\ell q} \\ R_{q\ell} & R_{qq} \end{bmatrix}; \quad \Sigma_{\ell\ell} = R_{\ell\ell}$$

$$c_{q|\ell} = R_{q\ell}^T R_{\ell\ell}^{-1} c_\ell$$

$$\Sigma_{q|\ell} = R_{qq} - R_{q\ell}^T R_{\ell\ell}^{-1} R_{\ell q}$$

D as in eq. (A.21) and

$$\varphi_\ell(c_\ell; \Sigma_{\ell\ell}) = (2\pi)^{\ell/2} (\det(\Sigma_{\ell\ell}))^{-1/2} \exp[-\tfrac{1}{2}(z^T \Sigma_{\ell\ell}^{-1} z)]$$

the ℓ–dimensional standard normal density.

STATISTICAL EXTREMES AS A TOOL FOR DESIGN

J. TIAGO DE OLIVEIRA
Academia das Ciências de Lisboa and
Faculdade de Ciências e Tecnologia, Universidade Nova de Lisboa

1. Introduction

Classical Statistical Extremes Theory describes the random behaviour of the largest or smallest values of independent and identically distributed (i.i.d.) samples: the theory may be extended to random interdependent sequences and thus it can be used in concrete decision or design problems where the crossing of some bounds can give rise to breakdown or disaster. Large waves, gusts of wind, floods, large insurance claims etc., are examples or maxima; droughts, fatigue, rupture, failures of nuclear reactors, etc., may be connected with minima; disaster can occur if some bounds are exceeded or not attained. The design of a breakwater, of a plane, of a high antenna or of a high tower, etc., must, take in to account the risks of (random) failure and/or disaster.

The study of these concrete problems should be dealt with, in a complete form, through the analysis of the extremes of a stochastic process.

The asymptotic behaviour of extremes of i.i.d. samples is a first and very important step, because, in general, the underlying distribution of the observed variables is completely unknown and, if known, still depends on unknown parameters. The analysis can be extended to dependent samples and also to time-series with correlations decaying with time; the asymptotic distributions possess some of the same properties as those for i.i.d. samples.

After presenting the i.i.d., asymptotic behaviour of extremes, we will refer to statistical decision for small samples of extremes.

The basic reference is Gumbel (1958). In this paper we will concentrate on the Gumbel distribution for maxima, the pivot distribution, and the Weibull distribution for minima. The last distribution without location parameter, can be made equivalent to the Gumbel distribution for maxima by an exponential transformation allowing thus statistical decisions to be made for design. A review paper by Muir and El-Sharawi

C. Guedes Soares (ed.), Probabilistic Methods for Structural Design, 213–225.

(1986) explains why Weibull distributions are used. The short comments in Gumbel (1958), section 6.3.4 and 6.3.8, are also of interest. Some remarks related to Structural Engineering applications can be found in Tiago de Oliveira (1979).

2. Asymptotic distribution of extremes

For an i.i.d. sample $(X_1, X_2, \ldots X_n)$ with distribution function $F(x) = \text{Prob }\{X \leq x\}$ the distribution function of the k-th order statistic X_k (i.e., the $(n-k+1)$ th maximum or k-th minimum) is

$$\text{Prob.}\{X_k \leq x\} = \{k \text{ or more } X_i \leq x\} = \sum_{j=k}^{n} \binom{n}{j} F(x)^j \left[1 - F(x)\right]^{n-j} =$$

$$= 1 - \sum_{j=0}^{n} \binom{n}{j} F(x)^j \left[1 - F(x)\right]^{n-j}$$

so that the distribution function of the maximum X_n is $F^n(x)$, and that of the minimum X_1 is $1 - [F(x)]^n$.

For maxima, it can be shown that, for a large family of distribution functions, there exist coefficients λ_n and $\delta_n (> 0)$ (not uniquely defined) such that, for $n \to \infty$,

$$F^n(\lambda_n + \delta_n z) \to L(z)$$

where the asymptotic distribution function $L(z)$ must be of one of the (reduced) forms:

$$\Lambda(z) = \exp(-e^{-z}), \ -\infty < z < +\infty, \qquad \text{Gumbel distribution;}$$

$$\begin{aligned} \Psi(z) \ &= \ 0 \text{ if } z < 0 \\ &= \ \exp(-z^{-\alpha}) \text{ if } z \geq 0, \ \alpha > 0, \end{aligned} \qquad \text{Fréchet distribution;}$$

$$\begin{aligned} \Psi_\alpha(z) \ &= \ \exp(-(-z)^\alpha) \text{ if } z < 0 \\ &= \ 1 \text{ if } z \geq 0, \ \alpha > 0, \end{aligned} \qquad \text{Weibull distribution.}$$

In applications, a location and a dispersion or scale parameter are introduced in these standardised forms by a transformation of the variable z.

The derivation of those asymptotic forms can be sketched as follows:

From

$$F''(\lambda_n + \delta_n z) \to L(z)$$

we get

$$F^{mn}(\lambda_{mn} + \delta_{mn}z) \to L(z)$$

or

$$F''(\lambda_{mn} + \delta_{mn}z) \to L^{1/m}(z)$$

and by Khintchine's theorem on the convergence of types (Gnedenko 1943) we know that

$$F''(\lambda_{mn} + \delta_{mn}z) \to L(\alpha_m + \beta_m z) \quad \text{where} \left(\beta_m > 0\right)$$

if $L(z)$ is proper and non-degenerate. Thus $L(z)$ must satisfy the functional equation:

$$L^{1/m}(z) = L(\alpha_m + \beta_m z)$$

for some α_m and $\beta_m(> 0)$, with m a positive integer. This functional equation is easily extended to m positive rational and, subsequently, to m positive real. The solutions of the functional equation are the three already given above.

The relation

$$F''(\lambda_n + \delta_n z) \to L(z), \left(0 < z < 1\right) \quad \text{is equivalent to the relation}$$

$$n\left[1 - F(\lambda_n + \delta_n z)\right] \to -\log L(z), \quad \text{wich} \quad \text{sometimes} \quad \text{is} \quad \text{easier} \quad \text{to}$$

manipulate.

The convergence is uniform, because the distribution functions $L(z)$ are continuous. By the stability relation $L^{1/m}(z) = L(\alpha_m + \beta_m z)$, and for large n we have $F''(\lambda_n + \delta_n z) \cong L(z)$. So we can approximate $F''(x)$ by $L[(x - \lambda)/\delta]$, with convenient location (λ) and dispersion (δ) parameters: Evidently, in the two last forms we have a shape parameter α. The graphs of the densities of the reduced random variables $\Lambda'(z), \Phi'_\alpha(z)$ and $\phi'_\alpha(z)$ are given in fig. in 1, 2 and 3.

If the distribution function of the maximum is $L(z)$, the asymptotic distribution of the k-th maximum i.e., the $(n-k+1)$-th order statistics from an i.i.d. sample is given by

$$L_k(z) = \lim_{n \to \infty} \sum_{n-k+1}^{n} F^j(\lambda_n + \delta_n z)\left[1 - F(\lambda_n + \delta_n z)\right]^{n-j} =$$

$$= \lim_{n \to \infty} \sum_{0}^{k-1} \binom{n}{n-j+1} F^{n-j+1}(\lambda_n + \delta_n z)[1 - F(\lambda_n + \delta_n z)] \, j - 1 =$$

$$= L(z) \sum_{0}^{k-1} \frac{[-\log L(z)]^j}{j!}$$

Note that $L_1(z) = L(z)$.

For practical purposes, the approximation, of the distribution function for the k-th maximum is, $L_k[(x - \lambda)/\delta]$. Note that $L(z)$ must be of one of the three forms listed earlier.

Because min $(X_i) = -\max(-X_i)$, we see that the asymptotic distribution functions for minima are $1-L(-x)$, and for the k-th minima. They are $1 - L_{n-k+1}(x)$. Note that if both the maximum and the minimum have asymptotic distributions (which does not always happen), they are not necessarily of the same types. For the exponential distribution, the asymptotic distribution of the maxima is of Gumbel type, but the asymptotic distribution for minima is exactly the exponential distribution itself $1 - \psi_1(-z)$.

We note, that if λ_n and δ_n are associated with the sample size n then under very general conditions,, the coefficients for the sample size N, satisfy the asymptotic relations

$$\lambda_N = \lambda_n + \delta_n \log(N/n), \; \delta_N = \delta_n \qquad \text{in the case of Gumbel distribution}$$

$$\lambda_N = \lambda_n = 0, \; \delta_N = \delta_n (N/n)^{1/\alpha} \qquad \text{in the case of Fréchet distribution and}$$

$$\lambda_N = \lambda_n = x_0, \; \delta_N = \delta_n (n/N)^{1/\alpha} \qquad \text{in the case of Weibull distribution}$$

Here x_0 denotes the right-end point, i.e., x_0 is such that $F(x_0) = 1$, but $F(x) < 1$ if $x < x_0$. This result is very important for pooling samples of different sizes. For details related to this section see Tiago de Oliveira (1972) and (1977).

These asymptotic distributions were obtained for i.i.d. samples. Similar results can be obtained under meaker conditions. For example, see Watson (1954), Newell (1964) and Loynes (1965) for details. From a practical point of view, we can then use one of the asymptotic distributions L(z), with adequate location and dispersion parameters, as approximate distributions of maxima in large samples even when the random observations are dependent (but not strongly) and the marginal distributions are not identical.

In fact we may use the limit

$$\lim_{\alpha \to \infty} \left(1 - \frac{z}{\alpha}\right)^\alpha = e^{-z}$$

to show that, with a linear change of variable depending on α, both $\Phi_\alpha(z)$ and $\Psi_\alpha(z)$ converge to the Gumbel distribution $\Lambda(z)$ for large α $(\alpha \geq 6)$.

From the relation between asymptotics of maxima and minima we see that the Weibull distribution for minima, is

$$W_\beta(z) = 1 - \Psi_\beta(-z) = 1 - \exp[(-z^\beta)] \quad \text{if} \quad z \geq 0, \quad \beta > 0.$$

Consider then the Weibull distribution $W_\beta(x|\theta)$ with a dispersion parameter β and zero (or known) location parameter. If X has the distribution $W_\beta(x|\theta)$, the new random variable $Y = -\log X$ has the distribution function

$$\text{Prob}\{Y \leq y\} = \text{Prob}\{X \geq e^{-y}\} = \exp\{-e^{-\beta(y+\log\theta)}\} = \Lambda\left[\frac{y - (-\log\theta)}{1/\beta}\right]$$

which is a Gumbel distribution with location parameter $\lambda = -\log\theta$ and dispersion parameter $\delta = \beta^{-1}$.

We are in general interested in the distribution of the largest or the smallest value of a sample; in case we could be interested in the 2nd, 3rd, ... largest or smallest value the asymptotic theory of k-th maxima or minima would then be in order.

3. Statistical decision for the Gumbel distribution

Some details about statistical decision for all asymptotic distributions of univariate maxima (estimation and testing) can be found in Tiago de Oliveira (1975).

For the Gumbel distribution, sometimes called the extreme distribution, with location and dispersion parameters, its density is:

$$\frac{1}{\delta} \cdot \Lambda'\left(\frac{x-\lambda}{\delta}\right) = \frac{1}{\delta} \cdot \exp-\left(\frac{x-\lambda}{\delta}\right) \cdot \exp\left[-\exp-\left(\frac{x-\lambda}{\delta}\right)\right]$$

The log-likelihood for an i.i.d. sample $(x_1, x_2 ... x_n)$ is then

$$\frac{1}{\delta^n} \cdot \exp\left(-\sum_{i=1}^{n}\frac{x_i - \lambda}{\delta}\right) \cdot \exp\left(-\sum_{i=1}^{n}\exp-\frac{x_i - \lambda}{\delta}\right)$$

so that the maximum likelihood equations lead to

$$\hat{\lambda} = -\hat{\delta} \log \left(\sum_{1}^{n} e^{-x_i/\hat{\delta}} / n \right)$$

$$\hat{\delta} = \bar{x} - \frac{\sum_{1}^{n} x_i e^{-X_i/\hat{\delta}}}{\sum_{1}^{n} e^{-X_i/\hat{\delta}}}$$

The second equation is easily solved, by computer, using the Newton-Raphson method, with few interations; once $\hat{\delta}$ is known, $\hat{\lambda}$ is immediately computed. The seed for the solution comes, in general, from the method of moments.

As the mean value and the variance of the reduced Gumbel random variable are γ (=.57722, ..., the Euler constant) and $\pi^2/6$, and, in the general case, the mean value and the variance are $\lambda + \gamma\delta$ and $\pi^2/6 \cdot \delta^2$, we can use the method of moments to obtain estimators (λ^*, δ^*) given by

$$\lambda^* + \gamma\delta^* = \bar{x}$$
$$\pi^2/(6 \cdot \delta^{*2}) = s^2.$$

In fact, only the second one will be used (δ^* as a seed to obtain $\hat{\delta}$). So the first equation $(\hat{\lambda} + \gamma\hat{\delta} = \bar{x})$ can be expected to be satisfied approximately, and an intuitive test of the values $(\hat{\lambda}, \hat{\delta})$ is the verification that $(\bar{x} - \hat{\lambda})/\hat{\delta} \approx \gamma$. It can be shown that it is asymptotically normal with mean values (λ, δ), variances $(1 + 6(1-\gamma)^2/\pi^2) \cdot \delta^2/n$ and $6/(\pi^2 \cdot \delta^2)/n$, covariance $6(1-\gamma)/(\pi^2 \cdot \delta^2)/n$, the asymptotic correlation coefficient being $\rho = (1 + \pi^2/6(1-\gamma)^2)^{-1/2} = .313$. For more details see Tiago de Oliveira (1972 and 1983).

The most important quantities to be estimated in Engineering are the larger or smaller values of a variable and the probability of crossing a level or a threshold.

As the I-quantile $(0 < p < 1)$ is the unique solution of the equation $\Lambda((x-\lambda)/\delta) = p$, the I-quantile has probability I of not being exceeded and $(1-I)$ of being exceeded. In practice, if we are dealing with execesses we choose I close to 1, and if we are dealing with lower bounds we choose I close to 0.

The quantiles for Gumbel distribution are $x_p = \lambda - \log(-\log p) \cdot \delta$ and so their natural estimators are $\hat{x}_p = \hat{\lambda} - \log(-\log p) \cdot \hat{\delta}$. But $(\hat{\lambda}, \hat{\delta})$ being asymptotically binormal, we see that X_p is also asymptotically normal with mean value x_p and variance $[1 + 6(1 - \gamma - \log(-\log p))^2/\pi^2]$.

Those results allow for the formation of confidence regions for (λ, δ) and confidence intervals for λ, δ and x_p; for the latter it is natural to form a one-sided confidence interval.

The computation of the mean square error MSE $(\hat{x}_p)$, of the bias $b(\hat{x}_p) = M\{\hat{x}_p\} - x_p$ and the variance $V(\hat{x}_p) \sim$ (const. /n) of the estimator $\hat{x}_p$ are easy to evaluate approximately; recall that MSE $(\hat{x}_p) = b^2(\hat{x}_p) + V(\hat{x}_p)$. The estimators given minimize the MSE either directly or approximately.

Lowering and Nash (1970) compare different methods of estimating (λ, δ), and conclude by favouring the maximum likelihood method.

To predict the average maximum in the next I observations one can take $p = \exp\left(-e^{-\gamma} / m\right)$ in the formulae for quantiles.

The probability of exceeding a level I, $P = 1 - \Lambda[(c - \lambda)/\delta]$, is estimated by $\hat{P} = 1 - \Lambda[(c - \hat{\lambda}/\hat{\delta})]$. The behaviour of this estimated probability is also asymptotically normal, as can be seen by the use of the δ-method (see Tiago de Oliveira, 1982).

Connected with this probability is the notion of return period. The return period for exceeding the level I is the mean value of the time interval between successive crossings of the level I in a sequence of independent experiments. It is given by $T = \{1 - \Lambda(x - \lambda)/\delta\}^{-1}$ and is approximated by $\overline{T} = -1/\log \Lambda[(x - \lambda)/\delta]$ $= \exp[(x - \lambda)/\delta]$, the approximation being on the safe side (underevaluation). It is estimated, if needed, by the usual substitution of $(\hat{\lambda}, \hat{\delta})$ for (λ, δ).

Let us consider then the question of long-range design, for the underlying Gumbel distribution. The probability that in the next I time units (e.g., years) the design level I will not be exceeded is:

$$\Lambda^T\left(\frac{a - \lambda}{\delta}\right);$$

If we want a design with probability $(1-\varepsilon)$ of being correct we should take:

$$\Lambda^T\left(\frac{a - \lambda}{\delta}\right) \geq 1 - \varepsilon$$

or $a_T(\varepsilon) \geq c_T(\varepsilon)$, where $c_T(\varepsilon)$ is given by:

$$\Lambda^T\left(\frac{c - \lambda}{\delta}\right) = 1 - \varepsilon;$$

or
$$c_T(\varepsilon) = \lambda + \delta\left[-\log\frac{-\log(1-\varepsilon)}{T}\right] < \lambda + \delta\log\left(\frac{T}{\varepsilon}\right)$$

The last expression is generally a good approximation. So to take $a_T(\varepsilon) = \lambda + \delta\log(T/\varepsilon)$ is a good and safe evaluation of the design level. For some θ, satisfying $0 < \theta < 1$, the design probability is $e^{-\varepsilon} = 1 - \varepsilon + \theta\cdot\varepsilon^2/2 > 1-\varepsilon$.

4. Statistical decision for the Weibull distribution

The most usual coefficients for the Weibull distribution (for minima) $W_\beta\left(x|\theta\right)$ are:

mean value $\qquad\qquad \theta\,\Gamma(1 + 1/\beta)$

mode $\qquad\qquad \theta\left(\dfrac{\beta-1}{\beta}\right)^{1/\beta}, (\beta > 1)$

median $\qquad\qquad \theta(\log 2)^{1/\beta}$

l-quantile $\qquad\qquad \theta[-\log(1-p)]^{1/\beta}$

variance $\quad \theta^2[\Gamma(1+2/\beta) - \Gamma^2(1+1/\beta)]$

The i.i.d. sample will be denoted by $(x_1, x_2 ... x_n)$.

We have noted that the transformation $Y = -\log X$ converte the Weibull distribution into the pivotal Gumbel distribution. The search of essential quantities (quantile estimator, point predictors, crossing probabilities, estimators, etc.), can either be done directly, or transforming the sample to $(y_1, y_2 ... y_n)$, where $y_i = -\log x_i$, and then using the methods and results given for the Gumbel distribution.

For example, to test a value $\alpha = \beta_o$ of the shape parameter is equivalent to testing the value $\delta = 1/\beta = \delta_o = 1/\beta_o$ of the transformed data. As $\sqrt{n}\,\dfrac{\pi}{\sqrt{6}}\dfrac{\delta - \delta_0}{\delta}$ or equivalently (see the δ-method) $\sqrt{n}\,\dfrac{\pi}{\sqrt{6}}\dfrac{\beta - \beta_o}{\beta_o}$ are asymptotically standard normally distributed an easy and asymptotically optimal test of $\beta = \beta_o$, which leads to the acceptance of α_o is to test if $\left|\dfrac{\beta - \beta_o}{\beta_o}\right| \le 1.96\dfrac{\sqrt{6}}{\pi\sqrt{n}}$ at the asymptotic significance level 5% (confidence coefficient 95%).

For the Weibull distribution $W_\beta(x|\theta)$ the test of $\beta=2$ (Rayleigh distribution) or of β =1 (exponential distribution) can be dealt with this way.

Dealing with the i.i.d. sample $(x_1, x_2 ... x_n)$ we find that the maximum likelihood estimators are given by:

$$\frac{n}{\hat{\beta}} - \sum_1^n \log x_i - n\frac{\sum\limits_1^n x_i^{\hat{\beta}} \log x_i}{\sum\limits_1^n x_i^{\hat{\beta}}} = 0$$

$$\hat{\theta} = \frac{1}{n} \sum_1^n x_i^{\hat{\beta}}$$

which correspond to those for the transformed parameters of the Gumbel distribution given above.

Once more the first equation can be solved by the Newton-Raphson method, using as a seed the corresponding seed for the Gumbel distribution (recall that $\delta = \frac{1}{\beta}$ for the $x_i = e^{-y_i}$). Alternatively since $x_{.75} / x_{.25} = 2^{1/\beta}$ than initial estimator is

$$\beta^* = \frac{\log 2}{\log x_{0.75}^* - \log x_{0.25}^*}$$

where x_p^* denotes the usual estimator of χ_p i.e., $x_p^* = X_{[np]+1}'$, is the $[np] +1$ order statistics, where $[np]$ denotes the integer part of np.

Also the random pair $(\hat{\theta}, \hat{\beta})$ has a binormal asymptotic behaviour with mean values (θ, β), variances $\left[1 + \frac{6}{\pi^2}(1-\gamma)^2\right]\theta^2 / n\beta^2$ and $6\beta^2 / n\pi^2$, covariance $6(1-\gamma)\theta / n\pi^2$, and correlation coefficient

$$\rho = \left(1 + \frac{\pi^2}{6(1-\lambda)^2}\right)^{-1/2} = \cdot 313.$$

As before, by the δ-method, we see that $\sqrt{n}\dfrac{\hat{\chi}_p - \chi_p}{\chi_p}$ is asymptotically equivalent to

$$\sqrt{n}\left[\frac{\hat{\theta}-\theta}{\theta} - \frac{\hat{\beta}-\beta}{\beta^2}\log[-\log(1-p)]\right]$$ and so we see that the (random) relative error

$\left(\dfrac{\hat{\chi}_p}{\chi_p} - 1\right)$ is asymptotically normal, with zero mean value, and variance

$$\left[1 + \frac{6}{\pi^2}(1 - \gamma - \log(-\log(1-p)))^2\right]n\beta^2$$

Similarly, the design level I such that, for I (time) units the probability that it will be not be reached, with probability smaller or equal to ε, is given by

$$1 - (1 - W_\beta(a/\theta))^T \le \varepsilon$$

or $$a_T(\varepsilon) \le c_T(\varepsilon)$$

where $$c_T(\varepsilon) = -\theta\frac{\log(1-\varepsilon)}{T} > \frac{\theta\varepsilon}{T}.$$

The last approximation ($\theta \varepsilon/T$) is a good one in general, the design probability being then $1 - e^{-\varepsilon}$, smaller than ε.

5. The check of modelling; test of fit

The crucial problem is, thus, the confirmation of the assumed model $W_\beta(x|\theta)$.

A graphical way is to plot $Y_i = -\log X_i$ in Gumbel probability paper if we suppose that the distribution is of Weibull type (for minima); see Gringorten (1963) and Cunnane (1978). By inspecting the linearity of the plots of the ordered $-\log x_i$ against the plotting positions we can check graphically whether the model is Weibull. It should be noted that the usual plotting positions are of the form $p_i = \dfrac{i-a}{n+1-2a}(0 \le a < 1)$; we have, in an immediate notation:

$$|p_i(a) - p_i(a')| = \frac{|a'-a||n+1-2i|}{(n+1-2a)(n+1-2a')} \le \frac{|a'-a|}{n-1} \le \frac{1}{n-1}$$

and so, for reasonable values of n(>30), the choices a=0, a=1/2, a=3/8, a=.44, for the platting positions. are irrelevant.

Analytically we can search for the best model by estimating the location and dispersion parameters for the Y_i, computing the Kolmogorov-Smirnov statistic

$$KS_n = \sup|S_n(x|x_i), -W_{\hat{\beta}}(x/\hat{\theta})| = \max_i \left\{ \max_i |\frac{i}{n} - W_{\hat{\beta}}(x_i'/\hat{\theta})|, \max_i |\frac{i-1}{n} - W_{\hat{\beta}}(x_i'/\hat{\theta})| \right\}$$

where x_i' is the order statistics of the sample $(x_1, x_2 ... x_n)$. We accept the Weibull model if $KS_n \leq 1.36/\sqrt{n}$ (at significance level 5%) or $KS_n \leq 1.63/\sqrt{n}$ (at significance level 1%); otherwere we reject it. If $\hat{\beta} \geq 6$ the estimate means that we may construct a good approximation using for x_i a Gumbel distribution for minima, or for $-x_i$ a Gumbel distribution for maxima with convenient location and dispersion parameters.

Recall that we have intuitively extended, but without solid theoretical foundation, the use of the Kolmogorov-Smirnov test as it is usually (and irregularly) done. There are other tests of fit, due to Cramer-von Mises, Stephens, Anderson-Darling, etc. which could also be used.

6. References

1. Cunnane, C.: "Unbiased Plotting Positions - a review", *J. Hydrol.*, 37, 205 222, 1978
2. Gnedenko, B.V.: "Sur la distribution limit du terme maximum d'une série aléatoire", *Ann. Math.*, **44**, 423-453, 1943
3. Gringorten, Irving I.: "A Plotting Rule for Extreme Probability Paper", Jeff. Res., **68**, 813-814, 1963
4. Gumbel, J.: *Statistics of Extremes*. Columbia University Press, 1958
5. Loynes, M.: "Extreme Values in Uniformly Mixing Stochastic Processes" *Ann. Math. Statist.*, **36**, 993-996, 1965.
6. Lowery and Nash, J.E.:, "A Comparison of Methods of Fitting the Double Exponential Distribution", *J. Hydrol.*, **10**, 259-275, 1970
7. Muir, Langley and El-Shaarawi, A.M.: "on the Calculation of Extreme Wave Heights: A Review", *ocean Engng.*, **13**, 93-118, 1986
8. Newell, F.: "Asymptotic Extremes for m-dependent Random Variables", *Ann. Math. Statist.*, **35**, 1322-1325, 1964
9. Oliveira, Tiago: "Statistics for Gumbel and Fréchet distributions" in, *Structural Safety and Reliability,* ed. by Freudenthal, 91-105, Pergamon Press, 1972
10. Oliveira, Tiago: "Statistical Decision for Extremes", *Trabaj. Estat. Inv. apart.,* XXVI, 453-471, 1975
11. Oliveira, Tiago: "Asymptotic Distributions for Univariate and Bivariate m-th Extremes", in *Recent Developments in Statistics,* ed. by Barra *et al.,* 613617, North Holland, 1977
12. Oliveira, Tiago: "Extreme Values and Applications", *Proc. 3rd. Intern. Confer. Appl. Statist. Probab. Soil Struct. Eng.,* I, 127-135, 1979
13. Oliveira, Tiago: "The δ-method for obtention of Asymptotic Distributions; Applications", *Publ. Inst. Statist. Univ. Paris,* XXVII, 49-70, 1982
14. Oliveira, Tiago: "Gumbel Distribution", in *Encyclopedia of Statistical Sciences, ed.* by N. L. Johnson and S. Kotz, III, 552-558, Wiley, 1983

15. Watson, S.: "Extreme Values in Samples for m-dependent Stationary Stochastic Processes", *Ann. Math. Statist.*, **25**, 798-800, 1954

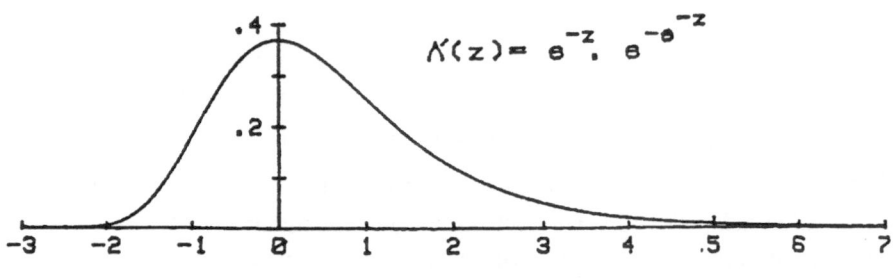

Fig 1 ; Gumbel distribution

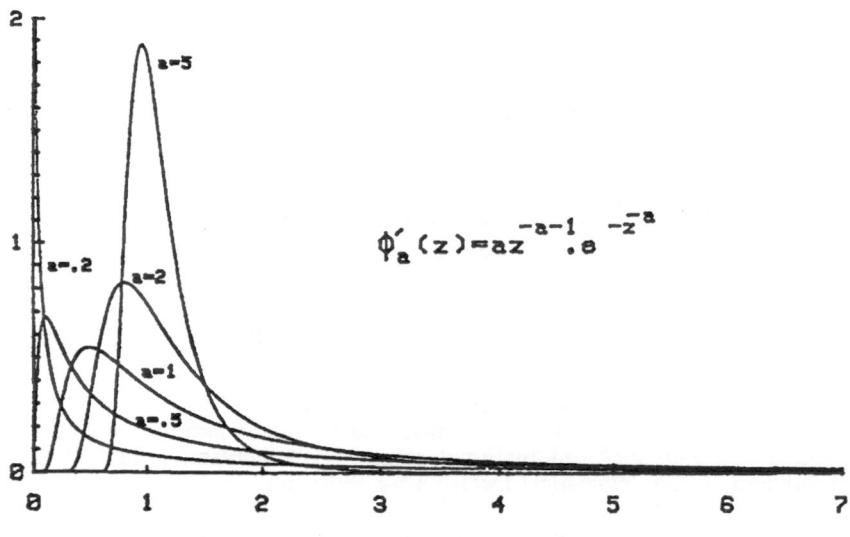

Fig 2 ; Fréchet distribution

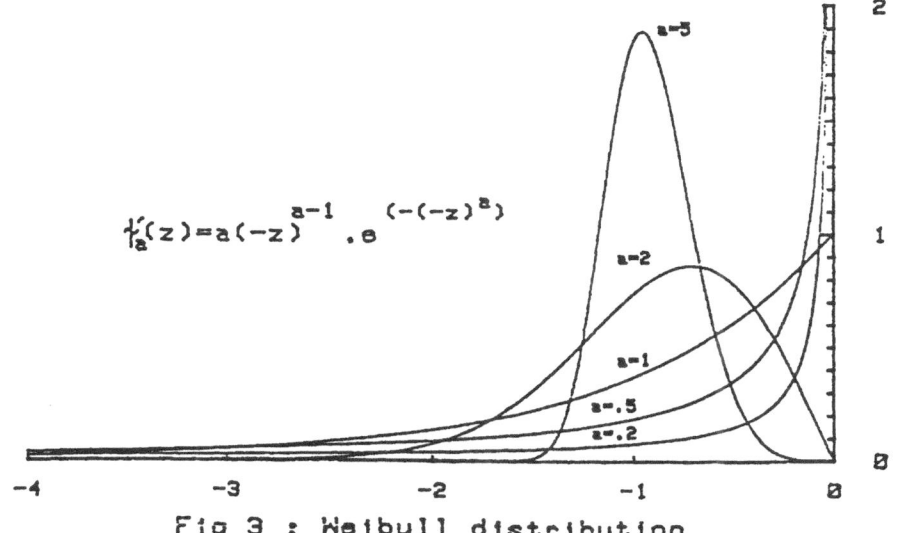

$$f'_a(z) = a(-z)^{a-1} \cdot e^{(-(-z)^a)}$$

Fig 3 ; Weibull distribution

STOCHASTIC MODELING OF LOAD COMBINATIONS

HENRIK O. MADSEN

Det Norske Veritas AS,
Tuborg Parkvej 8,
DK-2900 Hellerup,
DENMARK.

1. Introduction

When only one time-varying load acts on a structure, and failure is defined as the load process crossing some level, then the extreme value distribution of the load contains information which is sufficient for decisions about reliability. The theory of stochastic load combinations is applied in situations where a structure is subjected to two or more time-varying scalar loads acting simultaneously. The scalar loads can be components of the same load process or be components of different load processes. To evaluate the reliability of the structure, each load cannot be characterized by its extreme-value distribution alone; a more detailed characterization of the stochastic process is necessary. The reason is that the loads in general do not attain their extreme values at the same time.

We consider a structure subjected to loads defining a vector-valued load process $Q(t)$. Failure of the structure is assumed to occur at the time of the first exceedence of the deterministic function $\xi(t)$ by the random function $b(Q(t))$. Here $\xi(t)$ represents a strength threshold. The b-function transforms the load processes to the load effect process corresponding to the strength $\xi(t)$. A *linear load combination* is said to exist when the b-function is linear. Otherwise, the load combination is nonlinear. The failure event is illustrated geometrically in Fig. 1 for a combination of two loads and a constant threshold $\xi(t)$. The figure shows that failure can be thought of as either the first upcrossing of $\xi(t)$ by the process $b(Q(t))$ or as the first outcrossing of the set $B(t) = \{q \mid b(q) \leq \xi(t)\}$ by the vector process $Q(t)$. In both cases this is true under the condition that failure does not occur at time zero.

C. Guedes Soares (ed.), *Probabilistic Methods for Structural Design,* 227–243.

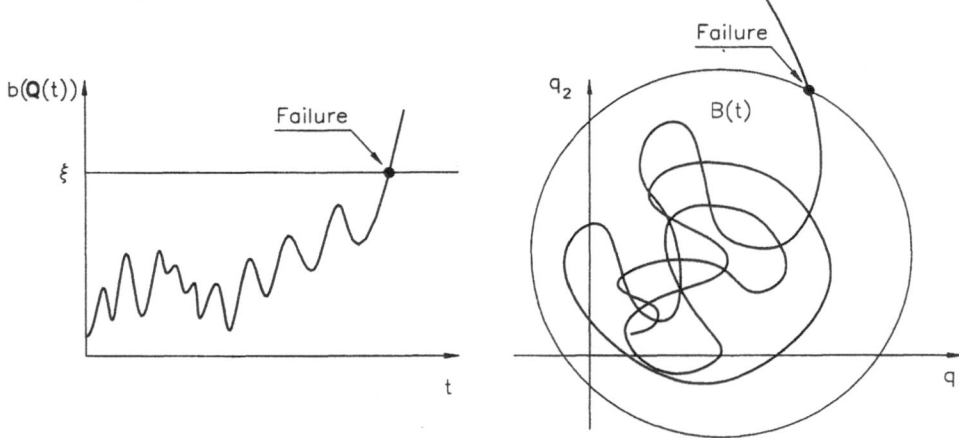

Figure 1.   Geometrical illustration of the failure event from combined loading.

2. Bounds on the extreme value distribution

A simple upper bound on the probability of failure P_F in the time interval $[0, T]$ is derived. Let $N_\xi(T)$ denote the number of upcrossings in $[0, T]$ of $\xi(t)$ by $b(Q(t))$ or, equivalently, the number of outcrossings in $[0, T]$ of the set $B(t)$ by $Q(t)$. The failure probability is expressed as

$$P_F = P(\text{failure at } t = 0) + P(N_\xi(T) \geq 1)$$
$$-P(\text{failure at } t = 0 \text{ and } N_\xi(T) \geq 1) \tag{1}$$

The negative term (last term), in the above equation, is numerically smaller than the smallest of the two positive terms in the equation. An upper bound on P_F is therefore

$$P_F \leq P_{F0} + P(N_\xi(T) \geq 1) \tag{2}$$

where $P_{F0} = P(\text{failure at } t = 0)$. This upper bound is a good approximation to P_F, at least if one of the terms on the right-hand side is much larger than the other term. The upper bound is further developed as

$$P_F \leq P_{F0} + \sum_{n=1}^{\infty} P(N_\xi(T) = n) \leq$$
$$P_{F0} + \sum_{n=1}^{\infty} n P(N_\xi(T) = n) = P_{F0} + \mathrm{E}\,[N_\xi(T)] \tag{3}$$

The right-hand side is a good approximation to P_F if Eq. (2) is a good approximation and if, further,

$$P(N_\xi(T) = 1) \gg \sum_{n=2}^{\infty} nP(N_\xi(T) = n) \qquad (4)$$

Eq. (4) will often be valid in practical situations with highly reliable structures when clustering of crossings can be neglected. The failure probability can be approximated by

$$P_F \approx 1 - (1 - P_{F0}) \exp\left[-\frac{E\left[N_\xi(T)\right]}{1 - P_{F0}}\right] \qquad (5)$$

When the probability distribution of $Q(0)$ is known, the probability of failure at $t = 0$ can be calculated by first- and second-order reliability methods described in (Madsen et al., 1986). The second term in Eq. (3) is written as

$$E\left[N_\xi(T)\right] = \int_{t=0}^{T} \nu(\xi, t)dt \qquad (6)$$

Here $\nu(\xi, t)$ is the mean-upcrossing rate of $\xi(t)$ or mean outcrossing rate of $B(t)$ at time t. In both situations $\nu(\xi, t)$ can be calculated by Rice's formula (Rice, 1944a; Rice, 1944b) or a generalization of it. If $\nu(\xi, t)$ is interpreted as the mean-upcrossing rate of $\xi(t)$, it follows directly from Rice's formula that

$$\nu(\xi, t) = \int_{\dot{s}=\xi}^{\infty} (\dot{s} - \dot{\xi}) f_{S\dot{S}}(\xi, \dot{s}, t) d\dot{s} \qquad (7)$$

Here a dot denotes a time derivative and the stochastic process $b(Q(t))$ is denoted by $S(t)$. If $\nu(\xi, t)$ is thought of as the mean-outcrossing rate of $B(t)$, it follows from a generalization of the arguments leading to Rice's formula that

$$\nu(\xi, t) = \int_{\partial B} \int_{\dot{q}_N=0}^{\infty} \dot{q}_N f_{Q\dot{Q}_N}(q, \dot{q}_N, t) d\dot{q}_N d(\partial B) \qquad (8)$$

It is here assumed that the set $B(t)$ is constant in time. $\dot{Q}_N$ denotes the projection of $\dot{Q}(t)$ on the outward normal to B at a point on the boundary ∂B. The inner integral in Eq. (8) can be viewed as a local outcrossing rate. A generalization of Eq. (8) to a time-varying set $B(t)$ is conceptually straightforward.

Very few closed-form results for $\nu(\xi, t)$ exist for general processes and general safe regions. Among these results can be mentioned results for Gaussian processes and different safe regions described in (Fuller, 1982; Veneziano et al., 1977; Ditlevsen, 1983a), and results for combinations of rectangular filtered Poisson processes described in (Breitung and Rackwitz, 1982; Waugh, 1977). Asymptotic results for Gaussian processes have been evaluated and are summarized in (Breitung, 1994).

Instead of using Rice's formula as in Eq. (7), it is proposed in (Madsen, 1992; Hagen and Tvedt, 1991) to calculate the mean up-crossing rate as a parallel system sensitivity measure available for time independent reliability analysis. $\nu(\xi, t)$ may be written as

$$
\begin{aligned}
\nu(\xi, t) &= \lim_{\Delta t \to 0^+} P[S(t) < \xi(t) \cap S(t + \Delta t) > \xi(t + \Delta t)] \\
&= \lim_{\Delta t \to 0^+} P[S(t) < \xi(t) \cap S(t) + \dot{S}(t)\Delta t > \xi(t) + \dot{\xi}(t)\Delta t] \\
&= \lim_{\Delta t \to 0^+} \{ P[\dot{S}(t) - \dot{\xi}(t) > 0 \cap S(t) + (\dot{S}(t) - \dot{\xi}(t))\Delta t > \xi(t)] \\
&\quad - P[\dot{S}(t) - \dot{\xi}(t) > 0 \cap S(t) > \xi(t)] \}
\end{aligned}
\tag{9}
$$

leading to

$$
\nu(\xi, t) = \frac{d}{d\theta} P[M_1(t) < 0 \cap M_2(t, \theta) < 0]
\tag{10}
$$

where the safety margins M_1 and M_2 are

$$
\begin{aligned}
M_1(t) &= \dot{\xi}(t) - \dot{S}(t) \\
M_2(t) &= \xi(t) - S(t) + (\dot{\xi}(t) - \dot{S}(t))\theta
\end{aligned}
\tag{11}
$$

Eq. (10) expresses $\nu(\xi, t)$ as a parametric sensitivity measure of the probability of an associated parallel system unsafe domain. Such sensitivity measures can be calculated by first-order reliability methods (Madsen, 1992) provided the joint distribution of $[Q(t), \dot{Q}(t)]$ is known and that the mapping of this vector into a set of independent standard Gaussain variables is possible at each time t in $[0, T]$. This is not a very restrictive condition.

With the notation $\nu(\xi, t) = \frac{d}{d\theta} P(t, \theta)$ in Eq. (10), Eq. (6) can be written as

$$
\begin{aligned}
E[N_\xi(T)] &= \int_0^T \frac{d}{d\theta} P(t, \theta) dt \\
&= T \frac{d}{d\theta} \int_0^T P(t, \theta) \frac{1}{T} dt
\end{aligned}
\tag{12}
$$

where the integeration and differentiation have been interchanged because the integrand is a continuous function of θ and t. Replacing the time t in Eq. (12) by the auxiliary random variable V uniformly distributed with probability density function $f_V(v)$ on the interval $[0, T]$, yields

$$
E[N_\xi(T)] = T \frac{d}{d\theta} \int_0^T P(v, \theta) f_V(v) dv = T\bar{\nu}
\tag{13}
$$

where the time averaged mean crossing rate $\bar{\nu}$ can be calculated by first order reliability methods applicable for calculation of ν, with one extra variable included (Hagen and Tvedt, 1991).

3. Linear load combination

In a general study for code purposes the case of stationary and mutually independent processes combined linearly is very important. The sum of two processes is first considered, so let

$$Q(t) = Q_1(t) + Q_2(t) \tag{14}$$

The expected rate of upcrossings of the constant level ξ, $\nu_Q(\xi)$, is calculated by Rice's formula:

$$\nu_Q(\xi) = \int_{\dot{q}=0}^{\infty} \dot{q} f_{Q\dot{Q}}(\xi, \dot{q}) d\dot{q} \tag{15}$$

The joint probability density function $f_{Q\dot{Q}}(\cdot, \cdot)$ is expressed in terms of the density functions $f_{Q_1\dot{Q}_1}(\cdot, \cdot)$ and $f_{Q_2\dot{Q}_2}(\cdot, \cdot)$ by the convolution integral

$$f_{Q\dot{Q}}(q, \dot{q}) = \int_{q_1=-\infty}^{\infty} \int_{\dot{q}_1=-\infty}^{\infty} f_{Q_1\dot{Q}_1}(q_1, \dot{q}_1) f_{Q_2\dot{Q}_2}(q - q_1, \dot{q} - \dot{q}_1) d\dot{q}_1 dq_1 \tag{16}$$

Inserting this result in Eq. (15) and substituting $\dot{q} = \dot{q}_1 + \dot{q}_2$ gives

$$
\begin{aligned}
\nu_Q(\xi) &= \int_{q=-\infty}^{\infty} \int_{\dot{q}_1=-\infty}^{\infty} \int_{\dot{q}_2=-\dot{q}_1}^{\infty} \dot{q}_1 f_{Q_1\dot{Q}_1}(q, \dot{q}_1) \times \\
&\qquad\qquad f_{Q_2\dot{Q}_2}(\xi - q, \dot{q}_2) d\dot{q}_2 d\dot{q}_1 dq \\
&+ \int_{q=-\infty}^{\infty} \int_{\dot{q}_1=-\infty}^{\infty} \int_{\dot{q}_2=-\dot{q}_1}^{\infty} \dot{q}_2 f_{Q_1\dot{Q}_1}(q, \dot{q}_1) \times \\
&\qquad\qquad f_{Q_2\dot{Q}_2}(\xi - q, \dot{q}_2) d\dot{q}_2 d\dot{q}_1 dq
\end{aligned} \tag{17}
$$

The two triple integrals can be evaluated analytically only in special cases. Simple upper and lower bounds on the mean-upcrossing rate can, however, be found by changing the area of integration in the $(\dot{q}_1, \dot{q}_2)$-plane for the two integrals in Eq. (17). Fig. 2a shows the common area of integration of the integrals. The vertical and horizontal hatching illustrate the integrations with the integrand of the first and second integral, respectively. Fig. 2b correspondingly illustrates the integrations in an upper bound and Fig. 2c the integrations in a lower bound.

The upper bound is

$$\nu_Q(\xi) \leq \int_{q=-\infty}^{\infty} \int_{\dot{q}_1=0}^{\infty} \int_{\dot{q}_2=-\infty}^{\infty} \dot{q}_1 f_{Q_1\dot{Q}_1}(q, \dot{q}_1) f_{Q_2\dot{Q}_2}(\xi - q, \dot{q}_2) d\dot{q}_2 d\dot{q}_1 dq$$

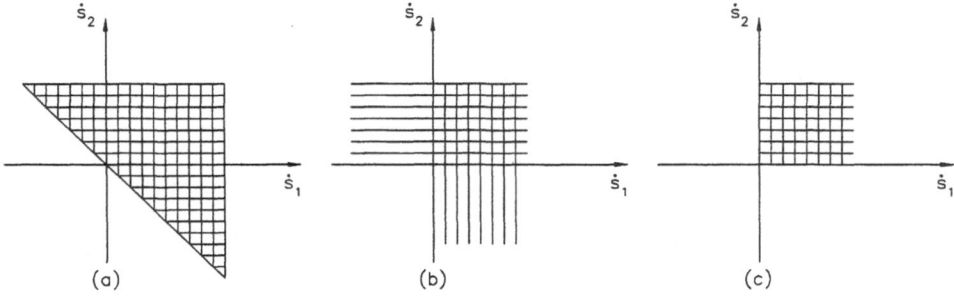

Figure 2. Areas of integration.

$$+ \int_{q=-\infty}^{\infty} \int_{\dot{q}_1=-\infty}^{\infty} \int_{\dot{q}_2=0}^{\infty} \dot{q}_2 f_{Q_1 \dot{Q}_1}(q, \dot{q}_1) f_{Q_2 \dot{Q}_2}(\xi - q, \dot{q}_2) d\dot{q}_2 d\dot{q}_1 dq$$

$$= \int_{q=-\infty}^{\infty} \nu_{Q_1}(q) f_{Q_2}(\xi - q) dq + \int_{q=-\infty}^{\infty} \nu_{Q_2}(q) f_{Q_1}(\xi - q) dq$$

$$= \nu_{Q_1} * f_{Q_2} + \nu_{Q_2} * f_{Q_1} \tag{18}$$

where * means convolution. The terms in the upper bound are generally called the *point crossing terms* (Larrabee and Cornell, 1981).

The lower bound is similarly

$$\nu_Q(\xi) \geq \int_{q=-\infty}^{\infty} \nu_{Q_1}(q) f_{Q_2}(\xi - q)(1 - F_{\dot{Q}_2|Q_2}(0, \xi - q)) dq$$

$$+ \int_{q=-\infty}^{\infty} \nu_{Q_2}(\xi - q) f_{Q_1}(q)(1 - F_{\dot{Q}_1|Q_1}(0, q)) dq \tag{19}$$

Here $\nu_{Q_1}(\cdot)$ is the mean-upcrossing rate function for $Q_1(t)$ and $f_{Q_1}(\cdot)$ is the marginal or arbitrary-point-in-time probability density function for $Q_1(t)$. The factor $1 - F_{\dot{Q}_1|Q_1}(0, q)$ gives the probability of a positive derivative $\dot{Q}_1(t)$ given that $Q_1(t) = q$. For a Gaussian process, this probability is 0.5, independent of q. For the sum of two Gaussian processes, the lower bound Eq. (19) is thus half the value of the upper bound Eq. (18).

The bounds are easily generalized to cover situations with nonstationary load processes or nonconstant thresholds (Madsen, 1982) and to situations where more than two time-varying loads are acting simultaneously. For the sum of three stationary and independent processes, $Q_1(t)$, $Q_2(t)$, and $Q_3(t)$, the upper bound on $\nu_Q(\xi)$ is

$$\nu_Q(\xi) \leq \int_{-\infty}^{\infty} \nu_{Q_1}(q) f_{Q_2+Q_3}(\xi - q) dq$$

$$+ \int_{-\infty}^{\infty} \nu_{Q_2}(q) f_{Q_1+Q_3}(\xi - q) dq$$

$$+ \int_{-\infty}^{\infty} \nu_{Q_3}(q) f_{Q_1+Q_2}(\xi - q) dq \qquad (20)$$

where, e.g.,

$$f_{Q_1+Q_2}(q) = \int_{-\infty}^{\infty} f_{Q_1}(x) f_{Q_2}(q - x) dx = f_{Q_1} * f_{Q_2} \qquad (21)$$

The upper bound can thus be written as

$$\nu_Q(\xi) \leq \nu_{Q_1} * f_{Q_2} * f_{Q_3} + \nu_{Q_2} * f_{Q_1} * f_{Q_3} + \nu_{Q_3} * f_{Q_1} * f_{Q_2} \qquad (22)$$

The upper bound on $\nu_Q(\xi)$ is thereby expressed solely in terms of convolution integrals of the mean-upcrossing rate function and the marginal probability density function for each process. In Table 1, these two functions are given for various Poisson pulse processes commonly used as load models. Additional results are given in (Madsen, 1979).

A more general treatment of the point-crossing-term idea for nonlinear load combinations and nonstationary processes is given in (Ditlevsen and Madsen, 1981; Ditlevsen, 1983b). The fact that the results of the point crossing method are in terms of convolution integrals makes a combination of the results with a first-order reliability methods very simple.

If the upper bound Eq. (18) on $\nu(\xi)$ is used in Eq. (3), a strict upper bound on the failure probability is maintained. This upper bound can still provide a very good approximation to the exact value if the upper bound on $\nu(\xi)$ is close to the exact value. Table 2 presents some exact results taken from (Larrabee and Cornell, 1981; Madsen, 1979; Madsen *et al.*, 1979) compared to the upper bound. The ratio of the exact result to the upper bound is in all cases close to unity, indicating that the upper bound is indeed a good approximation.

The analysis of linear load combinations of stationary and independent load processes can be summarized. It follows that if the extreme-value distribution of the combined load is well approximated in terms of the mean-upcrossing rate function, sufficiently accurate approximations to the extreme-value distribution can be computed from the mean-upcrossing rate function $\nu_Q(\xi)$ and the marginal probability density function $f_Q(q)$ for each process. It can thus be stated that the pair of functions (ν_Q, f_Q) provides sufficient information about each load process in a linear load combination. Several extensions of the bounding technique explained above to linear combinations of dependent processes are demonstrated in (Winterstein, 1980). An analysis of the clustering effect on failure rates of combined loads is presented in (Winterstein and Cornell, 1984).

The method using the point crossing terms in approximation of the failure probability is called the *point crossing method*. Another method developed by Y.K.Wen which is equally well developed is the *load coincidence*

TABLE 1. Mean-Upcrossing Rate Functions and Marginal Probability Density Functions for Poisson Pulse Processes (Intensity ν)

Process	$\nu_Q(q)$	$f_Q(q)$
Spike process	$\nu(1 - F_S(q))$	$\delta(q)$
Square-wave process	$\nu F_S(q)(1 - F_S(q))$	$f_S(q)$
Triangular pulse process	$\nu(1 - F_S(q))$	$\displaystyle\int_{s=q}^{\infty} \frac{1}{s} dF_S(s)$
Parabolic pulse process	$\nu(1 - F_S(q))$	$\displaystyle\int_{s=q}^{\infty} \frac{1}{2\sqrt{1 - \frac{q}{s}}} \frac{1}{s} dF_S(s)$

method see (Wen, 1977; Wen, 1981; Wen and Pearce, 1981; Wen and Pearce, 1983). This method is based on identification of load coincidences leading to a level crossing by the combined load. The mean rates of various types of load coincidences are computed together with the probability of a level crossing given that a load coincidence occurs.

4. Load combinations in codified structural design

An essential feature of structural design procedures is a set of requirements for load combination. Such formats provide a list of the combinations to be considered and a set of appropriate load factors to be applied to specified or

TABLE 2. Bounds on the Mean-Upcrossing Rate for Sums of Two Stationary Processes.

$$\Lambda = \int_{q=-\infty}^{\infty} \nu_{Q_1}(q) f_{Q_2}(\xi - q) dq + \int_{q=-\infty}^{\infty} \nu_{Q_2}(q) f_{Q_1}(\xi - q) dq$$

Processes combined	Bounds
Gaussian + Gaussian	$\Lambda/\sqrt{2} \leq \nu(\xi) \leq \Lambda$
Gaussian + unimodal renewal pulse process	$0.5\Lambda \leq \nu(\xi) \leq \Lambda$
Renewal spike process + arbitrary process	$\nu(\xi) = \Lambda$
Renewal rectangular pulse process + arbitrary process	$\nu(\xi) = \Lambda$
Unimodal Poisson pulse process + same	$0.75\Lambda \leq \nu(\xi) \leq \Lambda$
Filtered renewal rectangular pulse process + arbitrary process	$\nu(\xi) = \Lambda$

characteristic values of the individual loads. To provide for the many design situations that can arise, most codes have found it useful to categorize loads as either permanent (e.g., self-weight and prestressing forces) or variable. Variable loads can be further decomposed into long term (e.g., furniture loads, snow loads in some regions) and short term (e.g., wind and earthquake). For each type of load, codes specify characteristic or representative values, normally defined to have a specified probability of being exceeded in some specified period. As an example, the characteristic 50-year wind speed is the wind speed that is exceeded with a probability of 2% in one year.

To describe the basic design formats, permanent loads are denoted D and variable loads L, further decomposed into long-term components LL and short-term components LS. With this basic notation, a first subscript k is used to denote a characteristic value, and a second subscript, $j = 1, 2, ...,$ to denote a particular load type, such as wind or earthquake.

The format proposed by the (CEB, 1976; JCSS, 1978) is a family of total loads of the general form

$$\gamma_D D_k + \gamma_i L_{ki} + \sum_{j \neq i} \gamma_j \psi_{ij} L_{kj} \tag{23}$$

for the ultimate state, in which γ-values are load factors. The products $\gamma_j \psi_{ij} L_{kj}$ may be called companion values of the loads. The format involves the factored or design value of one load plus factored companion values of the others. There are at least as many such equations as there are load types. This combination method is called the *companion action format*.

Another basic design format in use by the National Building Code of

Canada is of the form

$$\gamma_D D_k + \phi \left(\sum_j \gamma_j L_{kj} \right) \tag{24}$$

in which ϕ is a probability factor to account for the fact that extreme values of different loads are unlikely to occur together. When only one load acts, $\phi = 1$. The American Concrete Institute uses a similar format. This combination format is called the *combination factor format*.

Russia has adopted a slightly different ultimate state format, which can be written

$$\gamma_D D_k + \gamma_{LL} LL_k + \phi \left(\sum_j \gamma_j LS_{kj} \right) \tag{25}$$

The long-term loads are considered at their full design values, and short-term actions are considered at reduced companion values by means of a common probability factor. Another closely related format is that in the proposed load and resistance factor design for steel structures, see e.g. (Ravaindra and Galambos, 1978).

The essential difference between the basic code formats is whether they multiply design loads $\gamma_i L_{ki}$ by combination factors after summation or before. In all cases, serviceability loads are obtained directly from the characteristic values.

Within a geographic region and a specific class of intended use, the physical effects of loads vary from structure to structure and between elements in a structure. The total variable load effect $S(t)$ in a linear or quasi-linear analysis can be written in the form

$$S(t) = c_1 \gamma_1 L_1(t) + c_2 \gamma_2 L_2(t) + c_3 \gamma_3 L_3(t) \tag{26}$$

in which $L_i(t)$ are the random time-dependent variable loads, c_i are deterministic influence coefficients, and γ_i are deterministic load factors. Within a single structure, c_i may be zero for one load type at one element and dominant at another element. The relative magnitudes of the random loads $L_i(t)$ depend on geography and intended use.

The fact that any load $L_i(t)$ in a combination can appear alone if $c_j = 0, j \neq i$ suggests the following statement of the design load combination problem: Establish a set of companion action factors ψ_{ij} in Eq. (23) or design probability factors ϕ in Eq. (24) such that the probability of exceeding design loads is approximately constant for all situations involving one or more loads, the spectrum of influence coefficients c_i, all geographic areas and intended structural uses, and all materials and types of structural form covered by a code. Given the scope of the problem definition and practical limitations on the number of factors permissible in any design procedure, it is evident that great precision cannot be expected.

Before proceeding to the determination of the load combination factors, it is of some interest to view the various load combination formats in the light of the results obtained for linear load combinations. In this context *Turkstra's rule* (Turkstra, 1970) plays a central role. The rule states that the maximum value of the sum of two independent random processes occurs when one of the processes has its maximum value. The rule is an approximation and corresponds to the assumption that the distribution functions of the two random variables

$$Z_1 = \max_{0 \leq t \leq T} (Q_1(t) + Q_2(t)) \tag{27}$$

and

$$Z_2 = \max \begin{cases} \max_{0 \leq t \leq T} Q_1(t) + Q_2(t) \\ Q_1(t) + \max_{0 \leq t \leq T} Q_2(t) \end{cases} \tag{28}$$

are the same.

For Z_2, the complementary cumulative distribution function is

$$P(Z_2 > \xi) = P(\max_{0 \leq t \leq T} Q_1(t) + Q_2(t) > \xi) + P(Q_1(t) + \max_{0 \leq t \leq T} Q_2(t) > \xi)$$
$$- P(\max_{0 \leq t \leq T} Q_1(t) + Q_2(t) > \xi \cap Q_1 + \max_{0 \leq t \leq T} Q_2(t) > \xi) \tag{29}$$

Neglect of the negative term and use of the bound Eq. (3) without P_{F0} leads to

$$P(Z_2 > \xi) \leq T \int_{q=-\infty}^{\infty} \nu_{Q_1}(q) f_{Q_2}(\xi - q) dq + T \int_{q=-\infty}^{\infty} \nu_{Q_2}(q) f_{Q_1}(\xi - q) dq \tag{30}$$

Use of the bound Eq. (3) without P_{F0}, together with the bound Eq. (18), leads to the same upper bound for $P(Z_1 > \xi)$. Based on these results, it can be concluded that when the upper bound Eq. (29) is a good approximation for both $P(Z_1 > \xi)$ and $P(Z_2 > \xi)$, Turkstra's rule is also a good approximation. The conditions for the applicability of the upper bound given here are, however, not necessary conditions for good accuracy of Turkstra's rule.

Turkstra's rule indicates that a natural code format for a combination of two loads is

$$\max \begin{cases} \gamma_1 q_{1k} + \gamma_2 \psi_2 q_{2k} \\ \gamma_1 \psi_1 q_{1k} + \gamma_2 q_{2k} \end{cases} \tag{31}$$

where the ψ-factors express the ratio between fractiles in the extreme-value distributions and the marginal distributions. There is thus a logic rationale for the format in Eq. (23) and studies such as those of (Turkstra and Madsen, 1980) also show that this format is superior to the others.

A comprehensive study aimed at determining the load combination factors ψ_{ij} in Eq. (23) has been presented in (Turkstra and Madsen, 1980). The analysis is restricted to cases where loads do not act in opposite senses, leading to stress reversal. The load combination factors are aimed at being the same for all materials and the criteria of probability levels of 0.01, 0.001, and 0.0001 for individual design loads and combinations are therefore adopted as objective. A linear combination as in Eq. (26) is used with the complete range of influence factors being covered. The major conclusions of the study are as follows:

  − The uncertainty in load models is of major importance in the study of individual loads. However, results for the combination of loads are relatively insensitive to the load models used.
  − Design combination rules depend on the probability level at which comparisons are made. In general, the more likely the exceedence of the design values of individual loads, the less important the combination problem.
  − Simple addition of design loads can lead to very conservative results. Ignoring load superposition can lead to extremely nonconservative results.
  − No combinations of transient live, earthquake, and wind loads need be made at the fractile levels used in conventional structural design.
  − The combination factor format leads to significant errors in a number of cases.
  − The companion action format, coupled with a simple model for the ψ-factors, leads to design values almost always within 10% and normally within 5% of theoretical values.

5. Load combination with random failure surface

In the previous sections, the level $\xi(t)$ or the safe set $B(t)$ was assumed to be deterministic. In general a reliability problem contains random variables in addition to the random processes. The random variables may describe uncertainty in ξ or B, or may describe uncertainty in the statistical parameters of the random load processes. This section therefore describes a general and efficient method for computing failure probabilities in design situations where uncertainties are represented by a vector of random variables and a vector of random processes. A formulation as a first-passage problem for a vector process outcrossing a safe set is first applied for a fixed value of the random variable vector. This gives a conditional failure probability, and a fast integration technique based on a first- or second-order reliability method is then applied to compute the unconditional failure probability.

The combination of Gaussian processes with random parameter uncertainty is very relevant for practical problems, where the response process is often assumed to be Gaussian or a function of a Gaussian process, e.g. a translation process (Grigoriu, 1984) or a Hermite transformation of a Gaussian process (Winterstein, 1988). Most response analyses within marine engineering, wind engineering and earthquake engineering make such an assumption.

The reliability of a structural element is generally analyzed with respect to one or more failure criteria. For one criterion the performance is described through the limit state function $G(\cdot)$

$$G(z) \begin{cases} < 0 & \text{for } z \text{ in failure set} \\ = 0 & \text{for } z \text{ on limit state surface} \\ > 0 & \text{for } z \text{ in safe set} \end{cases} \tag{32}$$

where z is a vector of basic variables including loading variables, material properties, geometrical variables, statistical estimates and model uncertainty parameters. Since some of the basic variables can be functions of time, the limit state function is written in the form

$$G(Z) = \min_{[0,T]} G(Z_1, Z_2(t)) \tag{33}$$

where $[0, T]$ is the considered time interval. Z_1 is modeled as a vector of random variables, with a continuous but otherwise arbitrary joint probability density. $Z_2(t)$ is modeled as a stationary vector process. The vector process is completely characterized by a set of parameters, some of which may be uncertain and included in Z_1. In the following it is therefore assumed that $Z_2(t)$ for a fixed value of Z_1 is a stationary and ergodic process.

The probability of failure, P_F, in a time period $[0, T]$ is

$$P_F = P(\min_{0 \le t \le T} G(Z_1, Z_2(t)) \le 0) = \int_{z_1} P_F(z_1) dz_1 \tag{34}$$

where the conditional failure probability $P_F(z_1)$ is defined as

$$P_F(z_1) = P(\min_{0 \le t \le T} G(z_1, Z_2(t)) \le 0) \tag{35}$$

The conditional failure probability is the probability that the vector process $Z_2(t)$ is in the failure set at the beginning or during the time interval $[0, T]$. The conditional failure probability is by Eq. (3) bounded in terms of the probability of failure at time $t = 0$, $P_{F0}(z_1)$, and the expected number of outcrossings into the failure set, $E[N(T, z_1)]$

$$P_F(z_1) \leq P_{F0}(z_1) + E[N(T, z_1)] = P(G(z_1, Z_2(0)) \leq 0) + \nu(z_1)T \quad (36)$$

where $\nu(z_1)$ is the mean outcrossing rate for the stationary process. Alternatively, the conditional failure probability can be approximated by, Eq. (5)

$$P_F(z_1) \approx 1 - (1 - P_{F0}(z_1)) \exp\left[-\frac{\nu(z_1)T}{1 - P_{F0}(z_1)}\right] \quad (37)$$

The term P_{F0} can be calculated by a method for time independent reliability analysis. The mean outcrossing rate is given by Rice's formula expressed as

$$
\begin{aligned}
\nu(z_1) &= \int_{\partial S(z_1)} \int_0^\infty \dot{z}_{2N} f_{Z_2, \dot{Z}_{2N}}(z_2, \dot{z}_{2N}) d\dot{z}_{2N} d(\partial S(z_1)) \\
&= \int_{\partial S(z_1)} f_{Z_2}(z_2) \int_0^\infty \dot{z}_{2N} f_{\dot{Z}_{2N}|Z_2}(z_2, \dot{z}_{2N}) d\dot{z}_{2N} d(\partial S(z_1)) \quad (38)
\end{aligned}
$$

where $\partial S(z_1)$ is the boundary of the safe set, and $\dot{Z}_{2N}$ is the component of the time derivative $\dot{Z}_2$ in the direction of the outward normal vector to the boundary. Alternatively $\nu(z_1)$ may be calculated as in Eq. (10).

To compute the unconditional failure probability, an auxiliary limit state function h may be defined as, (Wen and Chen, 1987)

$$h(z_1, u_{n+1}) = u_{n+1} - \Phi^{-1}(P_F(z_1)) \quad (39)$$

where Φ is the standard normal distribution function, and u_{n+1} is a standard normal variable. A FORM or SORM analysis for time independent reliability problems can be applied directly to this limit state function. The computational difficulty lies in evaluation of the conditional failure probability and its derivative with respect to elements in z_1. Efficient methods for calculation of these quantities when $Z_2(t)$ is a Gaussian process are described in (Madsen and Tvedt, 1990) and implemented in the computer program (PROBAN4, 1995).

As shown in section 4, Turkstra's rule often provides a good approximation for linear combination of independent processes. According to this rule, only the points in time where one of the processes is at its maximum value are considered. The extreme value distribution for the single load can often be determined or approximated, and the distribution of each of the companion load values is simply their marginal distribution. The failure probability is underestimated by Turkstra's rule, but the reliability analysis has been reduced to a time-independent analysis, and the error turns out to be small in most practical cases.

The principle behind Turkstra's rule can be easily extended to the time dependent problem which involves a non-linear combination of dependent Gaussian processes. With n_2 components in Z_2, n_2 time independent reliability analysis are defined, each combining the extreme value for one element in Z_2 with the companion values of the other elements. The distribution for the companion values is a joint Gaussian distribution.

6. Conclusions

The following conclusions can be stated:

1. A close upper bound to the extreme value distribution for a combination of time varying load processes is expressed in terms of the mean upcrossing rate of the combined process. The mean upcrossing rate is expressed as an extension of Rice's formula for scalar processes, or, alternatively is calculated as a parametric sensitivity factor of an associated parallel system unsafe domain.

2. For a linear combination, a close upper bound on the mean upcrossing rate for the combined process is expressed as a convolution integral of the mean upcrossing rates for one process and the random-point-in-time distributions for the other processes.

3. Various code format for load combinations have been reviewed and the companion factor format as used in Europe has been found to be most appropriate. This format is based on Turkstra's rule which is justified from results for linear combinations of independent load processes.

4. A general method for time dependent reliability analysis has been presented. The method extends the first-and second-order reliability methods for time independent reliability analysis. Uncertainty is described in terms of a vector of random variables with continuous but otherwise arbitrary joint distribution function and a stationary Gaussian vector process.

5. Turkstra's rule is considered for the nonlinear combination of dependent stationary Gaussian processes. Limited experience indicates that Turkstra's rule works well also in this case when the dependencies are properly accounted for. This reduces the reliability analysis to a time independent reliability analysis of a series system of component failure modes, corresponding to failure when one of the time varying parameters takes its extreme value.

References

K.W. Breitung and R. Rackwitz. Nonlinear combination of load processes. *Journal of Structural Mechanics*, 10(2):145–166, 1982.

K.W. Breitung. Asymptotic approximations for probability integrals. In *Lecture Notes in Mathematics*. Springer Verlag, 1994.

CEB (Comité Europeen du Beton). First order concepts for design codes. Technical report, Joint Committee on Structural Safety CEB-CECM-FIP-IABSE-IASS-RILEM, 1976. CEB Bulletin No. 112.

O. Ditlevsen. Gaussian outcrossing from safe convex polyhedrons. *Journal of Engineering Mechanics, ASCE*, 109:237-148, 1983.

O. Ditlevsen. Level crossings of random processes. In P. Thoft-Christensen, editor, *Reliability Theory and Its Applications in Structural and Soil Mechanics*, pages 57-83. Martinus Nijhoff, The Hague, 1983. NATO ASI Series E.

O. Ditlevsen and H.O. Madsen. Probabilistic modeling of man-made load processes and their individual and combined effects. In T. Moan and M. Shinouzuka, editors, *Structural Safety and Reliability*, pages 103-134, Trondheim, Norway, June 1981. Proceedings ICOSSAR '81.

J.R. Fuller. Boundary excursions for combined random loads. *AIAA Journal*, 20:130-1305, 1982.

M. Grigoriu. Crossing of non-gaussian translation processes. *Journal of Engineering Mechanics, ASCE*, 110(4):610-620, 1984.

Ø. Hagen and L. Tvedt. Vector process out-crossing as parallel system sensitivity measure. *Journal of Engineering Mechanics, ASCE*, 117(10):2201-2220, 1991.

H.O. Madsen, R. Kilcup and C.A. Cornell. Mean upcrossing rate for sums of pulse-type stochastic load processes. In *Proceedings of Specialty Conference on Probabilistic Mechanics and Structural Reliability*, pages 54-58, 1979. ASCE, Tucson, Arix.

JCSS (Joint Committee on Structural Safety). General principles of safety and serviceability regulations for structural design. Technical report, Lund Institute of Technology, Lund, Sweden, 1978. ed. L. Ostlund.

R.D. Larrabee and C.A. Cornell. Combination of various load processes. *Journal of the Structural Division, ASCE*, 197:223-239, 1981.

H.O. Madsen. Load models and load combinations. Technical Report Report No. R113, Department of Structural Engineering, Technical University of Denmark, February 1979.

H.O. Madsen. Reliability under combination of non stationary load processes. In *DIALOG 1-82*, pages 45-58. Danish Engineering Academy, Lyngby, Denmark, 1982.

H.O. Madsen, S. Krenk and N.C. Lind. *Methods of Structural Safety*. Prentice-Hall, Englewood Cliffs, N.J., 1986.

H.O. Madsen. Sensitivity factors for parallel systems. In G. Mohr, editor, *Miscellaneous Papers in Civil Engineering*. Danish Engineering Academy, Lyngby, Denmark, 1992. DIA 35'th Aniversary '92.

H.O. Madsen and L. Tvedt. Methods for time dependent reliability and sensitivity analysis. *Journal of Engineering Mechanics, ASCE*, 116(10):2118-2135, 1990.

PROBAN Version 4. Theoretical manual. Technical report, Det Norske Veritas Reseach, 1995.

M.K. Ravaindra and T.V. Galambos. Load and resistance factor design for steel. *Journal of the Structural Division, ASCE*, 104:1337-1353, 1978.

S.O. Rice. Mathematical analysis of random noise. *Bell System Technological Journal*, 23:282-332, 1944.

S.O. Rice. Mathematical analysis of random noise. *Bell System Technological Journal*, 24:46-156, 1944.

C.J. Turkstra. Theory of structural safety. Technical report, Solid Mechanics Division, University of Waterloo, Ontario, Canada, 1970. SM Study No. 2.

C.J. Turkstra and H.O. Madsen. Load combinations in codified structural design. *Journal of the Structural Division, ASCE*, 106:2527-2543, 1980.

D. Veneziano, M. Grigoriu and C.A. Cornell. Vector-process models for system reliability. *Journal of Engineering Mechanics, ASCE*, 103:441-460, 1977.

C.B. Waugh. Approximate models for stochastic load combinations. Technical Report

Report R77-1, Department of Civil Engineering, Massachusetts Institute of Technology, Cambridge, Mass., 1977.

Y.K. Wen. Statistical combination of extreme loads. *Journal of the Structural Division, ASCE*, 103:1079–1093, 1977.

Y.K. Wen. A clustering model for correlated load processes. *Journal of the Structural Division, ASCE*, 107:965–983, 1981.

Y.K. Wen and H.C. Chen. On fast integration for time variant structural reliability. *Probabilistic Engineering Mechanics*, 2(3):156–162, 1987.

Y.K. Wen and H.T. Pearce. Stochastic models for dependent load processes. Technical report, Civil Engineering Studies, University of Illinois, Urbana, Ill., 1981. Structural Research Series No. 489, UILU-ENG-81-2002.

Y.K. Wen and H.T. Pearce. Combined dynamic effects of correlated load processes. *Nuclear Engineering and Design*, 1983.

S.R. Winterstein. Combined dynamic response extreme and fatigue damage. Technical report, Department of Civil Eng., Massachusetts Institute of Technology, Cambridge, Mass., 1980. Report R80-46.

S.R. Winterstein. Non-linear vibration models for extremes and fatigue. *Journal of Engineering Mechanics, ASCE*, 114:1772–1790, 1988.

S.R. Winterstein and C.A. Cornell. Load combinations and clustering effects. *Journal of the Structural Division, ASCE*, 110:2690–2708, 1984.

TIME–VARIANT RELIABILITY FOR NON–STATIONARY PROCESSES
BY THE OUTCROSSING APPROACH

R. RACKWITZ
Technical University Munich
Arcisstr. 21, 80290 Munich, Germany

1. Introduction

Whereas theory and concepts for the computation of time–invariant reliability are now well–known and can be performed efficiently and reliably by various methods, much less theory is available for methods which are capable of handling time variant reliability problems. Time variant problems are usually present with time–variant environmental loading and possibly time–variant (deteriorating) structural properties. One needs to compute not primarily the probability that a structural system is in an adverse state at any given time. It is rather the probability that such an adverse state is reached for the first time in a given reference period. There are two important cases in which computation of so–called first passage probabilities is still possible with time–invariant methods. This is when the failure criterion is related to strictly increasing cumulative damage phenomena, for example in structural fatigue. Then, the probability of survival is equal to the probability that damage has not reached a critical value at a given time. The other case is when all variables are time–invariant except one which then can be replaced by its extreme value, but only in the stationary case.

In all other cases, virtually no closed form solution of sufficient generality and direct practical interest is known. In the following we will primarily discuss results which can be called exact in some sense and/or have found a practical, sufficiently general numerical solution. One of the first computationally efficient algorithms was designed for the special case of a combination of stationary random sequences with different change frequencies (see Ferry Borges and Castanheta, 1971, Rackwitz and Fießler, 1978). This scheme was restricted to the stationary case. Another well–known case is the Gaussian process and linear combinations thereof. In the first case, the so–called maximum approach has been employed, i.e. the vectorial problem is first reduced to a scalar problem for which the results for the maximum distribution of scalar random processes then become applicable. This approach is extremely difficult to generalize to other than random rectangular wave sequences. In the second case the so–called outcrossing approach is used, and this is the approach to be discussed below in more detail. Even for stationary Gaussian scalar processes exact results for the first passage times are extremely rare. But a well–known asymptotic result for the stationary case is due to Rice (1945) which is generalized to the non–stationary case in Cramer and Leadbetter (1967).

C. Guedes Soares (ed.), Probabilistic Methods for Structural Design, 245–260.

Computationally feasible approaches to such reliability problems at present are mostly of an asymptotic nature and, therefore, only furnish approximate solutions for either high or low reliability problems. They rest on the construction of a Poisson process for the exits of the structural state function into the failure domain. The intensity parameter of this Poisson process is determined via the outcrossing rates of the load effect processes through the possibly time variant limit state function. Given the outcrossing rates, the mean number of exits into the failure domain has to be determined by time integration. Whereas the calculation of the outcrossing rates in the general case is a non–trivial task, a second difficulty usually arises when assuring the Poisson nature (lack of memory) of the outcrossings under the presence of time invariant or at least non–ergodic basic variables.

Although various aspects have been recognized and formulated earlier mainly in the context of scalar Gaussian process theory (see Cramer and Leadbetter, 1967, Bolotin, 1981 and Leadbetter, et al., 1983), serious attempts to establish a sound theoretical basis and to design efficient numerical methods date back only a few years. First important steps for a practical solution of the problem have, among others, been made by Veneziano et al. (1977), Ditlevsen (1983) and Breitung (1988) who derived partly closed form solutions for the outcrossing rates of stationary Gaussian vector processes out of arbitrary domains. Breitung and Rackwitz (1982) derived a solution for stationary rectangular renewal wave processes (see also Breitung, 1994). Rackwitz (1993) extended this work to the non–stationary case. Breitung and Rackwitz (1979) published a solution for the superposition of filtered Poisson processes with rectangular marks. Important contributions to the combination of intermittent processes are due to Wen (1977, 1981, 1990), Shinozuka (1981), Winterstein (1981) and Schrupp and Rackwitz (1988). Further, on the basis of asymptotic concepts in Breitung (1988), Plantec and Rackwitz (1989) and Faber and Rackwitz (1990) derived solutions for the mean number of outcrossings of Gaussian (vector) processes under non–stationary conditions, by approximating the time integrals over the crossing rates by asymptotic analysis. In using the results for stationary Gaussian processes, Fujita et al. (1987) designed an algorithm to handle the time invariant and non–ergodic variables consistently. The approach will be discussed in more detail later. Schall et al. (1990), following Naess (1984), studied simple examples of the problem of the presence of time–invariant basic variables together with ergodic sequences which, for example, are used to model the sea states in marine engineering. It was found that an accurate treatment of those variables is, in fact, very important. In the following we will discuss more recent results which are primarily suitable for the non–stationary case and include ergodic sequences for the parameters of the processes as well as non–ergodic quantities.

2. General Concepts for Time–variant Reliability

Consider the general task of estimating the probability $P_f(t)$ that a realization $z(\tau)$ of a random state vector $Z(\tau)$ representative for a given problem, enters the failure domain $V = \{z(\tau) \mid g(z(\tau),\tau) \leq 0, 0 \leq \tau \leq t\}$. $g(\cdot)$ is the limit state function. $Z(\tau)$ may conveniently be separated into three components as

$$Z(\tau)^T = (R^T, Q(\tau)^T, S(\tau)^T) \tag{2.1}$$

where $\mathbf{R}$ is a vector of random variables independent of time t, $\mathbf{Q}(\tau)$ is a slowly varying random vector sequence and $\mathbf{S}(\tau)$ is a vector of not necessarily stationary, but sufficiently mixing random process variables having fast fluctuations as compared to $\mathbf{Q}(\tau)$. Typically, $\mathbf{Q}(\tau)$ characterizes slow variations of the parameters of the process $\mathbf{S}(\tau)$.

Consider first the case where only $\mathbf{S}(\tau)$ is present. If it can be assumed that the stream of crossings of the vector $\mathbf{S}(\tau)$ into the failure domain V is Poissonian, it is well known that the failure probability $P_f(t)$ can be estimated from (Cramer and Leadbetter, 1967)

$$P_f(t) \approx 1 - \exp(- E[N_{\mathbf{S}}^+(t)]) \leq E[N_{\mathbf{S}}^+(t)] \qquad (2.2)$$

for high reliability problems. $E[N_{\mathbf{S}}^+(t)]$ is the expected number of crossings of $\mathbf{S}(\tau)$ into the failure domain V in the considered time interval. It is assumed that there is negligible probability of failure at $\tau = 0$ or $\tau = t$. The upper bound is a strict upper bound and close to the exact result for small $P_f(t)$. The approximation in eq. (2.2) has found many applications in the past, not only because of its relative simplicity but also because there is no real practical alternative except in some special cases. The Poissonian character of exits into the failure domain will, in fact, be lost if, loosely speaking, the exits into the failure domain are neither rare nor independent. One possible route of investigation is along the following expansion given by Lange (1991) and Engelund et al. (1995)

$$P_f(t) = 1 - \exp\left\{- E[N_{\mathbf{S}}^+] \sum_{k=0}^{\infty} (-1)^k q_k(E[(N_{\mathbf{S}}^+-1)...(N_{\mathbf{S}}^+- k+1)])\right\} \qquad (2.3)$$

with certain Gram–Charlier–coefficients, which depend on the factorial moments $m_{(k)} = E[(N_{\mathbf{S}}^+-1)...(N_{\mathbf{S}}^+- k+1)]$. For simplicity, reference to the time interval is omitted. The Gram–Charlier–coefficients are given by

$$q_k = \sum_{i=1}^{k-1} (-1)^i \frac{m_1^i}{i!\,(k-i)!} (m_{(k-i)} - m_1^{k-i}) \text{ für } k > 0 \qquad (2.4)$$

with $q_0 = 1$. Another similar but less suitable "inclusion–exclusion" expansion has already been given by Rice (1945). The difficulty in using such improvements is the enormous effort reqired to compute the higher order factorial moments. Therefore, a different strategy must be followed to maintain as far as possible the Poissonian nature of the excursions.

For example, when both process variables $\mathbf{S}(\tau)$ and time invariant random variables $\mathbf{R}$ are present, the Poissonian nature of outcrossings is lost. Eq. (2.2) can furnish only conditional probabilities. The total failure probability must be obtained by integrating over the probabilities of all possible realizations of $\mathbf{R}$. Then the equivalent to eq. (2.2) is

$$P_f(t) \approx E_R[1-\exp(- E[N_S^*(t|R)])] \leq E_R[E[N_S^*(t|R)]] \tag{2.5}$$

In the general case, where all the different types of random variables R, $Q(\tau)$ and $S(\tau)$ are present, the failure probability $P_f(t)$ not only must be integrated over the time in—variant variables R, but an expectation operation must also be performed over the slowly varying variables $Q(\tau)$. In Schall et al. (1990) the following formula has been established, by making use of the ergodicity theorem

$$P_f(t) \approx 1 - E_R[\exp(- E_Q[E[N_S^*(t|R,Q)]])] \leq E_R[E_Q[E[N_S^*(t|R,Q)]]] \tag{2.6}$$

Eq. (2.6) is a rather good approximation for the stationary case but must be considered only as a first approximation whenever $S(\tau)$ is non—stationary or the limit state function exhibits strong dependence on τ. The bounds given in eqs. (2.5) and (2.6) again are strict but close to the exact result only for small failure probabilities.

3. Integration with Respect to Time–invariant Variables R

If there are time—invariant random vectors R present, several possibilities exist, the most straightforward being numerical integration using the upper bound solution. However, even for small dimensions of R, the computational effort can be considerable. The computational problem can be reformulated as (Fujita et al. 1987)

$$P_f(t) = \mathbb{P}(T(R) - t \leq 0) = \mathbb{P}(g(R,U_T) \leq 0) \tag{3.1}$$

where $T(R)$ is a random life time with realizations $t(r)$ given by the solution to the equation

$$E[N_S^*(t(r)|r)] + \ln(- \Phi(u_T)) = 0 \tag{3.2}$$

and u_T a realization of an auxiliary standard normal variable. Eq. (3.1) is appropriate for classical volume integration, for example by FORM/SORM, provided that the quantity $E[N_S^+(t(r)|r)]$ can be calculated. This formulation is always exact to the order of computation level (FORM or SORM). However for time varying limit state functions and/or non—stationary processes, the numerical solution of eq. (3.2), which must be performed at least twice in each iteration, becomes rather involved because of repeated calculations of $E[N_S^+(t(r)|r)]$, and the convergence of the algorithm may become slow if not unreliable.

Alternatively and to an arbitrary accuracy, the expectation operation in eq. (2.4) can be performed either by crude Monte Carlo integration or with importance sampling. Provided the important region for integration is known by r^* it is

$$E_R[1-\exp(-E_Q[E[N_S^+(t\,|\,Q,R)]])] = \int_{\mathbb{R}^n} \left[1-\exp(-E_Q[E[N_S^+(t\,|\,Q,r)]])\right] \frac{\varphi_R(r)}{h_R(r)}\, h_R(r)\, dr$$

(3.3)

where $h_R(r)$ is the sampling density. Then,

$$E_R[1 - \exp(- E_Q[E[N_S^+(t\,|\,Q,R)]])] =$$

$$\approx \frac{1}{N} \sum_{i=1}^{N} \left[1 - \exp(- E_Q[E[N_S^+(t\,|\,Q,r_i)]])\right] \frac{\varphi_R(r_i)}{h_R(r_i)}$$

(3.4)

The sampling density (standard space) can conveniently be chosen as the standard normal density with mean r^* determined from the upper bound solution and with covariance matrix I. A crude, usually conservative estimate is already obtained for $r_i = r^*$.

4. Relation between Mean Number of Crossing and Mean Number of Outcrossings

Frequently, the quantity $E[N_S^+(t)\,|\,r,q]$ cannot be determined directly. Instead, the expected number of crossings $E[N_S(t)\,|\,r,q]$ is determined first and then the expected number of outcrossings is obtained from the expected number of crossings by using the relation (see Cramer and Leadbetter, 1967, and Plantec and Rackwitz, 1989):

$$2\, E[N^+(t)] \approx E[N(t)]+\mathbb{P}(g(S;t) \le 0) - \mathbb{P}(g(S;0) \le 0)$$

(4.1)

This formula is derived as follows. In noting the obvious algebraic identity $a = (a + b)/2 + (a - b)/2$ one can also write

$$E[N^+(t)] = 1/2\, E[N(t)] + 1/2\, (E[N^+(t)] - E[N^-(t)])$$

The number of out– and incrossings can differ at most by one. If the process starts and finishes in the safe domain the number of out– and incrossings are equal. The same is true if the process starts and finishes in the failure domain. Therefore, the difference $N^+(t) - N^-(t)$ must be zero. The difference is $+1$ if the process starts in the safe domain and finishes in the failure domain and_is -1 in the opposite case. The probability of the first event (start in $\bar{V}$ and finish in V) is $\mathbb{P}[\{g(S;0) > 0\} \cap \{g(S;t) \le 0\}]$ whereas for the second event (start in V and finish in $\bar{V}$) it is $\mathbb{P}[\{g(S;0) \le 0\} \cap \{g(S;t) > 0\}]$. Therefore:

$$E[N^+(t)] - E[N^-(t)]$$
$$= \mathbb{P}[\{g(S;0) > 0\} \cap \{g(S;t) \le 0\}] - \mathbb{P}[\{g(S;0) \le 0\} \cap \{g(S;t) > 0\}]$$

(4.2)

According to the assumption of a sufficiently mixing process, the two events at time 0 and at time t, respectively, may be assumed independent for large t, and eq. (4.1) follows.

Eq. (4.2) can be simplified further with some loss of accuracy. Three cases with respect to the density of crossings in time must be distinguished. The stream of crossings is most dense either at $\tau^* = 0$, at $\tau^* = t$ or at some point in $0 \leq \tau^* \leq t$, respectively. Then

for $\tau^* = 0$:

$$2\,E[N^+(t)] \approx E[N(t)] - \mathbb{P}(g(\mathbf{S};0) \leq 0) \qquad (4.3a)$$

for $\tau^* = t$:

$$2\,E[N^+(t)] \approx E[N(t)] + \mathbb{P}(g(\mathbf{S};t) \leq 0) \qquad (4.3b)$$

for $0 < \tau^* < t$:

$$2\,E[N^+(t)] \approx E[N(t)] \qquad (4.3c)$$

5. Gaussian Processes

For convenience of notation, reference to the vectors r and q is now temporarily omitted. We first derive a general outcrossing formula. Let $V = \{g(\mathbf{s};\tau) \leq 0\}$ be the failure domain in the standard space with boundary ∂V varying in time. The latter is assumed to be at least locally twice differentiable in $\mathbf{x}$ and τ:

$$\partial V = \partial V(\mathbf{s};\tau) = \{\mathbf{s},\tau;\, g(\mathbf{s};\tau) = 0\} \qquad (5.1)$$

Following Belyaev (1972) and Bolotin (1981) the outcrossing rate of the process $S(\tau)$ through the hypersurface ∂V during a time interval $\Delta\tau$ can be defined as:

$$\nu^+(\partial V;\tau) = \lim_{\Delta\tau \to 0} \frac{1}{\Delta\tau} \mathbb{P}(N(\Delta\tau) = 1) = \lim_{\Delta\tau \to 0} \frac{1}{\Delta\tau} \mathbb{P}_1(\partial V;\Delta\tau) \qquad (5.2)$$

As usual, regularity of the point process of crossings is assumed, i.e. there is $\mathbb{P}(N(\Delta\tau) > 1) = o(\Delta\tau)$. $\mathbb{P}_1(\partial V;\Delta\tau)$ is the probability of a crossing of ∂V by the process $S(\tau)$ from the safe domain $V(\tau) = \{g(\mathbf{s},\tau) > 0\}$ into the failure domain $V = \{g(\mathbf{s},\tau) \leq 0\}$ during $\Delta\tau$. $\mathbb{P}_1(\partial V;\Delta\tau)$ can then be given by

$$P_1(\partial V;\Delta\tau) = \mathbb{P}(\{S(t) \in \Delta(\partial V)\} \cap \{\dot{S}_n(t) > \dot{\partial V}(\mathbf{x};t)\}) \qquad (5.3)$$

for $\tau \leq t \leq \tau + \Delta\tau$ and where $\dot{S}_n(t) = \mathbf{n}^T(\mathbf{s})\,\dot{S}(t)$ is the projection of $\dot{S}(t)$ on the normal $\mathbf{n}^T(\mathbf{x})$ of ∂V at the point $\mathbf{s}$, $\dot{\partial V}(\mathbf{s};t)$ is the time–variation of the surface ∂V at $\mathbf{s}$ and $\Delta(\partial V)$ is a thin layer enveloping ∂V. Introducing the joint probability density function $\varphi_{n+1}(\mathbf{s},\dot{s}_n;\tau)$ of S and $\dot{S}_n$ allows one to express $\mathbb{P}_1$ by the following integral:

$$P_1(\partial V(\tau);\Delta\tau) = \int_{\Delta(\partial V)} \int_{\dot{s}_n > \dot{\partial V}(\mathbf{s};\tau)} \varphi_{n+1}(\mathbf{s},\dot{s}_n;\tau)\,d\mathbf{s}\,d\dot{s}_n \qquad (5.4)$$

The integral over $\Delta(\partial V)$ can be transformed into a surface integral over ∂V. The layer $\Delta(\partial V)$ is understood as the sum of infinitely small cylinders with height $(\dot{s}_n(\tau) - \partial \dot{V}(s;\tau))\, \Delta\tau$ and base $ds(s)$. $ds(s)$ is a surface neighborhood of the crossing point. Hence, integrating over s yields:

$$P_1(\partial V; \Delta \tau) = \int_{\partial V} \int_{\dot{s}_n > \partial \dot{V}(s;\tau)} (\dot{s}_n - \partial \dot{V}(s;\tau))\, \varphi_{n+1}(s,\dot{s}_n;\tau)\, \Delta\tau\, d\dot{s}_n\, ds(s)$$

$$(5.5)$$

Introducing the density function of $\dot{S}_n$ conditional on $S = s$, proceeding to the limit according to eq. (5.2) and taking the integral over τ in $[0,t]$ as required in eq. (2.2) we find an integral for the mean number of outcrossings:

$$E[N^+(\partial V; t)] = \int_0^t \int_{\partial V} \int_{\dot{s}_n > \partial \dot{V}(s;\tau)} (\dot{s}_n - \partial \dot{V}(s;\tau))\, \varphi_1(\dot{s}_n | S(\tau) = s)\, \varphi_n(x)\, d\dot{s}_n\, ds(s)\, d\tau$$

$$(5.6)$$

Analogously, the expected number of incrossings can be determined. By combining the two contributions, we obtain the expected number of crossing:

$$E[N(\partial V; t)] = \int_0^t \int_{\partial V} \int_{\mathbb{R}} |\dot{s}_n - \partial \dot{V}(x;\tau)|\, \varphi_1(\dot{s}_n | S(\tau) = s)\, \varphi_n(s)\, d\dot{s}_n\, ds(s)\, d\tau$$

$$(5.7)$$

By considering the fact that the above is achieved by fixing the time τ, we note that the time–variation $\partial \dot{V}(s;\tau)$ of the surface ∂V corresponds to the time–variation of the function $g(s;\tau)$ and does not involve the time variation of its gradients. The conditional density function of $\dot{S}_n$ can be given explicitly by using the well known formulae for the conditional mean and variance of a Gaussian variable:

$$E[\dot{S}_n(\tau) | S(\tau) = s] = n^T(s;\tau)\, \dot{R}^T s = m(s;\tau) = m \qquad (5.8a)$$

$$\mathrm{Var}[\dot{S}_n(\tau) | S(\tau) = s] = n^T(s;\tau)\, [\bar{R} - \dot{R}^T \dot{R}]\, n(s;\tau) = \sigma_0^2 \qquad (5.8b)$$

Note that m depends explicitly on s. The complete formulation reads

$$P_f(t) \simeq 1 - E_R[\exp(-E_Q[N^+(t | q, r)])]$$

as outlined in section 2. The mean number of crossings also involving the slowly varying sequence $Q(\tau)$ is then

$$E_Q[E[N(t | q, r)]] =$$

$$= \int_{\mathbb{R}^{nq}} \int_{\partial V} \int_{\dot{s}_n > \partial \dot{V}} \int_0^t (\dot{s}_n - \partial \dot{V})\, f_{\dot{S}_n | S = s}(\dot{s}_n) f_S(s) f_Q(q)\, d\tau\, d\dot{s}_n\, ds(s)\, dq \quad (5.9)$$

with $\partial V = \{(\mathbf{q},\mathbf{s},\tau): g(\mathbf{r},\mathbf{q},\mathbf{s},\tau) = 0\}$ and $\partial \dot{V} = \frac{\partial g}{\partial \tau}\big|_{\mathbf{s}}$. If the upper bound solution is used in eq. (2.6), the vector $\mathbf{R}$ can be included in the vector $\mathbf{Q}$. This is the general formula for the mean number of crossings.

Scalar processes, which are by far the most important, can be dealt with easily in this framework. In particular, eq. (5.9) reduces to

$$E_{\mathbf{Q}}[E[N(t\,|\,\mathbf{q},\mathbf{r})]] = \int_{\mathbb{R}^{n_q}} \int_{\mathbb{R}} \int_0^t |\dot{s} - b_{\tau}(\tau,\mathbf{q})|\; f_{\dot{S}}(\dot{s})\; f_S(x)\; f_{\mathbf{Q}}(\mathbf{q})\; d\tau\; d\dot{s}\; d\mathbf{q}$$

$$(5.10)$$

where by introducing $z = \dot{s}/\sigma_{\dot{S}}$ into eq. (5.10) we find

$$E_{\mathbf{Q}}[E[N^+(t\,|\,\mathbf{q},\mathbf{r})]] = \int_{\mathbb{R}\times\mathbb{R}^{n_q}\times[0,t]} \frac{1}{(2\pi)^{1+n_q/2}}\, |k(\xi)|\; \exp(-\tfrac{1}{2}\,f(\xi))\; d\xi$$

$$(5.11)$$

with $\quad \xi = (z,\mathbf{q},\tau)^T$ $\hspace{6cm}$ (5.12a)

$\quad k(\xi) = \sigma_{\dot{S}}\, z - b_{\tau}(\tau,\mathbf{q})$ $\hspace{5cm}$ (5.12b)

$\quad f(\xi) = z^2 + \mathbf{q}^T\mathbf{q} + b^2(\tau,\mathbf{q})$ $\hspace{4.7cm}$ (5.12c)

$\quad S(\tau) = \dfrac{Y(\tau) - \mu(\tau)}{\sigma(\tau)}$ $\hspace{5cm}$ (5.12d)

$\quad b(\tau,\mathbf{r},\mathbf{q}) = \dfrac{a(\tau,\mathbf{r},\mathbf{q}) - \mu(\tau)}{\sigma(\tau)}$ $\hspace{4.3cm}$ (5.12e)

$\quad g(\tau,\mathbf{r},\mathbf{q},Y) = b(\tau,\mathbf{r},\mathbf{q}) - S(\tau)$ $\hspace{4cm}$ (5.12f)

$a(\tau,\mathbf{q},\mathbf{r})$ the original threshold function and $b_{\tau}(\tau,\mathbf{q})$ the time derivative of the normalized threshold. It is seen that the original process $Y(\tau)$ with mean value function $\mu(\tau)$ and standard deviation function $\sigma(\tau)$ is standardized as well as the threshold. $g(\tau,\mathbf{r},\mathbf{q},Y)$ is the usual state function. It is necessary to first locate the point $\xi^* = (z^*,\mathbf{q}^*,\tau^*)^T$ which minimizes $f(\xi)$ in $[0,t]$. Then expansions for functions f and k are used in the neighborhood of ξ^*.

If τ^* is an interior point of $[0,t]$, these expansions are

$$f(\xi) = (b^{*2} + \mathbf{q}^{*T}\mathbf{q}^*) + \tfrac{1}{2}\,\xi^T\, \mathbf{H}\, \xi \hspace{3cm} (5.13a)$$

$$k(\xi) = \mathbf{c}^T\, \xi \hspace{5.5cm} (5.13b)$$

$$\text{where:} \quad H = 2 \begin{bmatrix} 1 & \cdots 0 \cdots & 0 \\ \hline \vdots & & \\ 0 & I_{n_q} + b^* \overset{*}{B}_q + \nabla_q b^* (\nabla_q b^*)^T & b^* \overset{*}{b}_{i\tau} \\ \hline 0 & b^* \overset{*}{b}_{i\tau} & b^* \overset{*}{b}_{\tau\tau} \end{bmatrix} \quad (5.13c)$$

$$c^T = (\sigma_{\dot{S}}, - \overset{*}{b}_{1\tau}, \ldots, - \overset{*}{b}_{n_q\tau}, - \overset{*}{b}_{\tau\tau}) \tag{5.13d}$$

and $\overset{*}{B}_q$ denotes the Hessian matrix of b in the q–space at ξ^*,

$$\overset{*}{b}_{i\tau} = \frac{\partial^2 b}{\partial q_i \partial \tau}\Big|_{\xi^*} \text{ and } \overset{*}{b}_{\tau\tau} = \frac{\partial^2 b}{\partial \tau^2}\Big|_{\xi^*}$$

$\nabla_q b^*$ is the gradient of b in the q–space. Using the usual asymptotic arguments, one obtains by extending the integration over τ to the entire τ–axis, the expression

$$E_Q[E[N(t\,|\,q,r)]] \approx \sqrt{\frac{2}{\pi}} \exp(-\tfrac{1}{2}(b^{*2} + q^{*T}q^*)) \left[\frac{c^T H^{-1} c}{|\det(H)|}\right]^{1/2} \tag{5.14}$$

If τ^* is a boundary point of [0,t], the expansions are

$$f(\xi) = (b^{*2} + q^{*T}q^*) + 2b^* \overset{*}{b}_\tau \, \tau + \eta^T K \, \eta \tag{5.15a}$$

$$k(\xi) = \overset{*}{b}_\tau + \tilde{c}^T \eta \tag{5.15b}$$

with

$$K = \begin{bmatrix} 1 & \cdots 0 \cdots \\ \hline \vdots & \\ 0 & I_{n_q} + b^* \overset{*}{B}_q + \nabla_q b^* (\nabla_q b^*)^T \\ \vdots & \end{bmatrix} \tag{5.15c}$$

$$\tilde{c}^T = (\sigma_{\dot{S}}, - \overset{*}{b}_{1\tau}, \ldots, - \overset{*}{b}_{n_q\tau}) \tag{5.15d}$$

and $\eta^T = (z,q)$. One determines

$$E_Q[E[N(t\,|\,q,r)]] \approx \frac{1}{(2\pi)^{\frac{1}{2}}} \frac{\exp(-\tfrac{1}{2}(b^{*2} + q^{*T}q^*))}{|b^* \overset{*}{b}_\tau|} \left[\frac{\tilde{c}^T K^{-1} \tilde{c}}{|\det(K)|}\right]^{1/2} (2\varphi(a) - a + 2a\Phi(a)) \tag{5.16}$$

where:

$$a = \overset{*}{b}_\tau / (\tilde{c}^T K^{-1} \tilde{c})^{1/2} \tag{5.17}$$

For the stationary case the integration over time is trivial and the foregoing formulae can be reduced substantially.

Extensive numerical studies have shown that substantial improvements above the asymptotic solution can be achieved by treating the integration over time separately. In doing so, the second order interaction terms between q–variables and time τ are neglected, but integration over time of the first or second order expansion of $b(\tau)$ can be performed exactly (see also Hagen, 1992, for another route of improvement).

The formulae for the **interior point of [0,t]** are

$$E_Q[E[N(t\,|\,q,r)]] \approx 2\,\varphi(\beta_{qs}) \left[\frac{\omega_0^2 + c^T H^{-1} c}{|\det(H)|}\right]^{1/2} \left[\frac{\Phi(\sqrt{(h_{\tau\tau})}(t_2 - \tau^*) - \Phi(\sqrt{(h_{\tau\tau})}(t_1 - \tau^*)}{\sqrt{(h_{\tau\tau})}}\right]$$
(5.18)

now with c as in eq. (5.15d) except the last term, H as in eq. (5.15c) with the last column and row deleted, $h_{\tau\tau} = b(\tau^*)\,b_{\tau\tau}(\tau^*)$ and $\beta_{qs} = (b^{*2} + q^{*T}q^*)^{1/2}$. Further there is

$$\omega_0^2(\tau) = \frac{\partial^2}{\partial\tau_1\partial\tau_2}\,\rho_S(t_1,t_2)\big|_{\tau_1 = \tau_2 = \tau}$$
(5.19)

The stationary solution is readily obtained as

$$E_Q[E[N^+(t\,|\,q,r)]] \approx \frac{\omega_0}{\sqrt{(2\pi)}}\,\varphi(\beta_{qs})\left[\frac{1}{|\det(H)|}\right]^{1/2}|t_2 - t_1|$$
(5.20)

For the **boundary point of [0,t]**

$$E_Q[E[N^+(t\,|\,q,r)]] \approx \varphi(\beta_{qs})\left[\frac{\omega_0^2 + c^T K^{-1} c}{|\det(K)|}\right]^{1/2}\left[\frac{1 - \exp[-b^*|b_\tau^*|\,(t_2 - t_1)]}{b^*|b_\tau^*|}\right]$$
$$\times\,(2\,\varphi(a) - a + 2a\,\Phi(a))$$
(5.21)

with c, K and a as in eqs. (5.15d), (5.15c) and (5.17), respectively, is obtained. Note that $\Phi(-x) \approx \varphi(x)/x$ for large x. First order results with respect to the q–variables can be recovered by setting the terms involving the matrices H and K equal to unity. However, this is not recommended in view of the normally small dimension of the vector q. A first order result with respect to time integration can also be established; it is rather inaccurate.

6. Gaussian Vector Processes

The general vectorial case is not much more difficult as seen from eqs. (5.9). However, the dimension of $S(\tau)$ can be very large whereas the dimension of the sequence Q usually remains small. One may think of a vessel structure under Gaussian pressure loading and where the pressure field is reduced appropriately by local averaging to a vector process the components of which act at the nodal points of the structural finite element model. Therefore, due to the numerical effort for the determination of the Hessian in the q–s–space, a rigorous second order solution may require large numerical effort. But a first order approximation with respect to the vector $S(\tau)$ may be used while the integration with respect q is kept "exact" to second order. Thus, laborious evaluations of second order derivatives of the limit state function are avoided. However, certain errors may be expected, and no direct error quantification can be provided. At most, the potential error with respect to the s–integration can be crudely estimated theoretically by comparing the well known first order solution with the asymptotic solution for the stationary case given by Breitung (1989).

The conditional problem (the condition being the vector $\mathbf{r}$) is first formulated in the full, standard $\mathbf{q}$–$\mathbf{s}$–τ–space as usual, and the critical point is found. Then, the vector $\mathbf{S}(\tau)$ is replaced by a new standardized scalar process $W(\tau)$ according to

$$g(\mathbf{s}, \mathbf{q}, \mathbf{r}, \tau)$$
$$\approx \frac{g(\mathbf{s^*}, \mathbf{q}, \mathbf{r}, \tau^*) - \alpha_{\mathbf{s}}^T \mathbf{s^*}}{\|\alpha_{\mathbf{s}}\|} + \frac{\alpha_{\mathbf{s}}^T}{\|\alpha_{\mathbf{s}}\|} \mathbf{s} = \frac{g(\mathbf{s^*}, \mathbf{q}, \mathbf{r}, \tau^*) + \alpha_{\mathbf{s}}^T \mathbf{s^*}}{\|\alpha_{\mathbf{s}}\|} + w \quad (6.1)$$

where now $W(\tau)$ has correlation coefficient function

$$\rho_W(\tau_1, \tau_2) = \left[\frac{\alpha_{\mathbf{s}}}{\|\alpha_{\mathbf{s}}\|}\right]^T \rho_{\mathbf{S}}(\tau_1, \tau_2) \frac{\alpha_{\mathbf{s}}}{\|\alpha_{\mathbf{s}}\|} \quad (6.2)$$

and, thus, for $\tau = \tau^*$

$$\omega_0^2(\tau) = \frac{\partial^2}{\partial \tau_1 \partial \tau_2} \rho_W(t_1, t_2)\big|_{\tau_1 = \tau_2 = \tau} \quad (6.3)$$

Hence, the solution for the scalar case can be used with modified limit state function.

7. Renewal Rectangular Wave Processes

If the sequence of amplitudes $S_1, S_2, S_3, \ldots$ is an independent sequence (also independent of the jump times, of course) with distribution function $F_S(s; \mathbf{q}, \mathbf{r})$, then the mean number of exits into the failure domain $V = \{S \geq a\}$ is

$$E[N^+(t_1, t_2; \mathbf{q}, \mathbf{r})] = \lambda (t_2 - t_1) F_S(a; \mathbf{q}, \mathbf{r}) (1 - F_S(a; \mathbf{q}, \mathbf{r})) \quad (7.1)$$

Here, $\mathbf{q}$ is an ergodic sequence for the parameters of the distribution function of S and $\mathbf{r}$ is a possibly random, non–ergodic vector. For a n–dimensional renewal process with renewal rates λ_k and marks S_k with distribution function $F_{\mathbf{S}k}(\mathbf{x}; \mathbf{q}, \mathbf{r})$, it can be shown that (Breitung and Rackwitz, 1982)

$$E[N^+(t_1, t_2; \mathbf{q}, \mathbf{r})] = (t_2 - t_1) \sum_{i=1}^{n} \lambda_i \, \mathbb{P}(\{\mathbf{S}_i^- \in \overline{V}; \mathbf{q}, \mathbf{r}\} \cap \{\mathbf{S}_i^+ \in V; \mathbf{q}, \mathbf{r}\}) \quad (7.2)$$

where $\overline{V}$ and V are the safe and failure domain, respectively. $\mathbf{S}_i^+$ is the total load vector when the i–th component of the renewal process had a renewal. $\mathbf{S}_i^-$ denotes the total load vector just before the renewal. Therefore, $\mathbf{S}_i^-$ and $\mathbf{S}_i^+$ differ by the vectors $\mathbf{S}_i$ which are to be introduced as independent vectors in the first and the second set. Applying asymptotic concepts we can show, that (Breitung, 1995)

$$E[N^+(t_1,t_2;\mathbf{r})] \sim (t_2 - t_1) \sum_{i=1}^{n} \lambda_i \, \mathbb{P}(\{\mathbf{S} \in V(\mathbf{Q});\mathbf{r}\}) \qquad (7.3)$$

with $\mathbb{P}(\{\mathbf{S} \in V(\mathbf{Q});\mathbf{r}\})$ computed as a volume integral in the usual manner. This formula can be slightly improved for not small probabilities $\mathbb{P}(\{\mathbf{S} \in V(\mathbf{Q});\mathbf{r}\})$ by replacing the term $\mathbb{P}(\{\mathbf{S} \in V(\mathbf{Q});\mathbf{r}\})$ by $\mathbb{P}(\{\mathbf{S} \in V(\mathbf{Q});\mathbf{r}\}) - \mathbb{P}(\{\mathbf{S}_i^- \in V(\mathbf{Q}) \cap \{\mathbf{S}_i^+ \in V(\mathbf{Q});\mathbf{r}\})$. In numerous studies it was found that this correction remains usually insignificant. Note that integration with respect to $\mathbf{Q}$ is performed simultaneously with the integration with respect to $\mathbf{S}$.

The non–stationary case is not substantially more difficult. The renewal rates are assumed to vary slowly in time; they are denoted by $\lambda_k(\tau)$, $k = 1,2,...,n$. Also, the distribution function of $\mathbf{S}$ may now contain distribution parameters $\mathbf{r}(\tau)$, and the failure domain can be a function of time, i.e. $V = \{g(\mathbf{s},\mathbf{q},\mathbf{r},\tau) \leq 0\}$. Then, eq. (7.3) needs to be modified as

$$E[N^+(t_1,t_2;\mathbf{r})] \sim \int_{t_1}^{t_2} \sum_{i=1}^{n} \lambda_i(\tau) \, \mathbb{P}(\{\mathbf{S} \in V(\mathbf{Q},\tau);\mathbf{r}\}) \, d\tau \qquad (7.4)$$

The time–volume integral can be approximated in two steps. The probability $\mathbb{P}(\{\mathbf{S} \in V(\mathbf{Q},\tau);\mathbf{r}\})$ can be computed in the usual way for every time τ. At first, the critical point on the failure surface is located in the $\mathbf{s}$–$\mathbf{q}$–τ–space. At the critical point $(\mathbf{s}^*,\mathbf{q}^*,\tau^*)$ the probability $\mathbb{P}(\{\mathbf{S} \in V(\mathbf{Q},\tau);\mathbf{r}\})$ is estimated by

$$\mathbb{P}(\{\mathbf{S} \in V(\mathbf{Q},\tau);\mathbf{r}\}) = \Phi(-\beta(\tau^*)) \times C(\mathbf{s}^*,\mathbf{q}^*,\tau^*) \qquad (7.5)$$

with $C(\mathbf{s}^*,\mathbf{q}^*,\tau^*)$ the curvature correction term (in the $\mathbf{s}$–$\mathbf{q}$–space). Then, one can write

$$E[N^+(t_1,t_2;\mathbf{r})] \sim \int_{t_1}^{t_2} \sum_{i=1}^{n} \lambda_i(\tau) \, \Phi(-\beta(\tau)) \times C(\mathbf{s}^*,\mathbf{q}^*,\tau) \, d\tau$$

$$\approx C(\mathbf{s}^*,\mathbf{q}^*,\tau^*) \sum_{i=1}^{n} \lambda_i(\tau^*) \int_{t_1}^{t_2} \Phi(-\beta(\tau)) \, d\tau$$

$$= C(\mathbf{s}^*,\mathbf{q}^*,\tau^*) \sum_{i=1}^{n} \lambda_i(\tau^*) \int_{t_1}^{t_2} \exp[\ln[\Phi(-\beta(\tau))]] \, d\tau$$

$$= C(\mathbf{s}^*,\mathbf{q}^*,\tau^*) \sum_{i=1}^{n} \lambda_i(\tau^*) \int_{t_1}^{t_2} \exp[f(\tau)] \, d\tau \qquad (7.6)$$

where $f(\tau) = \ln[\Phi(-\beta(\tau))]$ with derivatives

$$f'(\tau) = -\frac{\varphi(-\beta(\tau))}{\Phi(-\beta(\tau))} \frac{\partial \beta(\tau)}{\partial \tau} \approx -\beta(\tau) \frac{\partial \beta(\tau)}{\partial \tau} \tag{7.7a}$$

$$f''(\tau) = \frac{\varphi(-\beta(\tau))}{\Phi(-\beta(\tau))} \left[\left[\frac{\partial \beta(\tau)}{\partial \tau} \right]^2 \left[\beta(\tau) + \frac{\varphi(-\beta(\tau))}{\Phi(-\beta(\tau))} \right] + \beta(\tau) \frac{\partial^2 \beta(\tau)}{\partial \tau^2} \right]$$

$$\approx \left[\left[\frac{\partial \beta(\tau)}{\partial \tau} \right]^2 2\beta^2(\tau) + \beta(\tau) \frac{\partial^2 \beta(\tau)}{\partial \tau^2} \right] \tag{7.7b}$$

While $\partial \beta / \partial \tau$ is directly obtained as a parametric sensitivity, the second derivative $\partial^2 \beta(\tau)/\partial \tau^2$ is be determined numerically by a simple forward difference scheme for the first order derivatives. As seen, the correction factor $C(s^*,q^*,\tau)$ is assumed to vary only slowly in time and therefore is replaced by $C(s^*,q^*,\tau^*)$. As for normal processes, we have separated integration over τ from the integration over s and q. Application of the method of asymptotic integral approximations then yields

$$\tau^* = t_1 \text{ or } \frac{\partial \beta(\tau)}{\partial \tau} > 0:$$

$$E[N^+(t_1,t_2;r)] \approx C(s^*,q^*,t_1) \sum_{i=1}^{n} \lambda_i(t_1)\, \Phi(-\beta(t_1)) \left\{ \frac{\exp[f'(t_1)\, t_2] - \exp[f'(t_1)\, t_1]}{\exp[f'(t_1)t_1]\, f'(t_1)} \right\} \tag{7.8}$$

$$\tau^* = t_2 \text{ or } \frac{\partial \beta(\tau)}{\partial \tau} < 0:$$

$$E[N^+(t_1,t_2;r)] \approx C(s^*,q^*,t_2) \sum_{i=1}^{n} \lambda_i(t_2)\, \Phi(-\beta(t_2)) \left\{ \frac{\exp[f'(t_2)\, t_2] - \exp[f'(t_2)\, t_1]}{\exp[f'(t_2)\, t_2]\, f'(t_2)} \right\} \tag{7.9}$$

$$t_1 < \tau^* < t_2 \text{ or } \frac{\partial \beta(\tau)}{\partial \tau} = 0 \text{ and } \frac{\partial^2 \beta(\tau)}{\partial \tau^2} > 0:$$

$$E[N^+(t_1,t_2;r)] \approx C(s^*,q^*,\tau^*) \sum_{i=1}^{n} \lambda_i(\tau^*)\, \Phi(-\beta(\tau^*)) \left[\frac{2\pi}{|f''(\tau^*)|} \right]^{1/2}$$

$$\times \left\{ \Phi(|f''(\tau^*)|^{1/2}(t_2 - \tau^*)) - \Phi(|f''(\tau^*)|^{1/2}(t_1 - \tau^*)) \right\} \tag{7.10}$$

As a generalization, the jumps can also be associated with vectors. In this case the summation over the terms in eqs. (7.8) to (7.10) extends only over all jump vectors.

8. Intermittent Processes

An important case is when the contributing processes are intermittent. Below we just refer to some results in the literature for easy reference. According to

Wen (1977), Winterstein (1980), Shinozuka (1981) the stationary outcrossing rate can be given by

$$
\nu^+(a) = \sum_{i=1}^{n} p_1^{(i)} \nu_{(i)} + \sum_{\substack{i=1 \\ i \neq j}}^{n} \sum_{j=1}^{n} p_2^{(i,j)} \nu_{(i,j)} + \sum_{\substack{i=1 \\ i \neq j \neq k}}^{n} \sum_{j=1}^{n} \sum_{k=1}^{n} p_3^{(i,j,k)} \nu_{(i,j,k)}
$$
$$
+ \dots + p_n \nu_{(1,2,\dots,n)} \tag{8.1}
$$

where the probabilities $p_m^{(M)}$ are the conditional probabilities of occurrence of exactly m members of the set M of actions which assumed non–zero values, and $\nu_{(m)}$ are the upcrossing rates if exactly the set M of load is "on". The possible occurrence and thus also exceedence events are subdivided into a exhaustive system of disjoint subevents. A multivariate Poisson renewal process for the occurrence events of the different actions is reasonable, which implies exponential distributions with parameters λ_i for the times between renewals. If it is further assumed that the durations of the action pulses are independently exponentially distributed with parameters μ_i, but truncated at the next renewal one can derive the occurrence probabilities by making use of a result in queuing theory. Note that duration will be truncated at new renewals. With $\rho_i = \lambda_i/\mu_i$ the stationary probability that no pulse is present is (Shinozuka, 1981)

$$
p_0 = 1/\prod_{m=1}^{n} (1 + \rho_m) \tag{8.2a}
$$

and that there are just one, two, three, ... pulses on

$$
p_1^{(i)} = \rho_i/\prod_{m=1}^{n} (1 + \rho_m) \tag{8.2b}
$$

$$
p_2^{(i,j)} = \rho_i \rho_j/\prod_{m=1}^{n} (1 + \rho_m); \; i \neq j \tag{8.2c}
$$

$$
p_2^{(i,j,k)} = \rho_i \rho_j \rho_k/\prod_{m=1}^{n} (1 + \rho_m); \; i \neq j, i \neq k, j \neq k \tag{8.2d}
$$

$$
p_n = \prod_{r=1}^{n} \rho_r/\prod_{m=1}^{n} (1 + \rho_m) \tag{8.2e}
$$

In eq. (8.1) the rates of threshold crossings for combined actions can be significantly larger than if only one action is present. On the other hand, the occurrence probabilities that more than one action is active simultaneously is rapidly decreasing with the number of such short–term actions. Combination of short–term actions with long–term action is just a limiting case $(\lambda_i/\mu_i \rightarrow 1)$. For the case of Erlang–distributed durations with $\rho_i = \lambda_i/(k_i\mu_i)$ and Poissonian arrival times, a similar result has been given by Shinozuka (1981). It can be shown that by   this more sophisticated model only little extra can be achieved, so that the classical Poissonian model should suffice in most practical cases. The mean duration of each load combination is simply t $p_m^{(M)}$ with t the reference period considered.

Slightly different and somewhat simplified models for load coincidences have been given by Winterstein (1980) and Wen (1978). They are especially useful for sparse load pulses. The above results can be generalized in various ways. For example, one may introduce certain dependencies between the various occurrence events. Clustering of events can be considered. This is modeled by a parent Poisson renewal process which generates, with some random delay, children processes with independent marks and independent durations. This increases the probability of overlap in time and thus the probabilities $p_m^{(M)}$ in eq. (8.1) (see Wen, 1990, and Schrupp and Rackwitz, 1988). It is important to note that the sequences of arrivals as well as of durations must be stationary. Within a load coincidence, the processes may, however, be non–stationary and thus may also have non–stationary crossing rates. The load coincidence method is no longer suitable if arrivals and durations form non–stationary sequences.

9. Summary

The theory of stationary crossings into failure domain is well established for a number of important stationary processes. This is not so for the non–stationary case. Two important cases, the scalar non–stationary Gaussian process and non–stationary rectangular renewal processes are discussed in more detail. A device is given to approximately reduce Gaussian vector processes to scalar processes. In both cases use is made of asymptotic concepts. From numerical studies it is found that some simplifying modifications are necessary. In particular, integrations over time need to be performed in a non–asymptotic sense.

10. References

Belyaev, Y.,K., On the Number of Exits across the Boundary of a Region by a Vector Stochastic Process, Theor. Probab. Appl., 1968, 13, pp. 320–324

Bolotin, V.V., Wahrscheinlichkeitsmethoden zur Berechnung von Konstruktionen, VEB Verlag für Bauwesen, Berlin, 1981

Breitung, K., Rackwitz, R., Upcrossing Rates for Rectangular Pulse Load Processes, Ber. Zuverlässigkeitstheorie d. Bauwerke, SFB 96, Technische Universität München, Heft 42, 1979

Breitung, K., Rackwitz, R., Nonlinear Combination of Load Processes, Journ. of Struct. Mech., Vol. 10, No.2, 1982, pp. 145–166

Breitung, K., Asymptotic Approximations for the Maximum of the Sum of Poisson Square Wave Processes, in: Berichte zur Zuverlässigkeitstheorie der Bauwerke, SFB 96, Technische Universität München, Heft 69, 1984, pp. 59–82

Breitung, K., Asymptotic Crossing Rates for Stationary Gaussian Vector Processes, Stochastic Processes and their Applications, 29, 1988, pp. 195–207

Bryla, P., Faber, M.H., Rackwitz, R., Second Order Methods in Time Variant Reliability Problems, Proc. OMAE '91, Stavanger, June, 1990

Cramer, H., Leadbetter, M.R., Stationary and Related Stochastic Processes. Wiley, New York, 1967

Ditlevsen, O., Gaussian Outcrossings from Safe Convex Polyhedrons, Journ. of the Eng. Mech. Div., ASCE, Vol. 109, 1983, pp. 127–148

Engelund, S., Rackwitz, R., C. Lange, Approximations of first—passage times for differentiable processes based on higher—order threshold crossings, Probabilistic Engineering Mechanics, Vol. 10, No.1, 1995, pp. 53—60

Faber, M., Rackwitz, R., Excursion Probabilities of Non—Homogeneous Gaussian Vector Fields based on Maxima Considerations, Proc. 2nd IFIP WG—7.5 "Reliability and Optimization of Structural Systems", London, September, 1988, pp. 117—134

Ferry Borges, J., Castanheta, M., Structural Safety, Laboratorio Nacional de Engenharia Civil, Lisbon, 1971

Fujita, M., Schall, G., Rackwitz,R., Time—Variant Component Reliabilities by FORM/SORM and Updating by Importance Sampling, Proc. ICASP 5, Vancouver, May, 1987, Vol. I, pp. 520—527

Hagen, O., Threshold Up—crossing by Second Order Methods, Probabilistic Engineering Mechanics, Vol. 7, No.4, 1992, pp. 235—241

Lange, C., First Excursion Probabilities for Low Threshold Levels by Differentiable Processes, Proc. 4th IFIP WG 7.5 Conference, München, September 1991, Springer—Verlag Berlin, 1992

Leadbetter, M.R., Lindgren, G., Rootzen, H., Extremes and Related Properties of Random Sequences and Processes, Springer—Verlag, New York, 1983

Madsen, H.O., Krenk, S., Lind, N.C., Methods of Structural Safety, Prentice—Hall, Englewood—Cliffs, 1986

Naess, A., On the Long—term Statistics of Extremes, Appl. Ocean Res., Vol. 6, 4, 1984, pp. 227—228

Plantec, J.—Y., Rackwitz, R., Structural Reliability under Non—Stationary Gaussian Vector Process Loads, Technische Universität München, Berichte zur Zuverlässigkeitstheorie der Bauwerke, Heft 85, 1989

Rackwitz, R., and Fiessler, B., Structural Reliability under Combined Random Load Sequences, Comp. & Struct., 1978, 9, 484—494

Rackwitz, R., On the Combination of Non—stationary Rectangular Wave Renewal Processes, Structural Safety, Vol. 13, 1+2, 1993, pp. 21—28

Rice, S.O., Mathematical Analysis of Random Noise, Bell System Tech. Journ., 32, 1944, pp. 282 and 25, 1945, pp. 46

Schall, G., Faber, M., Rackwitz, R., The Ergodicity Assumption for Sea States in the Reliability Assessment of Offshore Structures, Journal of Offshore Mechanics and Arctic Engineering, Transact. ASME,1991, Vol. 113, No. 3, pp. 241—246

Shinozuka, M., Stochastic Characterization of Loads and Load Combinations, Proc. 3rd ICOSSAR, Trondheim 32—25 June, 1981, Structural Safety and Reliability, T. Moan and M. Shinozuka (Eds.), Elsevier, Amsterdam, 1981

Veneziano, D., Grigoriu, M., Cornell, C.A., Vector—Process Models for System Reliability, Journ. of Eng. Mech. Div., ASCE, Vol. 103, EM 3, 1977, pp. 441—460

Wen, Y.K., Statistical Combination of Extreme Loads, Journ. of the Struct. Div., ASCE, Vol. 103, ST5, 1977, pp. 1079—1093

Wen, Y.K., A Clustering Model for Correlated Load Processes, Journ. of the Struct. Div., ASCE, Vol. 107, ST5, 1981, pp. 965—983

Wen, Y.—K., Structural Load Modeling and Combination for Performance and Safety Evaluation, Elsevier, Amsterdam, 1990

Winterstein, S.R., Stochastic Dynamic Response Combination, MIT, Cambridge, Ph.D. Thesis, 1981

SIMULATION OF STOCHASTIC PROCESSES AND FIELDS TO MODEL LOADING AND MATERIAL UNCERTAINTIES

G. DEODATIS

Department of Civil Engineering and Operations Research
Princeton University, Princeton, NJ 08544, USA

1. Introduction

Several methods are currently available to solve a large number of problems in mechanics involving uncertain quantities described by stochastic processes, fields or waves. At this time, however, Monte Carlo simulation appears to be the only universal method that can provide accurate solutions for certain problems in stochastic mechanics involving nonlinearity, system stochasticity, stochastic stability, parametric excitations, large variations of uncertain parameters, etc., and that can assess the accuracy of other approximate methods such as perturbation, statistical linearization, closure techniques, stochastic averaging, etc. The major advantage of Monte Carlo simulation is that accurate solutions can be obtained for any problem whose deterministic solution (either analytical or numerical) is known. The only disadvantage of Monte Carlo simulation is that it is usually time-consuming. It is the author's belief, however, that in the years to come, the continued evolution of digital computers will further enhance the usefulness of Monte Carlo simulation techniques in the area of engineering mechanics and structural engineering.

One of the most important parts of the Monte Carlo simulation methodology is the generation of sample functions of the stochastic processes, fields or waves involved in the problem. The generated sample functions must accurately describe the probabilistic characteristics of the corresponding stochastic processes, fields or waves that may be either stationary or non-stationary, homogeneous or non-homogeneous, one-dimensional or multi-dimensional, uni-variate or multi-variate, Gaussian or non-Gaussian. From the rich bibliography related to the various methods currently available for generating such sample functions, the following representative publications are mentioned here: Shinozuka and Jan 1972 (introduction of the spectral representation method), Gersch and Yonemoto 1977 (multi-variate ARMA model), Elishakoff 1979 (covariance decomposition method to simulate imperfections), Polhemus and Cakmak 1981 (ARMA model with frequency and amplitude modulation to account for non-stationarity in ground motion), Spanos

C. Guedes Soares (ed.), Probabilistic Methods for Structural Design, 261–288.

and Hansen 1981 (sea wave simulation by passing white noise through a recursive digital filter), Spanos 1983 (ARMA model for ocean waves), Elishakoff 1983 (book with chapter on simulation), Samaras et al. 1985 (multi-variate ARMA model), Mignolet and Spanos 1987a, 1987b (stability and invertibility aspects of AR to ARMA procedures for multi-variate random processes), Kozin 1988 (review paper on ARMA models), Elishakoff 1988 (covariance decomposition method for two-dimensional non-homogeneous fields), Hasofer 1989 (time- and frequency-domain hybrid simulation method), Yamazaki and Shinozuka 1990 (covariance decomposition method with statistical preconditioning), Fenton and Vanmarcke 1990 (simulation of stochastic fields by local average subdivision), Winterstein 1990 (simulation of stochastic processes using Fast Hartley Transform), Kareem and Li 1991 (simulation of non-stationary vector processes using an FFT-based approach), Li and Kareem 1993a (simulation of multi-variate processes using a hybrid discrete Fourier transform and digital filtering approach), Li and Kareem 1993b (ARMA, discrete convolution, discrete differentiation and discrete interpolation models for ocean wave processes), Soong and Grigoriu 1993 (book with chapter on simulation), Grigoriu 1993b (simulation of non-stationary processes by random trigonometric polynomials), Grigoriu and Balopoulou 1993 (simulation method based on sampling theorem), Ramadan and Novak 1993, 1994 (asymptotic and approximate spectral techniques to simulate multi-dimensional, multi-variate and anisotropic processes and fields), Gurley and Kareem 1995 (simulation of random processes using higher order spectra and wavelet transforms), Spanos and Zeldin 1995 (simulation of random fields using wavelet bases), Grigoriu 1995 (book with chapter on simulation of Gaussian and non-Gaussian processes).

Among the various methods mentioned above to generate sample functions of stochastic processes, fields and waves, the spectral representation method is one of the most widely used today. Although the concept of the method existed for some time (Rice 1954), it was Shinozuka (Shinozuka and Jan 1972, Shinozuka 1972) who first applied it for simulation purposes including multi-dimensional, multi-variate and non-stationary cases. Yang (1972, 1973) showed that the Fast Fourier Transform (FFT) technique can be used to dramatically improve the computational efficiency of the spectral representation algorithm, and proposed a formula to simulate random envelope processes. Shinozuka (1974) extended the application of the FFT technique to multi-dimensional cases. Recently, Deodatis and Shinozuka (1989) further extended the spectral representation method to simulate stochastic waves, Yamazaki and Shinozuka (1988) developed an iterative procedure to simulate non-Gaussian stochastic fields, and Grigoriu (1993a) compared two different spectral representation models. Three recent review papers on the subject of simulation using the spectral representation method were written by Shinozuka (1987) and Shinozuka and Deodatis (1988 and 1991).

In this paper, spectral representation algorithms are provided to simulate: (1) one-dimensional, uni-variate (1D-1V) stationary stochastic processes, (2) multi-dimensional, uni-variate (nD-1V) homogeneous stochastic fields, (3) one-dimensional, multi-variate (1D-mV) stationary stochastic processes, and (4) one-dimensional, multi-variate (1D-mV) non-stationary stochastic processes. Numerical examples are provided for three of the above four cases. Specifically, an example involving material properties (e.g. strength, elastic modulus, density, etc.) is provided for the second case, an example involving wind velocity at several locations is provided for the third case and an example involving non-stationary ground motion at several locations on the ground surface is provided for the fourth case.

2. Simulation of 1D-1V Stationary Stochastic Processes

Let $f_0(t)$ be a 1D-1V stationary stochastic process with mean value equal to zero, autocorrelation function $R_{f_0 f_0}(\tau)$ and two-sided power spectral density function $S_{f_0 f_0}(\omega)$. Then, the following relations hold:

$$\mathcal{E}[f_0(t)] = 0 \tag{1}$$

$$\mathcal{E}[f_0(t+\tau)f_0(t)] = R_{f_0 f_0}(\tau) \tag{2}$$

$$S_{f_0 f_0}(\omega) = \frac{1}{2\pi} \int_{-\infty}^{\infty} R_{f_0 f_0}(\tau)e^{-i\omega\tau} d\tau \tag{3}$$

$$R_{f_0 f_0}(\tau) = \int_{-\infty}^{\infty} S_{f_0 f_0}(\omega)e^{i\omega\tau} d\omega \tag{4}$$

where the last two equations constitute the well-known Wiener-Khintchine transform pair.

In the following, distinction will be made between the stochastic process $f_0(t)$ and its simulation $f(t)$. The stochastic process $f_0(t)$ can be simulated by the following series (Shinozuka and Deodatis 1991) as $N \rightarrow \infty$;

$$f(t) = \sqrt{2} \sum_{n=0}^{N-1} A_n \cdot \cos(\omega_n t + \Phi_n) \tag{5}$$

where:

$$A_n = \sqrt{2S_{f_0 f_0}(\omega_n)\Delta\omega} \quad ; \quad n = 0, 1, 2, \ldots, N-1 \tag{6}$$

$$\omega_n = n\Delta\omega \quad ; \quad n = 0, 1, 2, \ldots, N-1 \tag{7}$$

$$\Delta\omega = \frac{\omega_u}{N} \tag{8}$$

and:

$$A_0 = 0 \quad \text{or} \quad S_{f_0 f_0}(\omega_0 = 0) = 0 \tag{9}$$

In Eq. (8), ω_u represents an upper cut-off frequency beyond which the power spectral density function $S_{f_0 f_0}(\omega)$ may be assumed to be zero for either mathematical or physical reasons. As such, ω_u is a fixed value and hence $\Delta\omega \to 0$ as $N \to \infty$ so that $N\Delta\omega = \omega_u$. The following criterion is usually used to estimate the value of ω_u:

$$\int_0^{\omega_u} S_{f_0 f_0}(\omega)d\omega = (1 - \epsilon) \int_0^\infty S_{f_0 f_0}(\omega)d\omega \tag{10}$$

where $\epsilon << 1$ (e.g. $\epsilon = 0.01, 0.001$).

The $\Phi_0, \Phi_1, \Phi_2, \ldots, \Phi_{N-1}$ appearing in Eq. (5) are independent random phase angles distributed uniformly over the interval $[0, 2\pi]$.

The condition set in Eq. (9) is necessary (and must be forced if $S_{f_0 f_0}(0) \neq 0$) to guarantee that the temporal average and the temporal autocorrelation function of any sample function are identical to the corresponding targets, $\mathcal{E}[f_0(t)] = 0$ and $R_{f_0 f_0}(\tau)$, respectively (Shinozuka and Deodatis 1991). However, in order to avoid having to impose the condition shown in Eq. (9), the alternative of using the frequency shifting theorem (e.g. Papoulis 1962) was proposed by Zerva (1992), but with the side effect of doubling the period of the simulated field.

Under the condition of Eq. (9), it is easy to show that the simulated stochastic process $f(t)$ given by Eq. (5) is periodic with period T_0:

$$T_0 = \frac{2\pi}{\Delta\omega} \tag{11}$$

Equation (11) indicates that the smaller $\Delta\omega$, or equivalently the larger N under a specified upper cut-off frequency value ω_u, the longer the period of the simulated stochastic process.

Another very important point is that the simulated stochastic process $f(t)$ is asymptotically Gaussian as $N \to \infty$ because of the multi-variate central limit theorem (Shinozuka and Deodatis 1991).

At this point it should be noted that when generating sample functions of the simulated stochastic process according to Eq. (5), the time step Δt separating the generated values in the time domain has to obey the condition:

$$\Delta t \leq \frac{2\pi}{2\omega_u} \tag{12}$$

The condition set on Δt in Eq. (12) is necessary in order to avoid aliasing according to the sampling theorem (e.g. Bracewell 1986).

It can be shown (Shinozuka and Deodatis 1991) that the ensemble expected value $\mathcal{E}[f(t)]$ and the ensemble autocorrelation function $R_{ff}(\tau)$ of the simulated stochastic process $f(t)$ are identical to the corresponding targets, $\mathcal{E}[f_0(t)] = 0$ and $R_{f_0 f_0}(\tau)$, respectively.

Another very important property of the simulated stochastic process is the following: each and every sample function generated using Eq. (5) is ergodic in the mean value and in correlation. It can be shown (Shinozuka and Deodatis 1991) that the temporal average and the temporal autocorrelation function of any sample function are identical to the corresponding targets, $\mathcal{E}[f_0(t)] = 0$ and $R_{f_0 f_0}(\tau)$, respectively. In addition, it can be shown that these two identities are valid only when the length of the sample function is either equal to the period T_0 or when it approaches infinity.

Finally, it should be mentioned that the cost of digitally generating sample functions of the simulated stochastic process using Eq. (5) can be drastically reduced by using the FFT technique (Shinozuka and Deodatis 1991).

3. Simulation of nD-1V Homogeneous Stochastic Fields

In this section, the theory and the simulation algorithm for multi-dimensional stochastic fields are presented for the special case of two-dimensional fields. This is done for the sake of simplicity in the notation and because of space limitations. For the (straightforward) extensions to three-dimensional and multi-dimensional fields, the reader is referred to Shinozuka and Deodatis (1995).

Consider a 2D-1V homogeneous stochastic field $f_0(x_1, x_2)$ with mean value equal to zero, autocorrelation function $R_{f_0 f_0}(\xi_1, \xi_2)$ and power spectral density function $S_{f_0 f_0}(\kappa_1, \kappa_2)$.

For every 2D-1V homogeneous stochastic field, the following relations concerning the symmetry of $R_{f_0 f_0}(\xi_1, \xi_2)$ and $S_{f_0 f_0}(\kappa_1, \kappa_2)$ are valid:

$$R_{f_0 f_0}(\xi_1, \xi_2) = R_{f_0 f_0}(-\xi_1, -\xi_2) \tag{13}$$

$$S_{f_0 f_0}(\kappa_1, \kappa_2) = S_{f_0 f_0}(-\kappa_1, -\kappa_2) \tag{14}$$

For the special case of a "quadrant" 2D-1V homogeneous stochastic field (Vanmarcke 1983), the following relations concerning the symmetry of $R_{f_0 f_0}(\xi_1, \xi_2)$ and $S_{f_0 f_0}(\kappa_1, \kappa_2)$ are valid:

$$R_{f_0 f_0}(\xi_1, \xi_2) = R_{f_0 f_0}(\xi_1, -\xi_2) = R_{f_0 f_0}(-\xi_1, \xi_2) = R_{f_0 f_0}(-\xi_1, -\xi_2) \tag{15}$$

$$S_{f_0 f_0}(\kappa_1, \kappa_2) = S_{f_0 f_0}(\kappa_1, -\kappa_2) = S_{f_0 f_0}(-\kappa_1, \kappa_2) = S_{f_0 f_0}(-\kappa_1, -\kappa_2) \tag{16}$$

In the following, distinction will be made between the stochastic field $f_0(x_1, x_2)$ and its simulation $f(x_1, x_2)$.

It can be shown (Shinozuka and Deodatis 1995) that a stochastic field $f_0(x_1, x_2)$, with $R_{f_0 f_0}(\xi_1, \xi_2)$ and $S_{f_0 f_0}(\kappa_1, \kappa_2)$ following the symmetries described in Eqs. (13) and (14), can be simulated by the following series as $N_1, N_2 \to \infty$ simultaneously:

$$f(x_1, x_2) = \sqrt{2} \sum_{n_1=0}^{N_1-1} \sum_{n_2=0}^{N_2-1} \left[A_{n_1 n_2} \cos \left(\kappa_{1 n_1} x_1 + \kappa_{2 n_2} x_2 + \Phi_{n_1 n_2}^{(1)} \right) + \right.$$
$$\left. + \tilde{A}_{n_1 n_2} \cos \left(\kappa_{1 n_1} x_1 - \kappa_{2 n_2} x_2 + \Phi_{n_1 n_2}^{(2)} \right) \right] \tag{17}$$

where:

$$A_{n_1 n_2} = \sqrt{2 S_{f_0 f_0}(\kappa_{1n_1}, \kappa_{2n_2}) \Delta\kappa_1 \Delta\kappa_2} \tag{18}$$

$$\tilde{A}_{n_1 n_2} = \sqrt{2 S_{f_0 f_0}(\kappa_{1n_1}, -\kappa_{2n_2}) \Delta\kappa_1 \Delta\kappa_2} \tag{19}$$

$$\kappa_{1n_1} = n_1 \Delta\kappa_1 \quad ; \quad \kappa_{2n_2} = n_2 \Delta\kappa_2 \tag{20}$$

$$\Delta\kappa_1 = \frac{\kappa_{1u}}{N_1} \quad ; \quad \Delta\kappa_2 = \frac{\kappa_{2u}}{N_2} \tag{21}$$

and:

$$A_{0n_2} = A_{n_1 0} = 0 \quad \text{for} \quad n_1 = 0, 1, \ldots, N_1 - 1 \quad \text{and} \quad n_2 = 0, 1, \ldots, N_2 - 1 \tag{22}$$

$$\tilde{A}_{0n_2} = \tilde{A}_{n_1 0} = 0 \quad \text{for} \quad n_1 = 0, 1, \ldots, N_1 - 1 \quad \text{and} \quad n_2 = 0, 1, \ldots, N_2 - 1 \tag{23}$$

Equations (22) and (23) are equivalent to:

$$S_{f_0 f_0}(0, \kappa_2) = S_{f_0 f_0}(\kappa_1, 0) = 0 \quad \text{for} \quad -\infty < \kappa_1 < \infty \quad \text{and} \quad -\infty < \kappa_2 < \infty \tag{24}$$

In Eq. (21), κ_{1u} and κ_{2u} are the upper cut-off wave numbers corresponding to the x_1 and x_2 axes in the space domain, respectively (refer also to Section 2 about a similar discussion for 1D-1V processes). This implies that the power spectral density function $S_{f_0 f_0}(\kappa_1, \kappa_2)$ is assumed to be zero, for either mathematical or physical reasons, outside the region defined by:

$$-\kappa_{1u} \leq \kappa_1 \leq \kappa_{1u} \quad \text{and} \quad -\kappa_{2u} \leq \kappa_2 \leq \kappa_{2u} \tag{25}$$

The $\Phi_{n_1 n_2}^{(1)}$ and $\Phi_{n_1 n_2}^{(2)}$; $n_1 = 0, 1, \ldots, N_1 - 1$; $n_2 = 0, 1, \ldots, N_2 - 1$ appearing in Eq. (17) are independent random phase angles distributed uniformly over the interval $[0, 2\pi]$.

The conditions set in Eqs. (22) and (23) are necessary (and must be enforced if $S_{f_0 f_0}(0, \kappa_2) = S_{f_0 f_0}(\kappa_1, 0) \neq 0$) to guarantee that the spatial average and the spatial autocorrelation function of any sample function are identical to the corresponding targets, $\mathcal{E}[f_0(x_1, x_2)] = 0$ and $R_{f_0 f_0}(\xi_1, \xi_2)$, respectively (Shinozuka and Deodatis 1995). However, in order to avoid having to impose the conditions shown in Eqs. (22) and (23), the alternative of using the frequency shifting theorem (e.g. Papoulis 1962) was proposed by Zerva (1992), but with the side effect of quadrupling the two-dimensional period of the simulated field.

It is straightforward to show that the simulated stochastic field $f(x_1, x_2)$ given by Eq. (17) is periodic along the x_1 and x_2 axes with periods:

$$L_{x_1 0} = \frac{2\pi}{\Delta\kappa_1} \quad \text{along the } x_1 \text{ axis} \tag{26}$$

$$L_{x_2 0} = \frac{2\pi}{\Delta\kappa_2} \quad \text{along the } x_2 \text{ axis} \tag{27}$$

Equation (26) indicates that the smaller $\Delta\kappa_1$, or equivalently the larger N_1 under a specified upper cut-off wave number value κ_{1u}, the longer the period of the simulated stochastic field along the x_1 axis. A similar conclusion can be drawn from Eq. (27) concerning the period of the simulated stochastic field along the x_2 axis.

Another very important point is that the simulated stochastic field $f(x_1, x_2)$ is asymptotically Gaussian as $N_1, N_2 \to \infty$ simultaneously, because of the central limit theorem (Shinozuka and Deodatis 1991).

At this point it should be noted that when generating sample functions of the simulated stochastic field according to Eq. (17), the space increments Δx_1 and Δx_2 along the x_1 and x_2 axes respectively, separating the generated values in the two-dimensional space domain, have to obey the conditions:

$$\Delta x_1 \leq \frac{2\pi}{2\kappa_{1u}} \quad ; \quad \Delta x_2 \leq \frac{2\pi}{2\kappa_{2u}} \qquad (28)$$

The conditions set on Δx_1 and Δx_2 in Eq. (28) are necessary in order to avoid aliasing according to the sampling theorem (e.g. Bracewell 1986).

It can be shown (Shinozuka and Deodatis 1995) that the ensemble expected value $\mathcal{E}[f(x_1, x_2)]$ and the ensemble autocorrelation function $R_{ff}(\xi_1, \xi_2)$ of the simulated stochastic field $f(x_1, x_2)$ are identical to the corresponding targets, $\mathcal{E}[f_0(x_1, x_2)] = 0$ and $R_{f_0 f_0}(\xi_1, \xi_2)$, respectively.

Another very important property of the simulated stochastic field is the following: each and every sample function generated by Eq. (17) is ergodic in the mean value and in correlation. It can be shown (Shinozuka and Deodatis 1995) that the spatial average and the spatial autocorrelation function of any sample function are identical to the corresponding targets, $\mathcal{E}[f_0(x_1, x_2)] = 0$ and $R_{f_0 f_0}(\xi_1, \xi_2)$, respectively, only when the rectangular area $L_{x_1} \times L_{x_2}$ over which the sample function is simulated is either equal to the period $\left(L_{x_1} = L_{x_1 0} \text{ and } L_{x_2} = L_{x_2 0}\right)$ or when it approaches infinity $\left(L_{x_1} \to \infty \text{ and } L_{x_2} \to \infty\right)$.

Finally, it should be mentioned that the cost of digitally generating sample functions of the simulated stochastic field using Eq. (17) can be drastically reduced by using the FFT technique (Shinozuka and Deodatis 1995).

3.1 SPECIAL CASE – QUADRANT FIELDS

Consider a quadrant, 2D-1V homogeneous stochastic field $f_0(x_1, x_2)$ with mean value equal to zero and autocorrelation and power spectral density functions possessing the symmetries described in Eqs. (15) and (16), respectively. Such a quadrant field can be simulated by the following series as $N_1, N_2 \to \infty$ simultaneously:

$$f(x_1, x_2) = \sqrt{2} \sum_{n_1=0}^{N_1-1} \sum_{n_2=0}^{N_2-1} A_{n_1 n_2} \left[\cos\left(\kappa_{1n_1} x_1 + \kappa_{2n_2} x_2 + \Phi_{n_1 n_2}^{(1)} \right) \right.$$

$$+\cos\left(\kappa_{1n_1}x_1 - \kappa_{2n_2}x_2 + \Phi^{(2)}_{n_1n_2}\right)\right] \qquad (29)$$

Note that the simulation formula for quadrant fields (Eq. (29)) is slightly simpler than the corresponding formula for the general case (Eq. (17)), taking advantage of the symmetry of $S_{f_0f_0}(\kappa_1, \kappa_2)$ described in Eq. (16). All parameters appearing in Eq. (29) have already been defined with reference to Eq. (17).

3.2 NUMERICAL EXAMPLES

Consider a two-dimensional homogeneous stochastic field $f_0(x_1, x_2)$ with mean value equal to zero, autocorrelation function $R_{f_0f_0}(\xi_1, \xi_2)$ given by:

$$R_{f_0f_0}(\xi_1, \xi_2) = \sigma^2 \exp\left[-\left(\frac{\xi_1}{b_1}\right)^2 - \left(\frac{\xi_2}{b_2}\right)^2\right], \quad -\infty < \xi_1 < \infty \text{ and } -\infty < \xi_2 < \infty \tag{30}$$

and corresponding power spectral density function $S_{f_0f_0}(\kappa_1, \kappa_2)$ given by:

$$S_{f_0f_0}(\kappa_1, \kappa_2) = \sigma^2 \frac{b_1 b_2}{4\pi} \exp\left[-\left(\frac{b_1\kappa_1}{2}\right)^2 - \left(\frac{b_2\kappa_2}{2}\right)^2\right],$$
$$-\infty < \kappa_1 < \infty \text{ and } -\infty < \kappa_2 < \infty \tag{31}$$

In Eqs. (30) and (31), parameter σ is the standard deviation of stochastic field $f_0(x_1, x_2)$, while parameters b_1 and b_2 are proportional to the correlation distance of the stochastic field along the x_1 and x_2 axes, respectively.

At this juncture it should be mentioned that the expression for the power spectral density function shown in Eq. (31) has been used by Shinozuka and Lenoe (1976) to model the spatial variation of the (random) fluctuations of material strength around a constant mean value.

The following three cases are now selected for demonstration purposes:

Case 1 : $\sigma = 1$, $b_1 = 1$ m, $b_2 = 1$ m, $\kappa_{1u} = 5$ rad/m, $\kappa_{2u} = 5$ rad/m (32a)

Case 2 : $\sigma = 1$, $b_1 = 4$ m, $b_2 = 4$ m, $\kappa_{1u} = 1.25$ rad/m, $\kappa_{2u} = 1.25$ rad/m (32b)

Case 3 : $\sigma = 1$, $b_1 = 1$ m, $b_2 = 4$ m, $\kappa_{1u} = 5$ rad/m, $\kappa_{2u} = 1.25$ rad/m (32c)

with the upper cut-off wave numbers κ_{1u} and κ_{2u} having been defined in Eq. (25). If parameters N_1 and N_2 appearing in Eq. (21) are set equal to:

$$N_1 = 16 \quad ; \quad N_2 = 16 \tag{33}$$

then $\Delta\kappa_1$, $\Delta\kappa_2$ and L_{x_10}, L_{x_20} are computed from Eqs. (21), (26) and (27), respectively, as:

Case 1 : $\Delta\kappa_1 = \Delta\kappa_2 = 0.3125$ rad/m, $L_{x_10} = L_{x_20} = 20.1$ m (34a)

Case 2 : $\Delta\kappa_1 = \Delta\kappa_2 = 0.0781$ rad/m, $L_{x_10} = L_{x_20} = 80.4$ m (34b)

Case 3 : $\Delta\kappa_1 = 0.3125$ rad/m, $\Delta\kappa_2 = 0.0781$ rad/m, $L_{x_10} = 20.1$ m,

$$L_{x_20} = 80.4 \text{ m} \tag{34c}$$

One sample function of stochastic field $f_0(x_1, x_2)$ is generated for each one of the above three cases by applying the FFT technique (Shinozuka and Deodatis 1995) on Eq. (29). The generation is performed over an area equal to one period as specified in Eqs. (26), (27) and (34). The space increments Δx_1 and Δx_2 are set equal to:

$$\text{Cases 1, 2 and 3}: \quad \Delta x_1 = \pi/10 \text{ m}, \quad \Delta x_2 = \pi/10 \text{ m} \tag{35}$$

in order to satisfy the conditions shown in Eq. (28).

Figure 1 shows one sample function of stochastic field $f_0(x_1, x_2)$ for each one of the three cases defined in Eq. (32). Although each sample function is generated over a different area that is equal to the corresponding period (see Eq. (34)), they are all plotted over the same area of 20.1 m × 20.1 m in Fig. 1. This is done in order to demonstrate the effect of parameters b_1 and b_2 on the correlation structure of the stochastic field. Such a demonstration can only be achieved if each one of the three sample functions is plotted over the same area. Note that this 20.1 m × 20.1 m area is equal to one period for Case 1 (see Eq. (34a)), but it is less than one period for Cases 2 and 3 (see Eqs. (34b) and (34c), respectively). Note also that all three plots in Fig. 1 are using the same grid of 64×64 points.

4. Simulation of 1D-mV Stationary Stochastic Processes

In this section, the theory and the simulation algorithm for multi-variate stochastic processes are presented for the special case of a tri-variate vector process. This is done for the sake of simplicity in the notation and because of space limitations. The proposed algorithm can be extended in a straightforward fashion to any other dimension of the vector process.

Consider a 1D-3V (one-dimensional, tri-variate) stationary stochastic vector process with components $f_1^0(t)$, $f_2^0(t)$ and $f_3^0(t)$, having mean value equal to zero:

$$\mathcal{E}[f_j^0(t)] = 0 \quad ; \quad j = 1, 2, 3 \tag{36}$$

cross-correlation matrix given by:

$$\mathbf{R}^0(\tau) = \begin{bmatrix} R_{11}^0(\tau) & R_{12}^0(\tau) & R_{13}^0(\tau) \\ R_{21}^0(\tau) & R_{22}^0(\tau) & R_{23}^0(\tau) \\ R_{31}^0(\tau) & R_{32}^0(\tau) & R_{33}^0(\tau) \end{bmatrix} \tag{37}$$

and cross-spectral density matrix given by:

$$\mathbf{S}^0(\omega) = \begin{bmatrix} S_{11}^0(\omega) & S_{12}^0(\omega) & S_{13}^0(\omega) \\ S_{21}^0(\omega) & S_{22}^0(\omega) & S_{23}^0(\omega) \\ S_{31}^0(\omega) & S_{32}^0(\omega) & S_{33}^0(\omega) \end{bmatrix} \tag{38}$$

In Eq. (37), $R_{jj}^0(\tau)$; $j = 1, 2, 3$ are the auto-correlation functions of the three components $f_j^0(t)$; $j = 1, 2, 3$ of the process and $R_{jk}^0(\tau)$; $j = 1, 2, 3$; $k = 1, 2, 3$; $j \neq k$ are the corresponding cross-correlation functions. Due to the stationarity hypothesis, the following relations are valid:

$$R_{jj}^0(\tau) = R_{jj}^0(-\tau) \quad ; \quad j = 1, 2, 3 \tag{39}$$

$$R_{jk}^0(\tau) = R_{kj}^0(-\tau) \quad ; \quad j = 1, 2, 3; \ k = 1, 2, 3; \ j \neq k \tag{40}$$

The elements of the cross-correlation matrix are related to the corresponding elements of the cross-spectral density matrix through the Wiener-Khintchine transformation (τ is the time lag and ω is the frequency):

$$S_{jk}^0(\omega) = \frac{1}{2\pi} \int_{-\infty}^{\infty} R_{jk}^0(\tau) \, e^{-i\omega\tau} d\tau \quad ; \quad j, k = 1, 2, 3 \tag{41}$$

$$R_{jk}^0(\tau) = \int_{-\infty}^{\infty} S_{jk}^0(\omega) \, e^{i\omega\tau} d\omega \quad ; \quad j, k = 1, 2, 3 \tag{42}$$

In Eq. (38), $S_{jj}^0(\omega)$; $j = 1, 2, 3$ are the power spectral density functions of the three components of the process and $S_{jk}^0(\omega)$; $j = 1, 2, 3$; $k = 1, 2, 3$; $j \neq k$ are the corresponding cross-spectral density functions. It is noted that while the power spectral density function is a real and non-negative function of ω, the cross-spectral density function is a generally complex function of ω. Because of Eqs. (39)-(42), the following relations are valid:

$$S_{jj}^0(\omega) = S_{jj}^0(-\omega) \quad ; \quad j = 1, 2, 3 \tag{43}$$

$$S_{jk}^0(\omega) = S_{jk}^{0*}(-\omega) \quad ; \quad j = 1, 2, 3; \ k = 1, 2, 3; \ j \neq k \tag{44}$$

$$S_{jk}^0(\omega) = S_{kj}^{0*}(\omega) \quad ; \quad j = 1, 2, 3; \ k = 1, 2, 3; \ j \neq k \tag{45}$$

where the asterisk denotes the complex conjugate. Equation (45) indicates that the cross-spectral density matrix $\mathbf{S}^0(\omega)$ is Hermitian. It can also be shown (Shinozuka 1987) that matrix $\mathbf{S}^0(\omega)$ is non-negative definite.

In the following, distinction will be made between the stochastic vector process $f_j^0(t)$; $j = 1, 2, 3$ and its simulation $f_j(t)$; $j = 1, 2, 3$.

In order to simulate the 1D-3V stationary stochastic process $f_j^0(t)$; $j = 1, 2, 3$, its cross-spectral density matrix $\mathbf{S}^0(\omega)$ must be first decomposed into the following product:

$$\mathbf{S}^0(\omega) = \mathbf{H}(\omega) \, \mathbf{H}^{T*}(\omega) \tag{46}$$

where superscript ^{T} denotes the transpose of a matrix. This decomposition can be performed using Cholesky's method, in which case $\mathbf{H}(\omega)$ is a lower triangular matrix:

$$\mathbf{H}(\omega) = \begin{bmatrix} H_{11}(\omega) & 0 & 0 \\ H_{21}(\omega) & H_{22}(\omega) & 0 \\ H_{31}(\omega) & H_{32}(\omega) & H_{33}(\omega) \end{bmatrix} \tag{47}$$

whose diagonal elements are real and non-negative functions of ω and whose off-diagonal elements are generally complex functions of ω. The following relations are valid for the elements of matrix $\mathbf{H}(\omega)$:

$$H_{jj}(\omega) = H_{jj}(-\omega) \quad ; \quad j = 1, 2, 3 \tag{48}$$

$$H_{jk}(\omega) = H_{jk}^*(-\omega) \quad ; \quad j = 2, 3; \ k = 1, 2; \ j > k \tag{49}$$

If the off-diagonal elements $H_{jk}(\omega)$ are written in polar form as:

$$H_{jk}(\omega) = |H_{jk}(\omega)| \, e^{i\theta_{jk}(\omega)} \quad ; \quad j = 2, 3; \ k = 1, 2; \ j > k \tag{50}$$

where:

$$\theta_{jk}(\omega) = \tan^{-1} \left(\frac{\text{Im}[H_{jk}(\omega)]}{\text{Re}[H_{jk}(\omega)]} \right) \tag{51}$$

with Im and Re denoting the imaginary and the real part of a complex number, respectively, then Eq. (49) is written equivalently as:

$$|H_{jk}(\omega)| = |H_{jk}(-\omega)| \quad ; \quad j = 2, 3; \ k = 1, 2; \ j > k \tag{52}$$

$$\theta_{jk}(\omega) = -\theta_{jk}(-\omega) \quad ; \quad j = 2, 3; \ k = 1, 2; \ j > k \tag{53}$$

Once matrix $\mathbf{S}^0(\omega)$ is decomposed according to Eqs. (46)-(47), the stochastic process $f_j^0(t)$; $j = 1, 2, 3$ can be simulated by the following series (Deodatis 1995a) as $N \to \infty$

$$f_j(t) = 2 \sum_{m=1}^{3} \sum_{l=1}^{N} |H_{jm}(\omega_{ml})| \sqrt{\Delta\omega} \, \cos[\omega_{ml} \, t - \theta_{jm}(\omega_{ml}) + \Phi_{ml}] \quad ; \quad j = 1, 2, 3 \tag{54}$$

where:

$$\omega_{1l} = l\Delta\omega - \frac{2}{3}\Delta\omega \quad ; \quad l = 1, 2, \dots, N \tag{55a}$$

$$\omega_{2l} = l\Delta\omega - \frac{1}{3}\Delta\omega \quad ; \quad l = 1, 2, \dots, N \tag{55b}$$

$$\omega_{3l} = l\Delta\omega \quad\quad\quad\quad ; \quad l = 1, 2, \dots, N \tag{55c}$$

$$\Delta\omega = \frac{\omega_u}{N} \tag{56}$$

$$\theta_{jm}(\omega_{ml}) = \tan^{-1} \left(\frac{\text{Im}[H_{jm}(\omega_{ml})]}{\text{Re}[H_{jm}(\omega_{ml})]} \right) \tag{57}$$

In Eq. (56), ω_u represents an upper cut-off frequency beyond which the elements of the cross-spectral density matrix (Eq. (38)) may be assumed to be zero for either mathematical or physical reasons (refer also to section 2 about a similar discussion for 1D-1V processes).

The Φ_{1l}, Φ_{2l}, Φ_{3l}; $l = 1, 2, \ldots, N$ appearing in Eq. (54) are three sequences of independent random phase angles distributed uniformly over the interval $[0, 2\pi]$.

It is a straightforward task to show that the following three summations:

$$\sum_{l=1}^{N} A_{1l} \cos[\omega_{1l}\, t + \rho_{1l}] \quad ; \quad \sum_{l=1}^{N} A_{2l} \cos[\omega_{2l}\, t + \rho_{2l}] \quad ; \quad \sum_{l=1}^{N} A_{3l} \cos[\omega_{3l}\, t + \rho_{3l}] \tag{58}$$

where the A's and the ρ's denote amplitudes and phase angles respectively, are periodic with respective periods:

$$3\,\frac{2\pi}{\Delta\omega} \quad ; \quad 3\,\frac{2\pi}{\Delta\omega} \quad ; \quad \frac{2\pi}{\Delta\omega} \tag{59}$$

Following Eqs. (58)-(59), the simulated stochastic process $f_j(t)$; $j = 1, 2, 3$ is periodic with period T_0 given by:

$$T_0 = 3\,\frac{2\pi}{\Delta\omega} \tag{60}$$

Another very important point is that the simulated stochastic process $f_j(t)$; $j = 1, 2, 3$ is asymptotically Gaussian as $N \to \infty$ because of the central limit theorem (Shinozuka and Deodatis 1991).

At this point it should be noted that when generating sample functions of the simulated stochastic process according to Eq. (54), the time step Δt separating the generated values in the time domain has to obey the condition:

$$\Delta t \leq \frac{2\pi}{2\omega_u} \tag{61}$$

The condition set on Δt in Eq. (61) is necessary in order to avoid aliasing according to the sampling theorem (e.g. Bracewell 1986).

It can be shown (Deodatis 1995a) that the ensemble expected value $\mathcal{E}[f_j(t)]$; $j = 1, 2, 3$ and the ensemble auto-/cross-correlation function $R_{jk}(\tau)$; $j, k = 1, 2, 3$ of the simulated stochastic vector process $f_j(t)$ are identical to the corresponding targets, $\mathcal{E}[f_j^0(t)] = 0$; $j = 1, 2, 3$ and $R_{jk}^0(\tau)$; $j, k = 1, 2, 3$, respectively.

Another very important property of the simulated stochastic vector process is the following: each and every sample function generated by Eq. (54) is ergodic in the mean value and in correlation. It can be shown (Deodatis 1995a) that the temporal average and the temporal auto-/cross-correlation function of any sample function are identical to the corresponding targets, $\mathcal{E}[f_j^0(t)] = 0$; $j = 1, 2, 3$ and $R_{jk}^0(\tau)$; $j, k = 1, 2, 3$, respectively. In addition, it can be shown that these two identities are valid only when the length of the sample function is equal to the period T_0 given by Eq. (60). At this juncture it should be pointed out that although an algorithm based on the spectral representation method was available from the early seventies to simulate multi-variate stochastic processes (Shinozuka and Jan 1972), this algorithm was generating sample functions that were not ergodic. Several attempts have been made to modify the Shinozuka and Jan (1972) algorithm in order to achieve ergodicity. In one of the latest of these attempts, Shinozuka et al. (1989) introduced the concept of double-indexing the frequencies as indicated in Eq. (55), but the formula suggested in that paper was still generating sample functions that were not ergodic. It was Deodatis (1995a) that further modified the Shinozuka et al. (1989) simulation formula to generate ergodic sample functions of a multi-variate stochastic vector process for the first time.

Finally, it should be mentioned that the cost of digitally generating sample functions of the simulated stochastic vector process using Eq. (54) can be drastically reduced by using the FFT technique (Deodatis 1995a).

4.1 NUMERICAL EXAMPLES

Consider a tri-variate stationary stochastic process with components $f_1^0(t)$, $f_2^0(t)$ and $f_3^0(t)$, having mean value equal to zero and elements of the cross-spectral density matrix (Eq. (38)) defined as:

$$S_{11}^0(\omega) = \frac{38.3}{(1+6.19\omega)^{5/3}} \; ; \quad S_{22}^0(\omega) = \frac{43.3}{(1+6.98\omega)^{5/3}} \; ; \quad S_{33}^0(\omega) = \frac{135}{(1+21.8\omega)^{5/3}} \tag{62}$$

$$S_{jk}^0(\omega) = \sqrt{S_{jj}^0(\omega)S_{kk}^0(\omega)} \; \gamma_{jk}(\omega) \; ; \quad j,k = 1,2,3 \; ; \; j \neq k \tag{63}$$

with:

$$\gamma_{12}(\omega) = e^{-0.1757\omega} \; ; \quad \gamma_{13}(\omega) = e^{-3.478\omega} \; ; \quad \gamma_{23}(\omega) = e^{-3.292\omega} \tag{64}$$

According to Simiu and Scanlan (1986), the three components of the vector process defined in Eqs. (62)-(64) describe the longitudinal wind velocity fluctuations at heights $z = 35$ m (component 1), $z = 40$ m (component 2) and $z = 140$ m (component 3), assuming that the mean wind speed at $z = 35$ m is $U(35) = 45$ m/sec and that the surface roughness length is $z_0 = 0.001266$ m (corresponding shear velocity of the flow $u_* = 1.76$ m/sec). The mean wind speeds at $z = 40$ m and $z = 140$ m are computed using the logarithmic law as $U(40) = 45.6$ m/sec and $U(140) = 51.1$ m/sec. The expressions for the spectra shown in Eq. (62) were proposed by Kaimal et al. (1972), while the expressions for the coherence functions shown in Eq. (64) were proposed by Davenport (1968).

If the upper cut-off frequency ω_u and parameter N appearing in Eq. (56) are set equal to:

$$\omega_u = 4.0 \text{ rad/sec} \qquad \text{and} \qquad N = 128 \tag{65}$$

then $\Delta\omega$ and T_0 are computed from Eqs. (56) and (60), respectively, as:

$$\Delta\omega = 0.03125 \text{ rad/sec} \qquad \text{and} \qquad T_0 = 603.2 \text{ sec} \tag{66}$$

One sample function of the tri-variate stochastic process is now generated by applying the FFT technique (Deodatis 1995a) on Eq. (54). The generation is performed over one period as specified in Eqs. (60) and (66). The time step Δt is set equal to:

$$\Delta t = 0.1963 \text{ sec} \tag{67}$$

in order to satisfy the condition shown in Eq. (61).

Figure 2 shows one sample function of the tri-variate vector process with components $f_1^0(t)$, $f_2^0(t)$ and $f_3^0(t)$. The relatively stronger coherence between components $f_1^0(t)$ and $f_2^0(t)$, compared to the weaker coherence between components $f_1^0(t)$ and $f_3^0(t)$ is obvious in Fig. 2. Note that the time histories of the longitudinal wind velocity fluctuations shown in Fig. 2 consist of 3,072 points.

5. Simulation of 1D-mV Non-Stationary Stochastic Processes

Again as in Section 4, the theory and the simulation algorithm for multi-variate, non-stationary stochastic processes are presented for the special case of a tri-variate vector process. This is done for the sake of simplicity in the notation and because of space limitations. The proposed algorithm can be extended in a straightforward fashion to any other dimension of the vector process.

Consider a 1D-3V (one-dimensional, tri-variate) non-stationary stochastic vector process with components $f_1^0(t)$, $f_2^0(t)$ and $f_3^0(t)$, having mean value equal to zero:

$$\mathcal{E}[f_j^0(t)] = 0 \quad ; \quad j = 1, 2, 3 \tag{68}$$

cross-correlation matrix given by:

$$\mathbf{R}^0(t, t+\tau) = \begin{bmatrix} R_{11}^0(t, t+\tau) & R_{12}^0(t, t+\tau) & R_{13}^0(t, t+\tau) \\ R_{21}^0(t, t+\tau) & R_{22}^0(t, t+\tau) & R_{23}^0(t, t+\tau) \\ R_{31}^0(t, t+\tau) & R_{32}^0(t, t+\tau) & R_{33}^0(t, t+\tau) \end{bmatrix} \tag{69}$$

and cross-spectral density matrix given by:

$$\mathbf{S}^0(\omega, t) = \begin{bmatrix} S_{11}^0(\omega, t) & S_{12}^0(\omega, t) & S_{13}^0(\omega, t) \\ S_{21}^0(\omega, t) & S_{22}^0(\omega, t) & S_{23}^0(\omega, t) \\ S_{31}^0(\omega, t) & S_{32}^0(\omega, t) & S_{33}^0(\omega, t) \end{bmatrix} \tag{70}$$

Note that because of the non-stationarity of the vector process, the cross-correlation matrix is a function of two time instants: t and $t+\tau$ (t = time and τ = time lag), while the cross-spectral density matrix is a function of both frequency ω and time t.

Adopting the theory of evolutionary power spectra for non-stationary stochastic processes (Priestley 1965 and 1988), the elements of the cross-spectral density matrix are defined as:

$$S_{jj}^0(\omega, t) = |A_j(\omega, t)|^2 \, S_j(\omega) \quad ; \quad j = 1, 2, 3 \tag{71}$$

$$S_{jk}^0(\omega, t) = A_j(\omega, t) A_k(\omega, t) \sqrt{S_j(\omega) S_k(\omega)} \, \Gamma_{jk}(\omega) \quad ; \quad j, k = 1, 2, 3 \, ; \, j \neq k \tag{72}$$

where $A_j(\omega, t)$; $j = 1, 2, 3$ are the modulating functions of $f_1^0(t)$, $f_2^0(t)$ and $f_3^0(t)$, respectively, $S_j(\omega)$; $j = 1, 2, 3$ are the (stationary) power spectral density functions of $f_1^0(t)$, $f_2^0(t)$ and $f_3^0(t)$, respectively, and $\Gamma_{jk}(\omega)$; $j, k = 1, 2, 3$; $j \neq k$ are the complex coherence functions between $f_j^0(t)$ and $f_k^0(t)$.

It should be pointed out that Eqs. (71) and (72) imply that the modulating function $A_j(\omega, t)$ represents the change in the evolutionary power spectrum, relative to the (stationary) power spectral density function $S_j(\omega)$.

Consequently, for any time instant t, the diagonal elements of the cross-spectral density matrix are real and non-negative functions of ω satisfying:

$$S_{jj}^0(\omega, t) = S_{jj}^0(-\omega, t) \quad ; \quad j = 1, 2, 3 \quad \text{and for every } t \tag{73}$$

while the off-diagonal elements are generally complex functions of ω satisfying:

$$S_{jk}^0(\omega, t) = S_{jk}^{0*}(-\omega, t) \quad ; \quad j, k = 1, 2, 3 \, ; \, j \neq k \quad \text{and for every } t \tag{74}$$

$$S_{jk}^0(\omega, t) = S_{kj}^{0*}(\omega, t) \quad ; \quad j, k = 1, 2, 3 \, ; \, j \neq k \quad \text{and for every } t \tag{75}$$

where the asterisk denotes the complex conjugate. Equation (75) indicates that the cross-spectral density matrix $\mathbf{S}^0(\omega, t)$ is Hermitian for any value of t.

The elements of the cross-correlation matrix are related to the corresponding elements of the cross-spectral density matrix through the following transformations:

$$R_{jj}^0(t, t + \tau) = \int_{-\infty}^{\infty} A_j(\omega, t) A_j(\omega, t + \tau) \, e^{i\omega\tau} \, S_j(\omega) \, d\omega \quad ; \quad j = 1, 2, 3 \tag{76}$$

$$R_{jk}^0(t, t + \tau) = \int_{-\infty}^{\infty} A_j(\omega, t) A_k(\omega, t + \tau) \, e^{i\omega\tau} \, \sqrt{S_j(\omega) S_k(\omega)} \, \Gamma_{jk}(\omega) \, d\omega \quad ;$$
$$j, k = 1, 2, 3 \, ; \, j \neq k \tag{77}$$

For the special case of a uniformly modulated non-stationary stochastic vector process, the modulating functions $A_j(\omega, t)$; $j = 1, 2, 3$ are independent of the frequency ω:

$$A_j(\omega, t) = A_j(t) \quad ; \quad j = 1, 2, 3 \tag{78}$$

and Eqs. (76) and (77) reduce to:

$$R^0_{jj}(t, t+\tau) = A_j(t)A_j(t+\tau) \int_{-\infty}^{\infty} S_j(\omega)\, e^{i\omega\tau}\, d\omega \quad ; \quad j = 1, 2, 3 \qquad (79)$$

$$R^0_{jk}(t, t+\tau) = A_j(t)A_k(t+\tau) \int_{-\infty}^{\infty} \sqrt{S_j(\omega)S_k(\omega)}\; \Gamma_{jk}(\omega)\, e^{i\omega\tau}\, d\omega \quad ;$$
$$j, k = 1, 2, 3 \; ; \; j \neq k \qquad (80)$$

In such a case, the three components of the non-stationary stochastic vector process are expressed as:

$$f^0_j(t) = A_j(t)\, g^0_j(t) \quad ; \quad j = 1, 2, 3 \qquad (81)$$

where $g^0_j(t)$; $j = 1, 2, 3$ are the three components of a stationary stochastic vector process, having mean value equal to zero:

$$\mathcal{E}[g^0_j(t)] = 0 \quad ; \quad j = 1, 2, 3 \qquad (82)$$

and cross-spectral density matrix given by:

$$\mathbf{S}^0(\omega) = \begin{bmatrix} S_1(\omega) & \sqrt{S_1(\omega)S_2(\omega)}\Gamma_{12}(\omega) & \sqrt{S_1(\omega)S_3(\omega)}\Gamma_{13}(\omega) \\ \sqrt{S_2(\omega)S_1(\omega)}\Gamma_{21}(\omega) & S_2(\omega) & \sqrt{S_2(\omega)S_3(\omega)}\Gamma_{23}(\omega) \\ \sqrt{S_3(\omega)S_1(\omega)}\Gamma_{31}(\omega) & \sqrt{S_3(\omega)S_2(\omega)}\Gamma_{32}(\omega) & S_3(\omega) \end{bmatrix}$$
$$(83)$$

Note that the elements of matrix $\mathbf{S}^0(\omega)$ shown in Eq. (83) consist of terms that have been defined in Eqs. (71) and (72).

In the following, distinction will be made between the non-stationary stochastic vector process $f^0_j(t)$; $j = 1, 2, 3$ and its simulation $f_j(t)$; $j = 1, 2, 3$.

In order to simulate the 1D-3V non-stationary stochastic process $f^0_j(t)$; $j = 1, 2, 3$, its cross-spectral density matrix $\mathbf{S}^0(\omega, t)$ must be decomposed at every time instant t under consideration, into the following product:

$$\mathbf{S}^0(\omega, t) = \mathbf{H}(\omega, t)\, \mathbf{H}^{T*}(\omega, t) \qquad \text{for every } t \text{ under consideration} \qquad (84)$$

where superscript T denotes the transpose of a matrix. This decomposition can be performed using Cholesky's method, in which case $\mathbf{H}(\omega, t)$ is a lower triangular matrix:

$$\mathbf{H}(\omega, t) = \begin{bmatrix} H_{11}(\omega, t) & 0 & 0 \\ H_{21}(\omega, t) & H_{22}(\omega, t) & 0 \\ H_{31}(\omega, t) & H_{32}(\omega, t) & H_{33}(\omega, t) \end{bmatrix} \qquad (85)$$

whose diagonal elements are real and non-negative functions of ω and whose off-diagonal elements are generally complex functions of ω.

The following relation is satisfied by the diagonal elements of $\mathbf{H}(\omega,t)$:

$$H_{jj}(\omega,t) = H_{jj}(-\omega,t) \quad ; \quad j = 1,2,3 \quad \text{and for every } t \tag{86}$$

If the off-diagonal elements $H_{jk}(\omega,t)$ are written in polar form as:

$$H_{jk}(\omega,t) = |H_{jk}(\omega,t)| \, e^{i\theta_{jk}(\omega,t)} \quad ; \quad j = 2,3; \; k = 1,2; \; j > k \tag{87}$$

where:

$$\theta_{jk}(\omega,t) = \tan^{-1}\left(\frac{\text{Im}[H_{jk}(\omega,t)]}{\text{Re}[H_{jk}(\omega,t)]}\right) \tag{88}$$

with Im and Re denoting the imaginary and the real part of a complex number, respectively, then the following relations are satisfied:

$$|H_{jk}(\omega,t)| = |H_{jk}(-\omega,t)| \quad ; \quad j = 2,3; \; k = 1,2; \; j > k \quad \text{and for every } t \tag{89}$$

$$\theta_{jk}(\omega,t) = -\theta_{jk}(-\omega,t) \quad ; \quad j = 2,3; \; k = 1,2; \; j > k \quad \text{and for every } t \tag{90}$$

Once matrix $\mathbf{S}^0(\omega,t)$ is decomposed according to Eqs. (84)-(85), the non-stationary stochastic vector process $f_j^0(t)$; $j = 1,2,3$ can be simulated by the following series (Deodatis 1995b) as $N \to \infty$

$$f_j(t) = 2 \sum_{m=1}^{3} \sum_{l=1}^{N} |H_{jm}(\omega_l,t)|\sqrt{\Delta\omega} \, \cos[\omega_l t - \theta_{jm}(\omega_l,t) + \Phi_{ml}] \quad ; \quad j = 1,2,3 \tag{91}$$

where:

$$\omega_l = l\Delta\omega \quad ; \quad l = 1,2,\ldots,N \tag{92}$$

$$\Delta\omega = \frac{\omega_u}{N} \tag{93}$$

$$\theta_{jm}(\omega_l,t) = \tan^{-1}\left(\frac{\text{Im}[H_{jm}(\omega_l,t)]}{\text{Re}[H_{jm}(\omega_l,t)]}\right) \tag{94}$$

In Eq. (93), ω_u represents an upper cut-off frequency beyond which the elements of the cross-spectral density matrix (Eq. (70)) may be assumed to be zero for any time instant t (refer also to section 2 about a similar discussion for 1D-1V processes).

The Φ_{1l}, Φ_{2l}, Φ_{3l}; $l = 1,2,\ldots,N$ appearing in Eq. (91) are three sequences of independent random phase angles distributed uniformly over the interval $[0,2\pi]$.

It should be noted that the simulated non-stationary stochastic vector process $f_j(t)$; $j = 1,2,3$ is asymptotically Gaussian as $N \to \infty$ because of the central limit theorem (Shinozuka and Deodatis 1991).

It can be shown (Deodatis 1995b) that the ensemble expected value $\mathcal{E}[f_j(t)]$; $j = 1, 2, 3$ and the ensemble auto-/cross-correlation function $R_{jk}(t, t + \tau)$; $j, k = 1, 2, 3$ of the simulated non-stationary stochastic vector process $f_j(t)$ are identical to the corresponding targets, $\mathcal{E}[f_j^0(t)] = 0$; $j = 1, 2, 3$ and $R_{jk}^0(t, t + \tau)$; $j, k = 1, 2, 3$, respectively.

For the special case of a uniformly modulated non-stationary stochastic vector process, simulation can be performed on the basis of Eq. (81), instead of using Eq. (91). The simulation formula corresponding to Eq. (81) is:

$$f_j(t) = A_j(t)\, g_j(t) \quad ; \quad j = 1, 2, 3 \tag{95}$$

where $g_j(t)$ is the simulation of the stationary stochastic vector process $g_j^0(t)$ having mean value equal to zero and cross-spectral density matrix shown in Eq. (83). It should be mentioned that the simulation of stationary stochastic vector processes can be performed with great computational efficiency using the Fast Fourier Transform (FFT) technique, as described in Section 4.

At this juncture, it should be pointed out that it is not possible to take advantage of the FFT technique when using the (non-stationary) simulation formula shown in Eq. (91), in contrast to the corresponding formula for simulation of stationary stochastic vector processes (see Section 4). This is due to the fact that the coefficients $|H_{jm}(\omega_l, t)|\sqrt{\Delta\omega}$ in the double summation of Eq. (91) are now functions of both frequency and time. However, this shouldn't be of any great concern computationally, since in most cases of practical interest the non-stationary stochastic vector process $f_j^0(t)$; $j = 1, 2, 3$ is limited to relatively short durations by the modulating functions $A_j(\omega, t)$; $j = 1, 2, 3$ (e.g. ground motion acceleration time histories). The only case of non-stationary stochastic vector processes where the FFT technique can be used in the simulation formula is that of uniformly modulated processes (Eq. (95)).

5.1 NUMERICAL EXAMPLES

Consider that the acceleration time histories at three points on the ground surface (denoted by $f_1^0(t)$, $f_2^0(t)$ and $f_3^0(t)$, respectively) along the line of main wave propagation (see Fig. 3) are represented by a tri-variate, non-stationary stochastic vector process. The elements of the non-stationary cross-spectral density matrix with evolutionary power (see Eq. (70)) are defined now in the following way:

$$S_{jj}^0(\omega, t) = |A_j(\omega, t)|^2\, S_j(\omega) \quad ; \quad j = 1, 2, 3 \tag{96}$$

$$S_{jk}^0(\omega, t) = A_j(\omega, t) A_k(\omega, t)\, \sqrt{S_j(\omega) S_k(\omega)}\; \gamma_{jk}(\omega)\, \exp\left[-i\frac{\omega\, \xi_{jk}}{v}\right] \quad ;$$
$$j, k = 1, 2, 3 \; ; \; j \neq k \tag{97}$$

where $\gamma_{jk}(\omega)$ are the (stationary) coherence functions between $f_j^0(t)$ and $f_k^0(t)$ and $\exp[-i\omega\,\xi_{jk}/v]$ is the wave propagation term with ξ_{jk} being the distance between points j and k and v being the velocity of wave propagation.

It is pointed out that the expressions shown in Eqs. (96) and (97) describe earthquake ground motion that is non-homogeneous in space (since $S_1(\omega) \neq S_2(\omega) \neq S_3(\omega)$) and non-stationary in time.

The Clough-Penzien acceleration spectrum (Clough and Penzien 1975) is selected to model the (stationary) power spectral density functions $S_j(\omega)$; $j = 1, 2, 3$:

$$S_j(\omega) = S_{0j} \left[\frac{1 + 4\zeta_{gj}^2 \left[\frac{\omega}{\omega_{gj}}\right]^2}{\left\{1 - \left[\frac{\omega}{\omega_{gj}}\right]^2\right\}^2 + 4\zeta_{gj}^2 \left[\frac{\omega}{\omega_{gj}}\right]^2} \right] \left[\frac{\left[\frac{\omega}{\omega_{fj}}\right]^4}{\left\{1 - \left[\frac{\omega}{\omega_{fj}}\right]^2\right\}^2 + 4\zeta_{fj}^2 \left[\frac{\omega}{\omega_{fj}}\right]^2} \right]$$

(98)

where S_{0j} is a constant determining the intensity of acceleration at point j, ω_{gj} and ζ_{gj} can be thought of as some characteristic frequency and damping ratio of the ground at point j, and ω_{fj} and ζ_{fj} are filtering parameters for point j.

The Harichandran-Vanmarcke model (Harichandran and Vanmarcke 1986) is chosen to describe the (stationary) coherence functions $\gamma_{jk}(\omega)$; $j, k = 1, 2, 3$; $j \neq k$:

$$\gamma_{jk}(\omega) = A \exp\left[-\frac{2\xi_{jk}}{\alpha\,\theta(\omega)}(1 - A + \alpha A)\right] + (1 - A) \exp\left[-\frac{2\xi_{jk}}{\theta(\omega)}(1 - A + \alpha A)\right]$$

(99)

where $\theta(\omega)$ is the frequency-dependent correlation distance:

$$\theta(\omega) = k \left[1 + \left(\frac{\omega}{\omega_0}\right)^b\right]^{-1/2}$$

(100)

and A, α, k, ω_0 and b are model parameters.

The Bogdanoff-Goldberg-Bernard model (Bogdanoff, Goldberg and Bernard 1961) is used for the modulating functions $A_j(\omega, t)$; $j = 1, 2, 3$:

$$A_j(\omega, t) = A_j(t) = a_1 \left(t - \frac{\xi_{j1}}{v}\right) \cdot \exp\left[-a_2\left(t - \frac{\xi_{j1}}{v}\right)\right] \quad \text{for } t > \frac{\xi_{j1}}{v}; \; j = 1, 2, 3$$

(101)

where a_1 and a_2 are model parameters depending on such factors as earthquake magnitude and epicentral distance.

It should be pointed out that the models used for $S_j(\omega)$, $\gamma_{jk}(\omega)$ and $A_j(\omega, t)$ in Eqs. (98)-(101), were selected for demonstration purposes only. There are several other models in the literature that can be used for $S_j(\omega)$, $\gamma_{jk}(\omega)$ and $A_j(\omega, t)$.

The next step is to select numerical values for the parameters appearing in Eqs. (98)-(101). For ω_{gj} and ζ_{gj} in Eq. (98), the values suggested by Ellingwood and Batts (1982) for three different soil conditions are used in this study:

<u>Point 1</u>: Rock or stiff soil conditions: $\omega_{g1} = 8\pi$ rad/sec, $\zeta_{g1} = 0.60$ (102a)

<u>Point 2</u>: Deep cohesionless soils: $\omega_{g2} = 5\pi$ rad/sec, $\zeta_{g2} = 0.60$ (102b)

<u>Point 3</u>: Soft to medium clays and sands: $\omega_{g3} = 2.4\pi$ rad/sec, $\zeta_{g3} = 0.85$ (102c)

The filtering parameter ω_{fj} in Eq. (98) is set equal to 10% of the corresponding ω_{gj} value, while the other filtering parameter ζ_{fj} is set equal to the corresponding ζ_{gj} value, following the recommendation by Hindy and Novak (1980):

<u>Point 1</u>: Rock or stiff soil conditions: $\omega_{f1} = 0.8\pi$ rad/sec, $\zeta_{f1} = 0.60$ (103a)

<u>Point 2</u>: Deep cohesionless soils: $\omega_{f2} = 0.5\pi$ rad/sec, $\zeta_{f2} = 0.60$ (103b)

<u>Point 3</u>: Soft to medium clays and sands: $\omega_{f3} = 0.24\pi$ rad/sec, $\zeta_{f3} = 0.85$(103c)

The last parameter appearing in Eq. (98), S_{0j}, is computed so that the standard deviation of the Kanai-Tajimi part of the (stationary) power spectral density function is equal to 100 cm/sec^2 for all three points 1, 2 and 3:

$$S_{01} = 62.3 \text{ cm}^2/\text{sec}^3 \quad , \quad S_{02} = 99.7 \text{ cm}^2/\text{sec}^3 \quad , \quad S_{03} = 184.5 \text{ cm}^2/\text{sec}^3 \quad (104)$$

For the various parameters appearing in Eqs. (99) and (100), the values suggested by Harichandran and Wang (1990) are used in this study:

$$A = 0.626, \quad \alpha = 0.022, \quad k = 19,700 \text{ m}, \quad \omega_0 = 12.692 \text{ rad/sec}, \quad b = 3.47 \tag{105}$$

Finally, parameters a_1 and a_2 appearing in the expressions for the modulating functions (Eq. (101)) and the velocity of wave propagation v are set equal to:

$$a_1 = 0.906 \quad ; \quad a_2 = 1/3 \quad ; \quad v = 1,000 \text{ m/sec} \tag{106}$$

The simulation is performed at 3,072 time instants, with a time step $\Delta t = 6.14 \cdot 10^{-3}$ sec, over a length equal to $3,072 \cdot 6.14 \cdot 10^{-3} = 18.85$ sec. One generated sample function for the acceleration at points 1, 2 and 3, denoted by $f_1(t)$, $f_2(t)$ and $f_3(t)$, respectively, is displayed in Fig. 4. The non-stationarity and the different frequency contents of $f_1(t)$, $f_2(t)$ and $f_3(t)$ (specifically the frequency content of $f_1(t)$ is higher than that of $f_2(t)$ which is then higher than that of $f_3(t)$) can be easily identified in Fig. 4. Then, a segment of the acceleration time histories shown in Fig. 4 is magnified and displayed in Fig. 5. In this figure, the wave propagation effect and the loss of coherence are easily detected by following, for example, the movement and changing shape of peak A.

It is therefore obvious from Figs. 4 and 5 that the proposed algorithm is able to simulate non-stationary ground motion time histories that are spatially correlated according to a given coherence function, include the wave propagation effect and are non-homogeneous in space.

Acknowledgments

This work was supported by the National Science Foundation under Grant # BCS-9257900 with Dr. Clifford J. Astill as Program Director, by the NCEER Highway Project (FHWA Contracts DTFH61-92-C-00106 and DTFH61-92-C-00112) and by Kajima Corporation under Contract # AGMT 11/9/93.

References

Bogdanoff, J.L., Goldberg, J.E. and Bernard, M.C. (1961). "Response of a Simple Structure to a Random Earthquake-Type Disturbance," *Bulletin of the Seismological Society of America*, Vol. 51, No. 2, pp. 293-310.

Bracewell, R.N. (1986). *The Fourier Transform and its Applications*, McGraw-Hill.

Clough, R.W. and Penzien, J. (1975). *Dynamics of Structures*, McGraw-Hill.

Davenport, A.G. (1968). "The Dependence of Wind Load Upon Meteorological Parameters," *Proceedings of the International Research Seminar on Wind Effects on Buildings and Structures*, University of Toronto Press, pp. 19-82.

Deodatis, G. (1995a). "Simulation of Ergodic Multi-Variate Stochastic Processes," *Technical Report*, Department of Civil Engineering and Operations Research, Princeton University, (also submitted for publication).

Deodatis, G. (1995b). "Non-Stationary Stochastic Vector Processes: Seismic Ground Motion Applications," *Technical Report*, Department of Civil Engineering and Operations Research, Princeton University, (also submitted for publication).

Deodatis, G. and Shinozuka, M. (1989). "Simulation of Seismic Ground Motion Using Stochastic Waves," *Journal of Engineering Mechanics*, ASCE, Vol. 115, No. 12, pp. 2723-2737.

Elishakoff, I. (1979). "Buckling of a Stochastically Imperfect Finite Column on a Nonlinear Elastic Foundation," *Journal of Applied Mechanics*, ASME, Vol. 46, No. 2, pp. 411-416.

Elishakoff, I. (1983). *Probabilistic Methods in the Theory of Structures*, John Wiley.

Elishakoff, I. (1988). "Stochastic Simulation of an Initial Imperfection Data Bank for Isotropic Shells with General Imperfections," *Buckling of Structures: Theory and Experiment - The Josef Singer Anniversary Volume*, Elsevier, pp. 195-209.

Ellingwood, B.R. and Batts, M.E. (1982). "Characterization of Earthquake Forces for Probability-Based Design of Nuclear Structures," *Technical Report BNL-NUREG-51587, NUREG/CR-2945*, Department of Nuclear Energy, Brookhaven National Laboratory, Prepared for Office of Nuclear Regulatory

Research, U.S. Nuclear Regulatory Commission.

Fenton, G.A. and Vanmarcke, E. (1990). "Simulation of Random Fields Via Local Average Subdivision," *Journal of Engineering Mechanics*, ASCE, Vol. 116, No. 8, pp. 1733-1749.

Gersch, W. and Yonemoto, J. (1977). "Synthesis of Multi-Variate Random Vibration Systems: A Two-Stage Least Squares ARMA Model Approach," *Journal of Sound and Vibration*, Vol. 52, No. 4, pp. 553-565.

Grigoriu, M. (1993a). "On the Spectral Representation Method in Simulation," *Journal of Probabilistic Engineering Mechanics*, Vol. 8, No. 2, pp. 75-90.

Grigoriu, M. (1993b). "Simulation of Nonstationary Gaussian Processes by Random Trigonometric Polynomials," *Journal of Engineering Mechanics*, ASCE, Vol. 119, No. 2, pp. 328-343.

Grigoriu, M. (1995). *Applied Non-Gaussian Processes*, Prentice-Hall.

Grigoriu, M. and Balopoulou, S. (1993). "A Simulation Method for Stationary Gaussian Random Functions Based on the Sampling Theorem," *Journal of Probabilistic Engineering Mechanics*, Vol. 8, No. 3+4, pp. 239-254.

Gurley, K. and Kareem, A. (1995). "On the Analysis and Simulation of Random Processes Utilizing Higher Order Spectra and Wavelet Transforms," *Proceedings of Second International Conference on Computational Stochastic Mechanics*, Athens, Greece, Balkema, pp. 315-324.

Harichandran, R.S. and Vanmarcke, E.H. (1986). "Stochastic Variation of Earthquake Ground Motion in Space and Time," *Journal of Enginering Mechanics*, ASCE, Vol. 112, No. 2, pp. 154-174.

Harichandran, R.S. and Wang, W. (1990). "Effect of Spatially Varying Seismic Excitation on Surface Lifelines," *Proceedings of Fourth U.S. National Conference on Earthquake Engineering*, May 20-24, Palm Springs, pp. 885-894 (Vol. 1).

Hasofer, A.M. (1989). "Continuous Simulation of Gaussian Processes With Given Spectrum," *Proceedings of the 5th ICOSSAR*, (Eds. A.H-S. Ang, M. Shinozuka and G.I. Schueller), San Francisco, an ASCE publication, pp. 1201-1208.

Hindy, A. and Novak, M. (1980). "Pipeline Response to Random Ground Motion," *Journal of Engineering Mechanics Division*, ASCE, Vol. 106, No. EM2, pp. 339-360.

Kaimal, J.C. et al. (1972). "Spectral Characteristics of Surface-Layer Turbulence," *Journal of Royal Meteorological Society*, Vol. 98, pp. 563-589.

Kareem, A. and Li, Y. (1991). "Simulation of Multi-Variate Stationary and Non-Stationary Random Processes: A Recent Development," *Proceedings of First International Conference on Computational Stochastic Mechanics*, Corfu, Greece, Elsevier Applied Science, pp. 533-544.

Kozin, F. (1988). "Auto-Regressive Moving-Average Models of Earthquake Records," *Journal of Probabilistic Engineering Mechanics*, Vol. 3, No. 2, pp. 58-63.

Li, Y. and Kareem, A. (1993a). "Simulation of Multi-Variate Random Processes: Hybrid DFT and Digital Filtering Approach," *Journal of Engineering Mechanics*, ASCE, Vol. 119, No. 5, pp. 1078-1098.

Li, Y. and Kareem, A. (1993b). "Parametric Modelling of Stochastic Wave Effects on Offshore Platforms," *Applied Ocean Research*, Vol. 15, pp. 63-83.

Mignolet, M.P. and Spanos, P-T.D. (1987a). "Recursive Simulation of Stationary Multi-Variate Random Processes. Part I," *Journal of Applied Mechanics*, Vol. 54, No. 3, pp. 674-680.

Mignolet, M.P. and Spanos, P-T.D. (1987b). "Recursive Simulation of Stationary Multi-Variate Random Processes. Part II," *Journal of Applied Mechanics*, Vol. 54, No. 3, pp. 681-687.

Papoulis, A. (1962). *The Fourier Integral and Its Applications*, McGraw-Hill.

Polhemus, N.W. and Cakmak, A.S. (1981). "Simulation of Earthquake Ground Motions Using ARMA Models," *Earthquake Engineering and Structural Dynamics*, Vol. 9, No. 4, pp. 343-354.

Priestley, M.B. (1965). "Evolutionary Spectra and Non-Stationary Processes," *Journal of the Royal Statistical Society*, Series B, Vol. 27, pp. 204-237.

Priestley, M.B. (1988). *Non-Linear and Non-Stationary Time Series Analysis*, Academic Press.

Ramadan, O. and Novak, M. (1993). "Simulation of Spatially Incoherent Random Ground Motions," *Journal of Engineering Mechanics*, ASCE, Vol. 119, No. 5, pp. 997-1016.

Ramadan, O. and Novak, M. (1994). "Simulation of Multidimensional Anisotropic Ground Motions," *Journal of Engineering Mechanics*, ASCE, Vol. 120, No. 8, pp. 1773-1785.

Rice, S.O. (1954). "Mathematical Analysis of Random Noise," *Selected Papers on Noise and Stochastic Processes*, (Edited by Nelson Wax), Dover, pp. 133-294. (paper originally appeared in two parts in the Bell System Technical Journal in Vol. 23, July 1944 and in Vol. 24, January 1945)

Samaras, E., Shinozuka, M. and Tsurui, A. (1985). "ARMA Representation of Random Processes," *Journal of Engineering Mechanics*, ASCE, Vol. 111, No. 3, pp. 449-461.

Shinozuka, M. (1972). "Monte Carlo Solution of Structural Dynamics," *Computers and Structures*, Vol. 2, Nos. 5+6, pp. 855-874.

Shinozuka, M. (1974). "Digital Simulation of Random Processes in Engineering Mechanics with the Aid of FFT Technique," *Stochastic Problems in Mechanics*, (Eds. S. T. Ariaratnam and H. H. E. Leipholz), University of Waterloo Press, Waterloo, pp. 277-286.

Shinozuka, M. (1987). "Stochastic Fields and Their Digital Simulation," *Stochastic Methods in Structural Dynamics*, (Eds. G. I. Schuëller and M. Shinozuka), Martinus Nijhoff Publishers, Dordrecht, pp. 93-133.

Shinozuka, M. and Deodatis, G. (1988). "Stochastic Process Models for Earth-quake Ground Motion," *Journal of Probabilistic Engineering Mechanics*, Vol. 3, No. 3, pp. 114-123.

Shinozuka, M. and Deodatis, G. (1991). "Simulation of Stochastic Processes by Spectral Representation," *Applied Mechanics Reviews*, ASME, Vol. 44, No. 4, pp. 191-204.

Shinozuka, M. and Deodatis, G. (1995). "Simulation of Multi-Dimensional Stochastic Fields by Spectral Representation," *Technical Report*, Department of Civil Engineering and Operations Research, Princeton University, (also submitted for publication).

Shinozuka, M. and Jan, C-M. (1972). "Digital Simulation of Random Processes and its Applications," *Journal of Sound and Vibration*, Vol. 25, No. 1, pp. 111-128.

Shinozuka, M. and Lenoe, E. (1976). "A Probabilistic Model for Spatial Distribution of Material Properties," *Journal of Engineering Fracture Mechanics*, Vol. 8, pp. 217-227.

Shinozuka, M., Kamata, M. and Yun, C-B. (1989). "Simulation of Earthquake Ground Motion as Multi-Variate Stochastic Process," *Technical Report # 1989.5 from Princeton-Kajima Joint Research*, Department of Civil Engineering and Operations Research, Princeton University.

Simiu, E. and Scanlan, R.H. (1986). *Wind Effects on Structures*, John Wiley.

Soong, T.T. and Grigoriu, M. (1993). *Random Vibration of Mechanical and Structural Systems*, Prentice-Hall.

Spanos, P-T.D. (1983). "ARMA Algorithms for Ocean Spectral Modeling," *Journal of Energy Resources Technology*, ASME, Vol. 105, No. 3, pp. 300-309.

Spanos, P-T.D. and Hansen, J. (1981). "Linear Prediction Theory for Digital Simulations of Sea Waves," *Journal of Energy Resources Technology*, ASME, Vol. 103, No. 3, pp. 243-249.

Spanos, P-T.D. and Zeldin, B. (1995). "Random Field Simulation Using Wavelet Bases," *Proceedings of the ICASP-7 Conference*, Paris, France, July 10-13, pp. 1275-1283.

Vanmarcke, E. (1983). *Random Fields*, MIT Press.

Winterstein, S.R. (1990). "Random Process Simulation with Fast Hartley Transform," *Journal of Sound and Vibration*, Vol. 137, No. 3, pp. 527-531.

Yamazaki, F. and Shinozuka, M. (1988). "Digital Generation of Non-Gaussian Stochastic Fields," *Journal of Engineering Mechanics*, ASCE, Vol. 114, No. 7, pp. 1183-1197.

Yamazaki, F. and Shinozuka, M. (1990). "Simulation of Stochastic Fields by Statistical Preconditioning," *Journal of Engineering Mechanics*, ASCE, Vol. 116, No. 2, pp. 268-287.

Yang, J.-N. (1972). "Simulation of Random Envelope Processes," *Journal of Sound and Vibration*, Vol. 25, No. 1, pp. 73-85.

Yang, J.-N. (1973). "On the Normality and Accuracy of Simulated Random Processes," *Journal of Sound and Vibration*, Vol. 26, No. 3, pp. 417-428.

Zerva, A. (1992). "Seismic Ground Motion Simulations From a Class of Spatial Variability Models," *Earthquake Engineering and Structural Dynamics*, Vol. 21, No. 4, pp. 351-361.

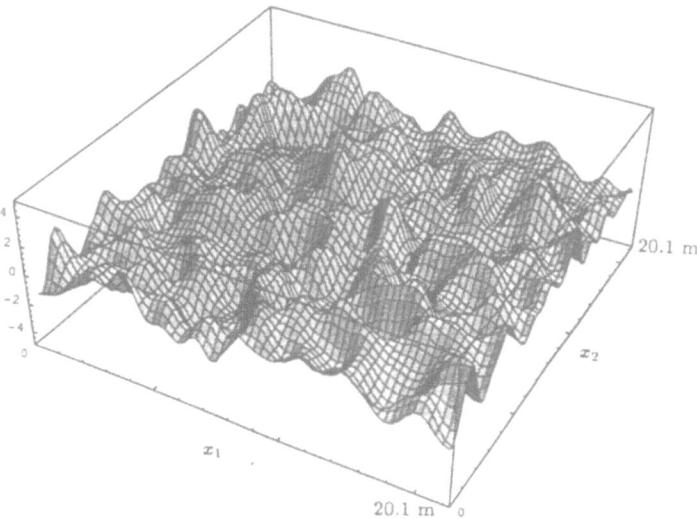

Figure 1a   Sample Function of Stochastic Field $f_0(x_1, x_2)$
for Case 1 ($b_1 = b_2 = 1.0$ m).

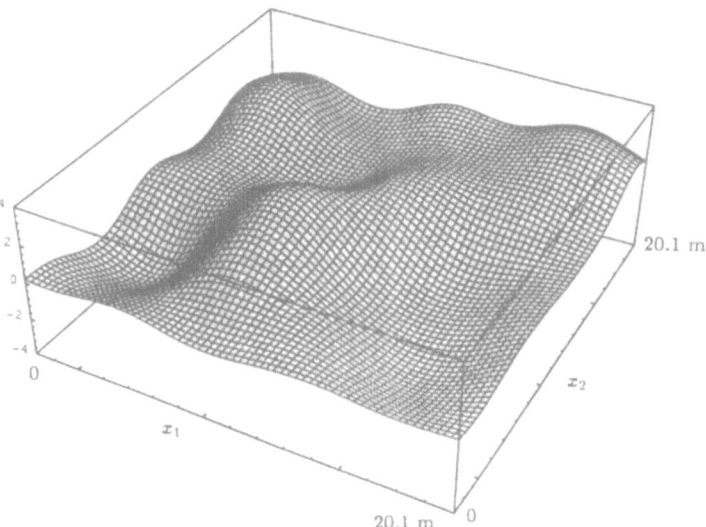

Figure 1b   Sample Function of Stochastic Field $f_0(x_1, x_2)$
for Case 2 ($b_1 = b_2 = 4.0$ m).

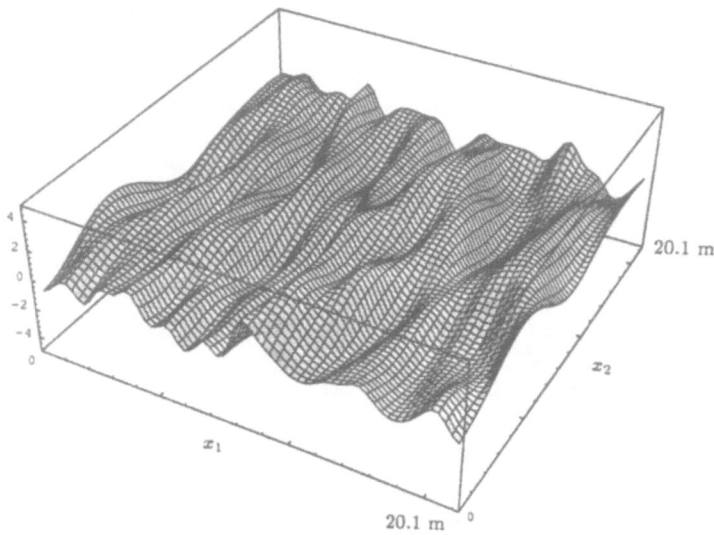

Figure 1c   Sample Function of Stochastic Field $f_0(x_1, x_2)$
for Case 3 ($b_1 = 1.0$ m, $b_2 = 4.0$ m).

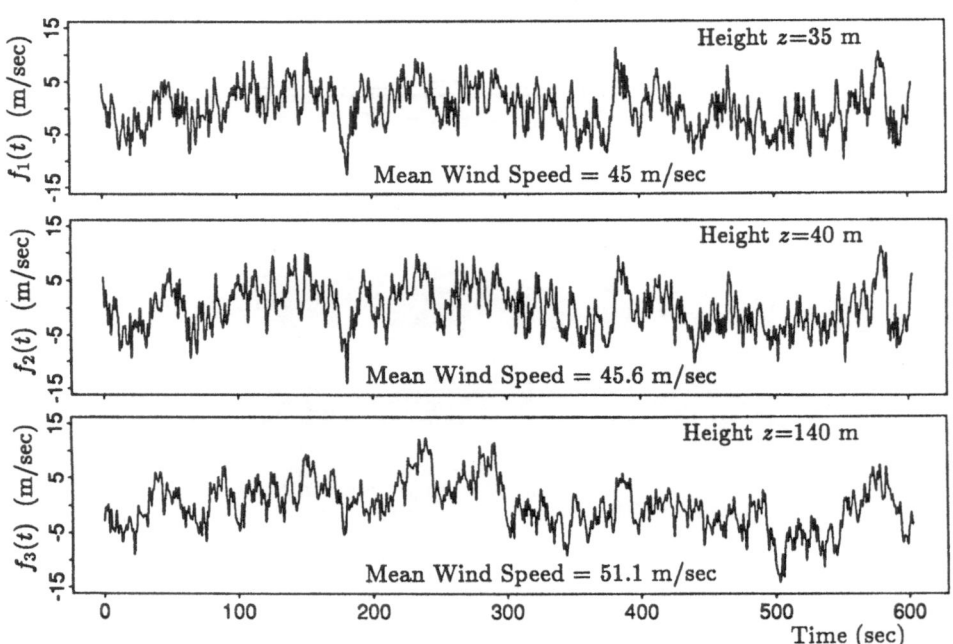

Figure 2   Sample Function for the Longitudinal Wind Velocity
Fluctuations at Three Different Heights.

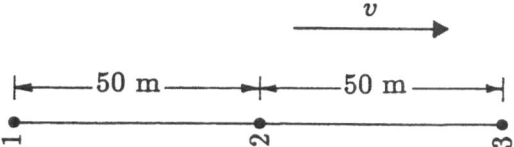

Figure 3   Configuration of Points 1, 2 and 3 on the Ground Surface.

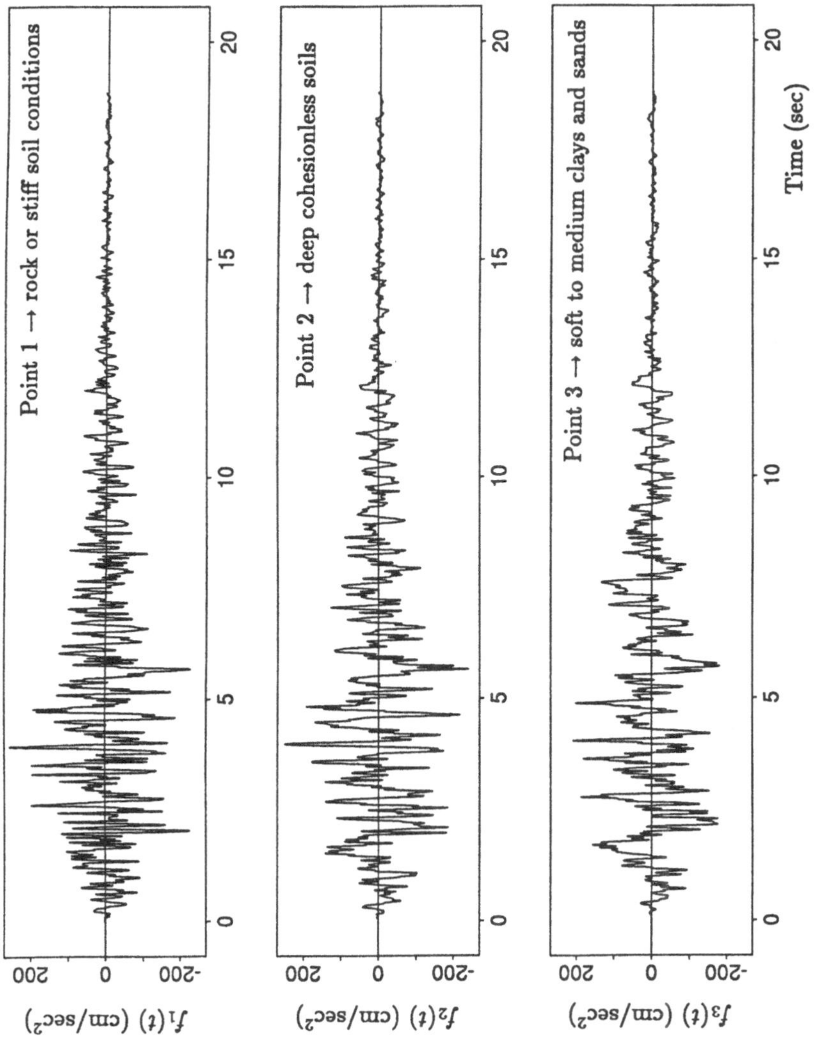

Figure 4   Sample Function for the Acceleration at Points 1, 2 and 3.

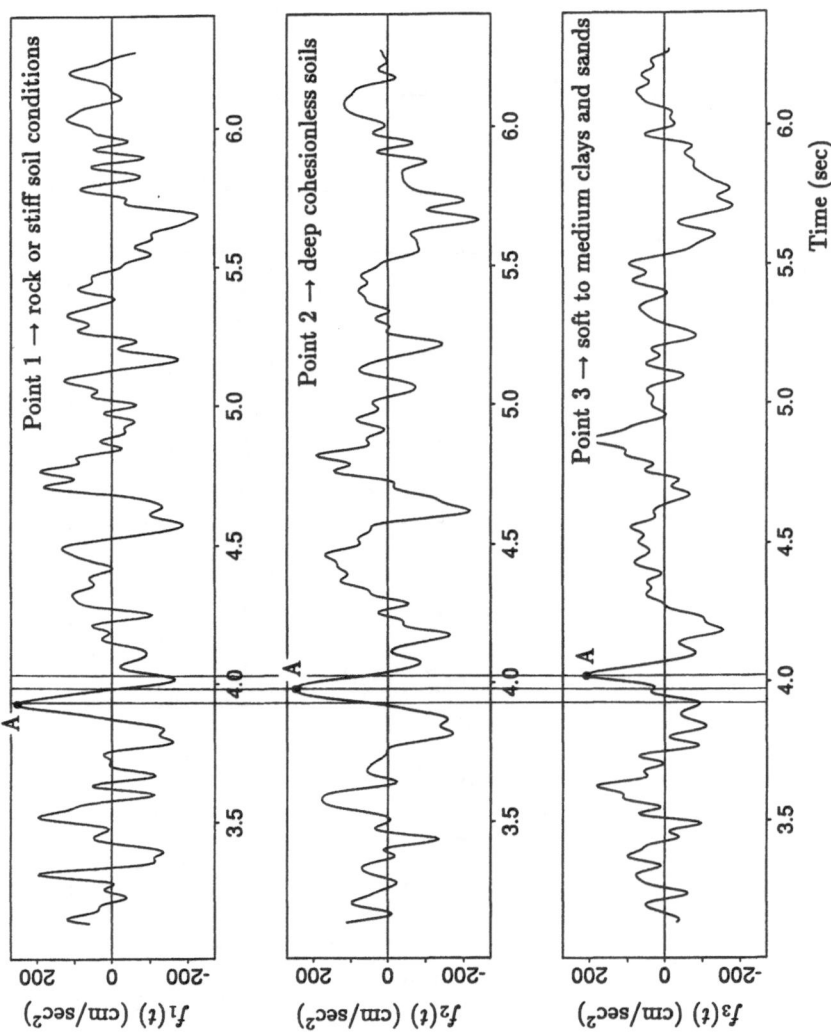

Figure 5   Magnified Sample Function for the Acceleration
at Points 1, 2 and 3.

A SPECTRAL FORMULATION OF STOCHASTIC FINITE ELEMENTS

R.G. GHANEM
The Johns Hopkins University
Baltimore, MD 21218.

AND

P.D. SPANOS
Rice University
Houston, TX 77251.

1. Introduction

Until recently, stochastic structural mechanics has addressed the issue of deterministic structures subjected to random loading. With the availability of more accurate analysis and design tools, however, quantifying the sensitivity of model predictions to uncertainty in the mechanical properties of structures has become possible. It has been observed, with the help of these tools, that this type of uncertainty can be more significant to the overall predictions of a particular structural model than the more traditional fluctuations attributed to external loads. In view of that, recent procedures have been developed for representing uncertainties in the parameters of a structural model, as well as, for propagating this uncertainty to obtain the associated uncertainty in the predicted response. The stochastic finite element method is a procedure for performing such an analysis, whereby the spatial extent of the structure has been represented within the context of the finite element method. This chapter describes a recent implementation of the stochastic finite element method that combines theoretical rigor with generality and efficiency of implementation. Specifically, the Spectral Stochastic Finite Element Method (SSFEM) presented in this chapter addresses the situation where the uncertain material properties are realizations of a spatially fluctuating random field. The formulation relies on discretizing the random processes using spectral expansions, thus eliminating the correlation between the requisite mesh size to meet energy-based convergence criteria, and the scales of fluctuation of the random

289

C. Guedes Soares (ed.), Probabilistic Methods for Structural Design, 289–312.

material properties involved. Moreover, reliability indicators obtained by various stochastic mechanics procedures have traditionally consisted of indices of reliability and second order moments. These descriptors feature condensed information about the uncertainty in the response process. The usefulness of these indicators are usually restricted by the levels of uncertainty in the structural parameters, as well as, by the level of complexity and nonlinearity of the structural model used. In the SSFEM presented herein, the outcome of the analysis is a complete probabilistic description of the response process that successfully eliminates the above restrictions, and therefore yields such quantities as the probability of failure of a structural system. In view of the computational implementation of the SSFEM, the outcome of the analysis can be adaptively refined, in a convergent fashion, as deemed necessary by the decision making context. This is analogous to mesh refinement in deterministic finite element procedures.

In the next section, a coherent mathematical framework is presented which is a natural setting for the analysis of systems with random parameters. Next, the theory of representation of stochastic processes is expanded with special emphasis on two spectral expansions, namely the Karhunen-Loeve and the Polynomial Chaos expansions. These are then used in the following section to develop the spectral stochastic finite element method.

2. The Mathematical Model

The class of problems dealt with in this study is not of the conventional engineering kind in that it involves concepts of a rather abstract and mathematical nature. It is both necessary and instructive to introduce at this point the mathematical concepts which are used in the sequel.

The Hilbert space of functions (Oden, 1979) defined over a domain $\mathbf{D}$, with values on the real line R, is denoted by $\mathbf{H}$. Let (Ω, Ψ, P) denote a probability space. By that is meant that Ω is a space of elementary events, Ψ is the σ-field generated by Ω, or loosely speaking, the space consisting of the various combinations of the elements of Ω, and finally, P is the probability measure defined on Ψ. Let $\mathbf{x}$ be an element of $\mathbf{D}$ and θ be an element of Ω. Then, the space of functions mapping Ω onto the real line is denoted by Θ. Each map $\Omega \rightarrow R$ defines a random variable.

The inner products over $\mathbf{H}$ and over Θ are defined using the Lebesgue measure and the probability measure, respectively. That is, for any two elements $h_i(\mathbf{x})$ and $h_j(\mathbf{x})$ in $\mathbf{H}$, their inner product $(\ h_i(\mathbf{x})\ ,\ h_j(\mathbf{x})\ )$ is defined as

$$(h_i(\mathbf{x}), h_j(\mathbf{x})) = \int_{\mathbf{D}} h_i(\mathbf{x}) h_j(\mathbf{x}) d\mathbf{x} . \tag{1}$$

The domain $\mathbf{D}$ represents the physical space over which the problem is defined. Similarly, given any two elements $\alpha(\theta)$ and $\beta(\theta)$ in Θ, their inner product is defined as

$$(\alpha(\theta), \beta(\theta)) = \int_\Omega \alpha(\theta)\beta(\theta)dP \qquad (2)$$

where dP is a probability measure. Under very general conditions, the integral in equation (2) is equivalent to the average of the integrand with respect to the probability measure dP, so that

$$(\alpha(\theta), \beta(\theta)) = <\alpha(\theta)\beta(\theta)> \qquad (3)$$

where $<.>$ denotes the operation of mathematical expectation. Any two elements of the Hilbert spaces defined above are said to be orthogonal if their inner product vanishes. A random process may then be described as a function defined on the product space $\mathbf{D} \times \Omega$. Viewed from this perspective a random process can be treated as a curve in either of $\mathbf{H}$ or Θ.

The physical model under consideration involves a medium whose properties exhibit random spatial fluctuations and which is subjected to a random external excitation. The mathematical representation of this problem involves an operator equation

$$\Lambda(\mathbf{x}, \theta)[u(\mathbf{x}, \theta)] = f(\mathbf{x}, \theta) \qquad (4)$$

where $\Lambda(\mathbf{x}, \theta)[.]$ is some operator defined on $\mathbf{H} \times \Theta$. In other words, Λ is a differential operator with coefficients exhibiting random fluctuations with respect to one or more of the independent variables. The aim then is to solve for the response $u(\mathbf{x}, \theta)$ as a function of both its arguments. With no loss of generality, Λ is assumed to be a differential operator whose random coefficients are restricted to being second order random processes. This is not a severe restriction for practical problems, since most physically measurable processes are of the second order type. Then, each one of these coefficients $a_k(\mathbf{x}, \theta)$ can be decomposed into a purely deterministic component and a purely random component in the form

$$a_k(\mathbf{x}, \theta) = \bar{a}_k(\mathbf{x}) + \alpha_k(\mathbf{x}, \theta) \qquad (5)$$

where $\bar{a}_k(\mathbf{x})$ is equal to the mathematical expectation of the process $a_k(\mathbf{x}, \theta)$, and $\alpha_k(\mathbf{x}, \theta)$ is a zero-mean random process, having the same covariance function as the process $a_k(\mathbf{x}, \theta)$. Equation (4) can then be written as

$$(\mathbf{L}(\mathbf{x}) + \mathbf{\Pi}(\mathbf{x}, \theta))[u(\mathbf{x}, \theta)] = f(\mathbf{x}, \theta), \qquad (6)$$

where $\mathbf{L}(\mathbf{x})[.]$ is a deterministic differential operator and $\mathbf{\Pi}(\mathbf{x}, \theta)[.]$ is a differential operator whose coefficients are zero-mean random processes.

Before a solution to equation (6) is sought, it is essential to clarify what is meant by such a solution.

It will prove instructive to start with the deterministic finite element method and see how the related concepts can be generalized. A finite element solution to a deterministic problem governed by a certain differential equation consists basically of computing the value of the dependent variables on a discrete mesh induced in the space spanned by the independent variables. This is probably the most widespread interpretation of a finite element solution; it has been crucial in disseminating the method as a powerful analysis and design tool into engineering practice. An alternative viewpoint which will prove to be more amenable to the required generalizations, is that a solution to a finite element problem consists in evaluating the value of the coefficients in the expansion of the solution along a certain basis in an appropriate functional space. The finite element procedure will consist in choosing a suitable basis and then computing optimal values of the coefficients with respect to this basis. From this perspective, the finite element mesh is naturally induced with specific choices of these bases. With other choices, however, the expansion coefficients do not necessarily carry an obvious physical interpretation. In the stochastic case, one of the independent variables spans the space of elementary events, which can only be discretized with respect to a probability measure, the result lacking any intuitive appeal. In this case, the appeal of the second interpretation of a finite element solution is obvious. The problem then becomes one of identifying a suitable basis in the space $\mathbf{H} \times \Theta$ over which the solution is defined, and of determining a meaningful optimality criterion for computing the coefficients in the associated expansion. Obviously, the basis functions in this case will be random. By simulating realizations of these functions, corresponding realizations of the solution process can be obtained. Alternatively, by defining a suitable inner product over the space of random variables, various statistics or, equivalently, norms of the solution process may be evaluated.

3. Representation of Stochastic Processes

Similarly to the case of the deterministic finite element method, whereby functions are represented by a denumerable set of parameters consisting of the values of the function and its derivatives at the nodal points, the problem encountered in the stochastic case is that of representing a random process by a denumerable set of random variables, thereby discretizing the process.

In the deterministic case discretization of the domain has a physical appeal. The domain in the stochastic case does not, however, have a physical meaning that permits a sensible discretization. In this context the functional analysis foundation of the finite element method becomes useful as it can be extended to deal with random functions. Two of the most useful expansions for random processes are the Karhunen-Loeve expansion, and the Polynomial Chaos expansion. The first requires knowledge of the covariance structure of the process under consideration, while the second one is more general. The difference between these two expansions can be loosely compared to that between a modal expansion and a Fourier-type expansion of a system response. Although the former has better convergence properties, the latter is more general and does not require knowledge of the properties of the system. These two expansions are discussed next.

3.1. KARHUNEN-LOEVE EXPANSION

The major conceptual difficulty from the viewpoint of the class of problems considered herein, involves the treatment of functions defined on these abstract spaces, namely random variables defined on the σ-field of random events. The most widely used method, the Monte Carlo simulation, consists of sampling these functions at randomly chosen elements of this σ-field, in a random, collocation-like, scheme. Obviously, a quite large number of points must be sampled if a good approximation is to be achieved. Alternatively, these functions could be expanded in a Fourier-type series as

$$w(\mathbf{x}, \theta) = \sum_{n=1}^{\infty} \sqrt{\lambda_n} \xi_n(\theta)\, f_n(\mathbf{x}) , \qquad (7)$$

where $\{\xi_n(\theta)\}$ is a set of random variables to be determined, λ_n is some constant, and $\{f_n(\mathbf{x})\}$ is an orthonormal set of deterministic functions. This is exactly what the Karhunen-Loeve expansion achieves. The expansion was derived independently by a number of investigators (Karhunen, 1947; Loeve, 1948; Kac and Siegert, 1947).

Let $w(\mathbf{x}, \theta)$ be a random process, function of the position vector $\mathbf{x}$ defined over the domain $\mathbf{D}$, with θ belonging to the space of random events Ω. Let $\bar{w}(\mathbf{x})$ denote the expected value of $w(\mathbf{x}, \theta)$ over all possible realizations of the process, and $C(\mathbf{x}_1, \mathbf{x}_2)$ denote its covariance function. By definition of the covariance function, it is bounded, symmetric and positive definite. Thus, it has the spectral decomposition (Courant and Hilbert, 1953)

$$C(\mathbf{x}_1, \mathbf{x}_2) = \sum_{n=1}^{\infty} \lambda_n\, f_n(\mathbf{x}_1)\, f_n(\mathbf{x}_2) \qquad (8)$$

where λ_n and $f_n(\mathbf{x})$ are the eigenvalue and the normalized eigenvector of the covariance kernel, respectively. That is, they are the solution to the integral equation

$$\int_{\mathbf{D}} C(\mathbf{x}_1, \mathbf{x}_2)\, f_n(\mathbf{x})\, d\mathbf{x}_1 = \lambda_n\, f_n(\mathbf{x}_2)\,. \tag{9}$$

Due to the symmetry and the positive definiteness of the covariance kernel (Loève, 1977), its eigenfunctions are orthogonal and form a complete set. They have further been normalized so that the following equation holds,

$$\int_{\mathbf{D}} f_n(\mathbf{x})\, f_m(\mathbf{x})\, d\mathbf{x} = \delta_{nm}\,, \tag{10}$$

where δ_{nm} is the Kronecker delta. Clearly, $w(\mathbf{x}, \theta)$ can be written as

$$w(\mathbf{x}, \theta) = \bar{w}(\mathbf{x}) + \alpha(\mathbf{x}, \theta)\,, \tag{11}$$

where $\alpha(\mathbf{x}, \theta)$ is a process with zero mean and covariance function $C(\mathbf{x}_1, \mathbf{x}_2)$. The process $\alpha(\mathbf{x}, \theta)$ can be expanded in terms of the eigenfunctions $f_n(\mathbf{x})$ as

$$\alpha(\mathbf{x}, \theta) = \sum_{n=1}^{\infty} \xi_n(\theta)\, \sqrt{\lambda_n}\, f_n(\mathbf{x})\,. \tag{12}$$

Second order properties of the random variables ξ_n can be determined by multiplying both sides of equation (12) by $\alpha(\mathbf{x}_2\,,\, \theta)$ and taking the expectation on both sides. Specifically, it is found that

$$\begin{aligned} C(\mathbf{x}_1, \mathbf{x}_2) &= <\alpha(\mathbf{x}_1, \theta)\, \alpha(\mathbf{x}_2, \theta)> \\ &= \sum_{n=1}^{\infty} \sum_{m=1}^{\infty} <\xi_n(\theta)\, \xi_m(\theta)> \sqrt{\lambda_n\, \lambda_m}\, f_n(\mathbf{x}_1)\, f_m(\mathbf{x}_2)\,. \end{aligned} \tag{13}$$

Then, multiplying both sides of equation (14) by $f_k(\mathbf{x}_2)$, integrating over the domain $\mathbf{D}$, and making use of the orthogonality of the eigenfunctions, yields

$$\begin{aligned} \int_{\mathbf{D}} C(\mathbf{x}_1, \mathbf{x}_2)\, f_k(\mathbf{x}_2)\, d\mathbf{x}_2 &= \lambda_k\, f_k(\mathbf{x}_1) \\ &= \sum_{n=1}^{\infty} <\xi_n(\theta)\, \xi_k(\theta)> \sqrt{\lambda_n \lambda_k}\, f_n(\mathbf{x}_1)\,. \end{aligned} \tag{14}$$

Multiplying once more by $f_l(\mathbf{x}_1)$ and integrating over $\mathbf{D}$, gives

$$\lambda_k \int_{\mathbf{D}} f_k(\mathbf{x}_1)\, f_l(\mathbf{x}_1)\, d\mathbf{x}_1 = \sum_{n=1}^{\infty} E<\xi_n(\theta)\, \xi_k(\theta)> \sqrt{\lambda_n \lambda_k}\, \delta_{nl}. \tag{15}$$

Then, using equation (10) leads to

$$\lambda_k \, \delta_{kl} \; = \; \sqrt{\lambda_k \, \lambda_l} \; <\xi_k(\theta) \, \xi_l(\theta) > \, . \tag{16}$$

Equation (16) can be rearranged to give

$$<\xi_k(\theta) \, \xi_l(\theta) > \; = \; \delta_{kl} \, . \tag{17}$$

Thus, the random process $w(\mathbf{x}, \theta)$ can be written as

$$w(\mathbf{x}, \theta) \; = \; \bar{w}(\mathbf{x}) \; + \; \sum_{n=1}^{\infty} \xi_n(\theta) \, \sqrt{\lambda_n} \, f_n(\mathbf{x}) \, . \tag{18}$$

where,

$$<\xi_n(\theta) > \; = \; 0 \quad , \quad <\xi_n(\theta) \, \xi_m(\theta)> \; = \; \delta_{nm} \, , \tag{19}$$

and λ_n, $f_n(\mathbf{x})$ are solution to equation (9). Truncating the series in equation (18) at the Mth term, gives

$$w(\mathbf{x}, \theta) \; = \; \bar{w}(\mathbf{x}) \; + \; \sum_{n=0}^{M} \xi_n(\theta) \, \sqrt{\lambda_n} \, f_n(\mathbf{x}) \, . \tag{20}$$

An explicit expression for $\xi_n(\theta)$ can be obtained by multiplying equation (12) by $f_n(\mathbf{x})$ and integrating over the domain $\mathbf{D}$. That is,

$$\xi_n(\theta) \; = \; \frac{1}{\lambda_n} \int_{\mathbf{D}} \alpha(\mathbf{x}, \theta) \, f_n(\mathbf{x}) \, d\mathbf{x} \, . \tag{21}$$

It is well known from functional analysis that the steeper a bilinear form decays to zero as a function of one of its arguments, the more terms are needed in its spectral representation in order to reach a preset accuracy. Noting that the Fourier transform operator is a spectral representation, it may be concluded that the faster the autocorrelation function tends to zero, the broader is the corresponding spectral density, and the greater the number of requisite terms to represent the underlying random process by the Karhunen-Loeve expansion.

For the special case of a random process possessing a rational spectrum, the integral eigenvalue problem can be replaced by an equivalent differential equation that is easier to solve (Van Trees, 1968). In the same context, it is reminded that a necessary and sufficient condition for a process to have a finite dimensional Markov realization is that its spectrum be rational

(Kree and Soize, 1986). Further, note that analytical solutions for the integral equation (10) are obtainable for some quite important and practical forms of the kernel $C(\mathbf{x}_1, \mathbf{x}_2)$ (Juncosa, 1945; Slepian and Pollak, 1961; Van Trees, 1968). In the general case, however, the integral equation must be solved numerically. Various techniques are available to this end (Ghanem and Spanos, 1991).

3.2. HOMOGENEOUS CHAOS

It is clear from the preceding discussion that the implementation of the Karhunen-Loeve expansion requires knowledge of the covariance function of the process being expanded. As far as the system under consideration is concerned, this implies that the expansion can be used for the random coefficients in the operator equation. However, it cannot be implemented for the solution process, since its covariance function and therefore the corresponding eigenfunctions are not known. An alternative expansion is clearly needed which circumvents this problem. Such an expansion could involve a basis of known random functions with deterministic coefficients to be found by minimizing some norm of the error resulting from a finite representation. This should be construed as similar to the Fourier series solution of deterministic differential equations, whereby the series coefficients are determined so as to satisfy some optimality criterion. To clarify this important idea further, a general functional form of the solution process is written as

$$u = h\ [\xi_i(\theta), x] \tag{22}$$

where $h[.]$ is a nonlinear functional of its arguments. In equation (22), the random processes involved have all been replaced by their corresponding Karhunen-Loeve representations. It is clear now that what is required is a nonlinear expansion of $h[.]$ in terms of the set of random variables $\xi_i(\theta)$. If the processes defining the operator are Gaussian, this set is a sampled derivative of the Wiener process (Doob, 1953). In this case, equation (22) involves functionals of the Brownian motion. This is exactly what the concept of Homogeneous Chaos provides. This concept was first introduced by Wiener (1938) and consists of an extension of Volterra's work on the generalization of Taylor series to functionals (Volterra, 1913). Wiener's contributions were the result of his investigations of nonlinear functionals of the Brownian motion. Based on Wiener's ideas, Cameron and Martin (1947) constructed an orthogonal basis for nonlinear functionals in terms of Fourier-Hermite functionals.

3.2.1. *Definitions and Properties*

Let $\{\xi_i(\theta)\}_{i=1}^{\infty}$ be a set of orthonormal Gaussian random variables. Consider the space $\hat{\Gamma}_p$ of all polynomials in $\{\xi_i(\theta)\}_{i=1}^{\infty}$ of degree not exceeding p. Let Γ_p represent the set of all polynomials in $\hat{\Gamma}_p$ orthogonal to $\hat{\Gamma}_{p-1}$. Finally, let $\bar{\Gamma}_p$ be the space spanned by Γ_p. Then, the subspace $\bar{\Gamma}_p$ of Θ is called the pth Homogeneous Chaos, and Γ_p is called the Polynomial Chaos of order p.

Based on the above definitions, the Polynomial Chaoses of any order p consist of all orthogonal polynomials of order p involving any combination of the random variables $\{\xi_i(\theta)\}_{i=1}^{\infty}$. It is clear, then, that the number of Polynomial Chaoses of order p, which involve a specific random variable out of the set $\{\xi_i(\theta)\}_{i=1}^{\infty}$ increases with p. This fact plays an important role in connection with the finite dimensional Polynomial Chaoses to be introduced in the sequel. Furthermore, since random variables are themselves functions, it becomes clear that Polynomial Chaoses are functions of functions and are therefore functionals.

The set of Polynomial Chaoses is a linear subspace of the space of square-integrable random variables Θ, and is a ring with respect to the functional multiplication $\Gamma_p\Gamma_l(\omega) = \Gamma_p(\omega)\Gamma_l(\omega)$. In this context, square integrability must be construed to be with respect to the probability measure defining the random variables. Denoting the Hilbert space spanned by the set $\{\xi_i(\theta)\}$ by $\Theta(\xi)$, the resulting ring is denoted by $\Phi_{\Theta(\xi)}$, and is called the ring of functions generated by $\Theta(\xi)$. Then, it can be shown that under some general conditions, the ring $\Phi_{\Theta(\xi)}$ is dense in the space Θ (Kakutani, 1961). This means that any square-integrable random function $(\Omega \rightarrow R)$ can be approximated as closely as desired by elements from $\Phi_{\Theta(\xi)}$. Thus, any element $\mu(\theta)$ from the space Θ admits the following representation,

$$\mu(\theta) = \sum_{p \geq 0} \sum_{n_1+\ldots+n_r=p} \sum_{\rho_1,\ldots,\rho_r} a_{\rho_1\ldots\rho_r}^{n_1\ldots n_r} \, \Gamma_p(\xi_{\rho_1}(\theta),\ldots,\xi_{\rho_r}(\theta)) \,, \qquad (23)$$

where $\Gamma_p(.)$ is the Polynomial Chaos of order p. The superscript n_i refers to the number of occurrences of $\xi_{\rho_i}(\theta)$ in the argument list for $\Gamma_p(.)$. Also, the double subscript provides for the possibility of repeated arguments in the argument list of the Polynomial Chaoses, thus preserving the generality of the representation given by equation (23). Briefly stated, the Polynomial Chaos appearing in equation (23) involves r distinct random variables out of the set $\{\xi_i(\theta)\}_{i=1}^{\infty}$, with the kth random variable $\xi_k(\theta)$ having multiplicity n_k, and such that the total number of random variables involved is equal to the order p of the Polynomial Chaos. The Polynomial Chaoses of any order will be assumed to be symmetric with respect to their arguments. Such a symmetrization is always possible. Indeed, a symmetric polynomial can be obtained from a non-symmetric one by taking the average of the polynomial over all permutations of its arguments. The form of the coefficients appearing in equation (23) can then be simplified, resulting in the following expanded expression for the representation of random variables,

$$\mu(\theta) = a_0 \, \Gamma_0 + \sum_{i_1=1}^{\infty} a_{i_1} \Gamma_1(\xi_{i_1}(\theta)) \tag{24}$$

$$+ \sum_{i_1=1}^{\infty} \sum_{i_2=1}^{i_1} a_{i_1 i_2} \Gamma_2(\xi_{i_1}(\theta), \xi_{i_2}(\theta))$$

$$+ \sum_{i_1=1}^{\infty} \sum_{i_2=1}^{i_1} \sum_{i_3=1}^{i_2} a_{i_1 i_2 i_3} \Gamma_3(\xi_{i_1}(\theta), \xi_{i_2}(\theta), \xi_{i_3}(\theta))$$

$$+ \sum_{i_1=1}^{\infty} \sum_{i_2=1}^{i_1} \sum_{i_3=1}^{i_2} \sum_{i_4=1}^{i_3} a_{i_1 i_2 i_3 i_4} \Gamma_4(\xi_{i_1}(\theta), \xi_{i_2}(\theta), \xi_{i_3}(\theta), \xi_{i_4}(\theta)) + \ldots,$$

where $\Gamma_p(.)$ are successive Polynomial Chaoses of their arguments, the expansion being convergent in the mean-square sense. The upper limits on the summations in equation (24) reflect the symmetry of the Polynomial Chaoses with respect to their arguments, as discussed above. The Polynomial Chaoses of order greater than one have mean zero. Polynomials of different order are orthogonal to each other; so are same order polynomials with different argument list. At times in the ensuing developments, it will prove notationally expedient to rewrite equation (24) in the form

$$\mu(\theta) = \sum_{j=0}^{\infty} \hat{a}_j \, \Psi_j[\boldsymbol{\xi}(\theta)], \tag{25}$$

where there is a one-to-one correspondence between the functionals $\Psi[.]$ and $\Gamma[.]$, and also between the coefficients $\hat{a}_j$ and $a_{i_1 \ldots i_r}$ appearing in equation (24). Implicit in equation (24) is the assumption that the expansion (24) is carried out in the order indicated by that equation. In other words, the contribution of polynomials of lower order is accounted for first.

Throughout the previous theoretical development, the symbol θ has been used to emphasize the random character of the quantities involved. It will be deleted in the ensuing development whenever the random nature of a certain quantity is obvious from the context.

As defined above, each Polynomial Chaos is a function of the infinite set $\{\xi_i\}$, and is therefore an infinite dimensional polynomial. In a computational setting, however, this infinite set has to be replaced by a finite one. In view of that, it seems logical to introduce the concept of a finite dimensional Polynomial Chaos. Specifically, the n-dimensional Polynomial

Chaos of order p is the subset of the Polynomial Chaos of order p, as defined above, which is a function of only n of the uncorrelated random variables ξ_i. As $n \to \infty$, the Polynomial Chaos as defined previously is recovered. Obviously, the convergence properties of a representation based on the n-dimensional Polynomial Chaoses depend on n as well as on the choice of the subset $\{\xi_{\lambda_i}\}_{i=1}^n$ out of the infinite set. In the ensuing analysis, this choice will be based on the Karhunen-Loeve expansion of an appropriate random process. Since the finite dimensional Polynomial Chaos is a subset of the (infinite-dimensional) Polynomial Chaos, the same symbol will be used for both, with the dimension being specified. Note that for this case, the infinite upper limit on the summations in equation (24) is replaced by a number equal to the dimension of the Polynomials involved. For clarity, the two-dimensional counterpart of equation (24) is rewritten, in a fully expanded form, as

$$
\begin{aligned}
\mu(\theta) = \ & a_0\,\Gamma_0 \ + \ a_1\,\Gamma_1(\xi_1) \ + \ a_2\,\Gamma_1(\xi_2) \\
& + \ a_{11}\,\Gamma_2(\xi_1,\xi_1) \ + \ a_{12}\Gamma_2(\xi_2,\xi_1) \ + \ a_{22}\Gamma_2(\xi_2,\xi_2) \\
& + \ a_{111}\,\Gamma_3(\xi_1,\xi_1,\xi_1) \ + \ a_{211}\,\Gamma_3(\xi_2,\xi_1,\xi_1) \ + \ a_{221}\,\Gamma_3(\xi_2,\xi_2,\xi_1) \\
& + \ a_{222}\,\Gamma_3(\xi_2,\xi_2,\xi_2) \ \dots \ .
\end{aligned}
\tag{26}
$$

In view of this last equation, it becomes clear that, except for a different indexing convention, the functionals $\Psi[.]$ and $\Gamma[.]$ are identical. In this regard, equation (26) can be recast in terms of $\Psi_j[.]$ as follows

$$
\begin{aligned}
\mu(\theta) = \ & \hat{a}_0\Psi_0 + \hat{a}_2\Psi_2 + \hat{a}_3\Psi_3 + \hat{a}_4\Psi_4 + \hat{a}_5\Psi_5 \\
& + \hat{a}_6\Psi_6 + \hat{a}_7\Psi_7 + \hat{a}_8\Psi_8 + \hat{a}_9\Psi_9 + \dots \,,
\end{aligned}
\tag{27}
$$

from which the correspondence between $\Psi[.]$ and $\Gamma[.]$ is evident. For example, the term $a_{211}\Gamma_3(\xi_2,\xi_1,\xi_1)$ of equation (26) is identified with the term $\hat{a}_7\Psi_7$ of equation (27).

3.2.2. *Construction of the Polynomial Chaos*

A direct approach to construct the successive Polynomial Chaoses is to start with the set of homogeneous polynomials in $\{\xi_i(\theta)\}$ and to proceed, through a sequence of orthogonalization procedures. The zeroth order polynomial is a constant and it can be chosen to be 1. That is

$$
\Gamma_0 \ = \ 1 \, .
\tag{28}
$$

The first order polynomial has to be chosen so that it is orthogonal to all zeroth order polynomials. In this context, orthogonality is understood to be with respect to the inner-product defined by equation (2). Since the set $\{\xi_i\}$ consists of zero-mean elements, the orthogonality condition implies

$$
\Gamma_1(\xi_i) \ = \ \xi_i \, .
\tag{29}
$$

The second order Polynomial Chaos consists of second order polynomials in $\{\xi_i\}$ that are orthogonal to both constants and first order polynomials. Formally, a second order polynomial can be written as

$$\Gamma_2(\xi_{i_1}, \xi_{i_2}) = a_0 + a_{i_1}\xi_{i_1} + a_{i_2}\xi_{i_2} + a_{i_1 i_2}\xi_{i_1}\xi_{i_2} , \tag{30}$$

where the constants are so chosen as to satisfy the orthogonality conditions. The second of these requires that

$$<\Gamma_2(\xi_{i_1}, \xi_{i_2})\,\xi_{i_3}> \, = \, 0 . \tag{31}$$

This leads to the following equation

$$a_{i_1}\delta_{i_1 i_3} + a_{i_2}\delta_{i_2 i_3} = 0 . \tag{32}$$

Allowing i_3 to be equal to i_1 and i_2 successively, permits the evaluation of the coefficients a_{i_1} and a_{i_2} as

$$a_{i_1} = 0 , \quad a_{i_2} = 0 . \tag{33}$$

The first orthogonality condition yields

$$a_0 + a_{i_1 i_2}\delta_{i_1 i_2} = 0 . \tag{34}$$

Equation (34) can be normalized by requiring that

$$a_{i_1 i_2} = 1 . \tag{35}$$

This leads to

$$a_0 = -\,\delta_{i_1 i_2} . \tag{36}$$

Thus, the second Polynomial Chaos can be expressed as

$$\Gamma_2(\xi_{i_1}, \xi_{i_2}) = \xi_{i_1}\xi_{i_2} - \delta_{i_1 i_2} . \tag{37}$$

In a similar manner, the third order Polynomial Chaos has the general form

$$\begin{aligned}\Gamma_3(\xi_{i_1}, \xi_{i_2}, \xi_{i_3}) = &\, a_0 + a_{i_1}\xi_{i_1} + a_{i_2}\xi_{i_2} + a_{i_3}\xi_{i_3} + a_{i_1 i_2}\xi_{i_1}\xi_{i_2} \\ &+ a_{i_1 i_3}\xi_{i_1}\xi_{i_3} + a_{i_2 i_3}\xi_{i_2}\xi_{i_3} + a_{i_1 i_2 i_3}\xi_{i_1}\xi_{i_2}\xi_{i_3},\end{aligned} \tag{38}$$

with conditions of being orthogonal to all constants, first order polynomials, and second order polynomials. The first of these conditions implies that

$$<\Gamma_3(\xi_{i_1}, \xi_{i_2}, \xi_{i_3})> \, = \, 0 . \tag{39}$$

That is,

$$a_0 + a_{i_1 i_2} \delta_{i_1 i_2} + a_{i_1 i_3} \delta_{i_1 i_3} + a_{i_2 i_3} \delta_{i_2 i_3} = 0 . \tag{40}$$

The second condition implies that

$$<\Gamma_3(\xi_{i_1}, \xi_{i_2}, \xi_{i_3}) \, \xi_{i_4}> = 0 , \tag{41}$$

which leads to

$$a_{i_1} \delta_{i_1 i_4} + a_{i_2} \delta_{i_2 i_4} + a_{i_3} \delta_{i_3 i_4} + a_{i_1 i_2 i_3} <\xi_{i_1} \xi_{i_2} \xi_{i_3} \xi_{i_4}> . \tag{42}$$

The last orthogonality condition is equivalent to

$$<\Gamma_3(\xi_{i_1}, \xi_{i_2}, \xi_{i_3}) \, \xi_{i_4} \, \xi_{i_5}> = 0 , \tag{43}$$

which gives

$$a_0 \, \delta_{i_4 i_5} \, a_{i_1 i_2} <\xi_{i_1} \xi_{i_2} \xi_{i_4} \xi_{i_5}> + a_{i_1 i_3} <\xi_{i_1} \xi_{i_3} \xi_{i_4} \xi_{i_5}>$$
$$+ a_{i_2 i_3} <\xi_{i_2} \xi_{i_3} \xi_{i_4} \xi_{i_5}> = 0. \tag{44}$$

The above equations can be normalized by requiring that

$$a_{i_1 i_2 i_3} = 1 . \tag{45}$$

Then equation (42) becomes

$$a_{i_1} \delta_{i_1 i_4} + a_{i_2} \delta_{i_2 i_4} + a_{i_3} \delta_{i_3 i_4} + <\xi_{i_1} \xi_{i_2} \xi_{i_3} \xi_{i_4}> = 0 \tag{46}$$

Due to the Gaussian property of the set $\{\xi_i\}$, the following equation holds

$$<\xi_{i_1} \xi_{i_2} \xi_{i_3} \xi_{i_4}> = \delta_{i_1 i_2} \delta_{i_3 i_4} + \delta_{i_1 i_3} \delta_{i_2 i_4} + \delta_{i_1 i_4} \delta_{i_2 i_3} . \tag{47}$$

Substituting for the expectations in equations (46) and (44) yields

$$a_{i_1} \delta_{i_1 i_4} + a_{i_2} \delta_{i_2 i_4} + a_{i_3} \delta_{i_3 i_4}$$
$$+ \delta_{i_1 i_2} \delta_{i_3 i_4} + \delta_{i_1 i_3} \delta_{i_2 i_4} + \delta_{i_1 i_4} \delta_{i_2 i_3} = 0 , \tag{48}$$

and

$$a_0 \, \delta_{i_4 i_5} + a_{i_1 i_2} [\, \delta_{i_1 i_2} \delta_{i_4 i_5} + \delta_{i_1 i_4} \delta_{i_2 i_5} + \delta_{i_1 i_5} \delta_{i_2 i_4} \,]$$
$$+ a_{i_1 i_3} [\, \delta_{i_1 i_3} \delta_{i_4 i_5} + \delta_{i_1 i_4} \delta_{i_3 i_5} + \delta_{i_1 i_5} \delta_{i_3 i_4} \,]$$
$$+ a_{i_2 i_3} [\, \delta_{i_2 i_3} \delta_{i_4 i_5} + \delta_{i_2 i_4} \delta_{i_3 i_5} + \delta_{i_2 i_5} \delta_{i_3 i_4} \,] = 0. \tag{49}$$

Substituting for a_0 from equation (40), equation (49) can be rewritten as

$$a_{i_1 i_2} \left[\delta_{i_1 i_4} \delta_{i_2 i_5} + \delta_{i_1 i_5} \delta_{i_2 i_4} \right] + a_{i_1 i_3} \left[\delta_{i_1 i_4} \delta_{i_3 i_5} + \delta_{i_1 i_5} \delta_{i_3 i_4} \right]$$
$$+ a_{i_2 i_3} \left[\delta_{i_2 i_4} \delta_{i_3 i_5} + \delta_{i_2 i_5} \delta_{i_3 i_4} \right] = 0 . \tag{50}$$

From equation (50), the coefficients $a_{i_1 i_2}$, $a_{i_1 i_3}$, and $a_{i_2 i_3}$ can be evaluated as

$$
\begin{aligned}
a_{i_1 i_2} &= 0 \\
a_{i_1 i_3} &= 0 \\
a_{i_2 i_3} &= 0 .
\end{aligned}
\tag{51}
$$

Using equation (49) again, it is found that

$$a_0 = 0 . \tag{52}$$

Equation (48) can be rewritten as

$$\delta_{i_1 i_4}(a_{i_1} + \delta_{i_2 i_3}) + \delta_{i_1 i_4}(a_{i_1} + \delta_{i_2 i_3}) + \delta_{i_1 i_4}(a_{i_1} + \delta_{i_2 i_3}) = 0, \tag{53}$$

from which the coefficients a_{i_1}, a_{i_2}, and a_{i_3} are found to be,

$$
\begin{aligned}
a_{i_1} &= -\delta_{i_2 i_3} \\
a_{i_2} &= -\delta_{i_1 i_3} \\
a_{i_3} &= -\delta_{i_1 i_2} .
\end{aligned}
\tag{54}
$$

The third order Polynomial Chaos can then be written as

$$\Gamma_3(\xi_{i_1}, \xi_{i_2}, \xi_{i_3}) = \xi_{i_1} \, \xi_{i_2} \, \xi_{i_3} - \xi_{i_1} \, \delta_{i_2 i_3} - \xi_{i_2} \, \delta_{i_1 i_3} - \xi_{i_3} \, \delta_{i_1 i_2} . \tag{55}$$

After laborious algebraic manipulations, the fourth order Polynomial Chaos can be expressed as

$$
\begin{aligned}
\Gamma_4(\xi_{i_1}, \xi_{i_2}, \xi_{i_3}, \xi_{i_4}) =\ & \xi_{i_1} \, \xi_{i_2} \, \xi_{i_3} \, \xi_{i_4} \\
& - \xi_{i_1} \, \xi_{i_2} \, \delta_{i_3 i_4} - \xi_{i_1} \, \xi_{i_3} \, \delta_{i_2 i_4} - \xi_{i_1} \, \xi_{i_4} \, \delta_{i_2 i_3} \\
& - \xi_{i_2} \, \xi_{i_3} \, \delta_{i_1 i_4} - \xi_{i_2} \, \xi_{i_4} \, \delta_{i_1 i_3} - \xi_{i_3} \, \xi_{i_4} \, \delta_{i_1 i_2} \\
& + \delta_{i_1 i_2} \delta_{i_3 i_4} + \delta_{i_1 i_3} \delta_{i_2 i_4} + \delta_{i_1 i_4} \delta_{i_2 i_3} .
\end{aligned}
\tag{56}
$$

It is readily seen that, in general, the nth order Polynomial Chaos can be written as

$$
\Gamma_p(\xi_{i_1}, ..., \xi_{i_n}) =
\begin{cases}
\displaystyle\sum_{\substack{r=n \\ r \text{ even}}}^{0} (-1)^r \sum_{\pi(i_1,...,i_n)} \prod_{k=1}^{r} \xi_{i_k} < \prod_{l=r+1}^{n} \xi_{i_l} > \\
\\
\qquad\qquad n \text{ even} \\
\\
\displaystyle\sum_{\substack{r=n \\ r \text{ even}}}^{0} (-1)^{r-1} \sum_{\pi(i_1,...,i_n)} \prod_{k=1}^{r} \xi_{i_k} < \prod_{l=r+1}^{n} \xi_{i_l} > \\
\\
\qquad\qquad n \text{ odd}
\end{cases}
\tag{57}
$$

where $\pi(.)$ denotes a permutation of its arguments, and the summation is over all such permutations such that the sets $\{\xi_{i_1}, ..., \xi_{i_r}\}$ is modified by the permutation.

Note that the Polynomial Chaoses as obtained in equations (28), (29), (37), (55) and (56) are orthogonal with respect to the Gaussian probability measure, which makes them identical with the corresponding multidimensional Hermite polynomials (Grad, 1949). These polynomials have been used extensively in relation to problems in turbulence theory (Imamura et.al, 1965a-b). This equivalence is implied by the orthogonality of the Polynomial Chaoses with respect to the inner product defined by equation (2) where dP is the Gaussian measure $e^{-\frac{1}{2}\xi^T \xi} d\xi$, where ξ denotes the vector of n random variables $\{\xi_{i_k}\}_{k=1}^{n}$. This measure is exactly the weighing function with respect to which the Hermite polynomials are orthogonal in the L_2 sense (Oden, 1979). This fact suggests another method for constructing the Polynomial Chaoses, namely from the generating function of the Hermite polynomials. Specifically, the Polynomial Chaos of order n can be obtained as

$$
\Gamma_n(\xi_{i_1}, ..., \xi_{i_n}) = (-1)^n \frac{\partial^n}{\partial \xi_{i_1} ... \partial \xi_{i_n}} e^{-\frac{1}{2}\xi^T \xi}
\tag{58}
$$

The first two terms in equation (25) represent the Gaussian component of the function $\mu(\theta)$. Therefore, for a Gaussian process, this expansion reduces to a single summation, the coefficients a_{i_1} being the coefficients in the Karhunen-Loeve expansion of the process. Note that equation (25) is a convergent series representation for the functional operator $h[.]$ appearing in equation (23). For a given non-Gaussian process defined by its probability distribution function, a representation in the form given by equation (25) can be obtained by projecting the process on the successive Homogeneous Chaoses. This can be achieved by using the inner product defined by equation (2) to determine the requisite coefficients. This concept has been successfully applied in devising efficient variance reduction techniques to be coupled with the Monte Carlo simulation method (Chorin, 1971; Maltz and Hitzl, 1979).

4. Projection on the Homogeneous Chaos

In this section the Karhunen-Loeve expansion and the Polynomial Chaos expansion presented earlier are implemented into a stochastic finite element method which features a number of similarities with the deterministic finite element method. Specifically, the geometric interpretation of the finite element method as a projection in function space is preserved.

Equation (6) constitute the starting point. Assuming that

$$\mathbf{\Pi}(\mathbf{x}, \omega)[.] = \alpha(\mathbf{x}, \theta) \, \mathbf{R}(\mathbf{x})[.] , \tag{59}$$

and expanding $\alpha(\mathbf{x}, \theta)$ in a Karhunen-Loeve series gives

$$\left(\mathbf{L}(\mathbf{x}) + \sum_{n=1}^{M} \xi_n \, a_n(\mathbf{x}) \, \mathbf{R}(\mathbf{x}) \right) [u(\mathbf{x}, \theta)] = f(\mathbf{x}, \theta) . \tag{60}$$

Assuming, without loss of generality, that $u(\mathbf{x}, \theta)$ is a second order process, it lends itself to a Karhunen-Loeve expansion of the form

$$u(\mathbf{x}, \theta) = \sum_{j=1}^{L} e_j \, \chi_j(\theta) \, b_j(\mathbf{x}) , \tag{61}$$

where

$$\int_{\mathbf{D}} C_{uu} (\mathbf{x_1} , \mathbf{x_2}) \, b_j(\mathbf{x_2}) \, d\mathbf{x_2} = e_j \, b_j(\mathbf{x_1}) , \tag{62}$$

and

$$\chi_j(\theta) = \frac{1}{e_j} \int_{\mathbf{D}} u(\mathbf{x}, \theta) \, b_j(\mathbf{x}) \, d\mathbf{x} . \tag{63}$$

Obviously, the covariance function $C_{uu}(\mathbf{x_1}, \mathbf{x_2})$ of the response process is not known at this stage. Thus, e_j and $b_j(\mathbf{x})$ are also not known. Further, $u(\mathbf{x}, \theta)$, not being a Gaussian process, the set $\chi_j(\theta)$ is not a Gaussian vector. Therefore, equation (61) is of little use in its present form. Relying on the discussion concerning the Homogeneous Chaos, the second order random variables $\chi_j(\theta)$ can be represented by the mean-square convergent expansion

$$\chi_j(\theta) = a_{i_0}^{(j)} \, \Gamma_0 + \sum_{i_1=1}^{\infty} a_{i_1}^{(j)} \, \Gamma_1(\xi_{i_1})$$

$$+ \sum_{i_1=1}^{\infty} \sum_{i_2=1}^{i_1} a_{i_1,i_2}^{(j)} \ \Gamma_2(\xi_{i_1},\xi_{i_2}) + \sum_{i_1=1}^{\infty} \sum_{i_2=1}^{i_1} \sum_{i_3=1}^{i_2} a_{i_1 i_2 i_3}^{(j)} \ \Gamma_3(\xi_{i_1},\xi_{i_2},\xi_{i_3})$$

$$+ \sum_{i_1=1}^{\infty} \sum_{i_2=1}^{i_1} \sum_{i_3=1}^{i_2} \sum_{i_4=1}^{i_3} a_{i_1 i_2 i_3 i_4}^{(j)} \ \Gamma_4(\xi_{i_1} \ , \ \xi_{i_2} \ , \ \xi_{i_3} \ , \ \xi_{i_4}) + \cdots, \tag{64}$$

where $a_{i_1,\ldots i_p}^{(j)}$ are deterministic constants independent of θ and $\Gamma_p(\xi_{i_1}, \ldots, \xi_{i_p})$ is the pth order Homogeneous Chaos. Equation (64) is truncated after the Pth polynomial and is rewritten for convenience, as discussed in equation (27), in the following form,

$$\chi_j(\theta) \ = \ \sum_{i=0}^{P} x_i^{(j)} \ \Psi_i[\{\xi_r\}] \ , \tag{65}$$

where $x_i^{(j)}$ and $\Psi_i[\{\xi_r\}]$ are identical to $a_{i_1\ldots i_p}^{(j)}$ and $\Gamma_p(\xi_{i_1},\ldots\xi_{i_p})$, respectively. In equation (65), P denotes the total number of Polynomial Chaoses used in the expansion, excluding the *zeroth* order term. Given the number M of terms used in the Karhunen-Loeve expansion, and the order p of Homogeneous Chaos used, P may be determined by the equation

$$P \ = \ 1 \ + \ \sum_{s=1}^{p} \frac{1}{s!} \prod_{r=0}^{s-1} (M+r) \ . \tag{66}$$

Substituting equation (65) for $\chi_j(\theta)$, equation (61) becomes

$$u(\mathbf{x},\theta) \ = \ \sum_{j=1}^{L} \sum_{i=0}^{P} x_i^{(j)} \ \Psi_i[\{\xi_r\}] \ c_j(\mathbf{x}) \ , \tag{67}$$

where

$$c_j(\mathbf{x}) \ = \ e_j \ b_j(\mathbf{x}) \ . \tag{68}$$

Changing the order of summation in equation (67) gives

$$u(\mathbf{x},\theta) \ = \ \sum_{i=0}^{P} \Psi_i[\{\xi_r\}] \sum_{j=1}^{L} x_i^{(j)} \ c_j(\mathbf{x})$$

$$= \ \sum_{i=0}^{P} \Psi_i[\{\xi_r\}] \ d_i(\mathbf{x}) \ , \tag{69}$$

where,

$$d_i(\mathbf{x}) \ = \ \sum_{k=1}^{L} x_i^{(j)} \ c_j(\mathbf{x}) \ . \tag{70}$$

Substituting equation (69) for $u(\mathbf{x}, \theta)$, equation (60) becomes

$$\left(\mathbf{L}(\mathbf{x}) + \sum_{n=1}^{M} \xi_n \, a_n(\mathbf{x}) \, \mathbf{R}(\mathbf{x}) \right) \left[\sum_{j=0}^{P} \Psi_j[\{\xi_r\}] \, d_j(\mathbf{x}) \right] = f(\mathbf{x}) , \quad (71)$$

where reference to the parameter θ was eliminated for notational simplicity. The response $u(\mathbf{x}, \theta)$ can be completely determined once the functions $d_i(\mathbf{x})$ are known. In terms of the eigenfunctions $b_j(\mathbf{x})$ of the covariance function of $u(\mathbf{x}, \theta)$, $d_i(\mathbf{x})$ can be expressed as

$$d_i(\mathbf{x}) = \sum_{j=1}^{L} x_i^{(j)} \, e_j \, b_j(\mathbf{x})$$

$$= \sum_{j=1}^{L} y_i^{(j)} \, b_j(\mathbf{x}) . \quad (72)$$

Equation (71) may be written in an alternative form

$$\sum_{j=0}^{P} \Psi_j[\{\xi_j\}] \, \mathbf{L}(\mathbf{x}) \, [d_j(\mathbf{x})] + \sum_{j=0}^{P} \sum_{i=1}^{M} \xi_i \, \Psi_j[\{\xi_r\}] \, \mathbf{R}(\mathbf{x}) \, [d_j(\mathbf{x})] = f(\mathbf{x}). \quad (73)$$

This form of the equation shows that $d_j(\mathbf{x})$ belongs to the intersection of the domains of $\mathbf{R}(\mathbf{x})[.]$ and $\mathbf{L}(\mathbf{x})[.]$. Then, following the standard deterministic finite element method, the function $d_j(\mathbf{x})$ may be expanded in an appropriate function space as

$$d_j(\mathbf{x}) = \sum_{k=1}^{N} d_{kj} \, g_k(\mathbf{x}) . \quad (74)$$

Then, equation (73) becomes

$$\sum_{j=0}^{P} \sum_{k=1}^{N} d_{kj} \, \Psi_j[\{\xi_r\}] \, \mathbf{L}(\mathbf{x}) \, [g_k(\mathbf{x})] \quad (75)$$

$$+ \sum_{j=0}^{P} \sum_{i=1}^{M} \xi_i(\theta) \, \Psi_j[\{\xi_r\}] \sum_{k=1}^{N} d_{kj} \, \mathbf{R}(\mathbf{x}) \, [g_k(\mathbf{x})] = f(\mathbf{x}) .$$

Equation (75) may be rearranged to give

$$\sum_{j=0}^{P} \sum_{k=1}^{M} d_{kj} \, [\, \Psi_j[\{\xi_r\}] \, \mathbf{L}(\mathbf{x}) \, [g_k(\mathbf{x})]$$

$$+ \sum_{i=1}^{M} \xi_i(\theta) \ \Psi_j[\{\xi_r\}] \ \mathbf{R}(\mathbf{x}) \ [g_k(\mathbf{x})] \Bigg] = f(\mathbf{x}) . \qquad (76)$$

Multiplying both sides of equation (76) by $g_l(\mathbf{x})$ and integrating throughout yields

$$\sum_{j=0}^{P} \sum_{k=1}^{M} d_{kj} \ \Bigg[\ \Psi_j[\{\xi_r\}] \int_{\mathbf{D}} \mathbf{L}(\mathbf{x}) \ [g_k(\mathbf{x})] \ g_l(\mathbf{x}) \ d\mathbf{x}$$

$$+ \sum_{i=1}^{M} \xi_i \ \Psi_j[\{\xi_r\}] \int_{\mathbf{D}} \mathbf{R}(\mathbf{x}) \ [g_k(\mathbf{x})] \ g_l(\mathbf{x}) \ d\mathbf{x} \ \Bigg]$$

$$= \int_{\mathbf{D}} f(\mathbf{x}) \ g_l(\mathbf{x}) \ d\mathbf{x} \ , \ l = 1 \ ,..., \ N \ . \qquad (77)$$

Setting

$$\mathbf{L}_{kl} = \int_{\mathbf{D}} \mathbf{L}(\mathbf{x}) \ [g_k(\mathbf{x})] \ g_l(\mathbf{x}) \ d\mathbf{x} \qquad (78)$$

$$\mathbf{R}_{ikl} = \int_{\mathbf{D}} \mathbf{R}(\mathbf{x}) \ [g_k(\mathbf{x})] \ g_l(\mathbf{x}) \ a_i(\mathbf{x}) \ d\mathbf{x} \qquad (79)$$

$$\mathbf{f}_l = \int_{\mathbf{D}} f(\mathbf{x}) \ g_l(\mathbf{x}) \ d\mathbf{x} \ , \qquad (80)$$

equation (77) becomes

$$\sum_{j=0}^{P} \sum_{k=1}^{N} \ \Bigg[\ \Psi_j[\{\xi_r\}] \ \mathbf{L}_{kl} + \sum_{i=1}^{M} \xi_i(\theta) \ \Psi_j[\{\xi_r\}] \ \mathbf{R}_{ikl} \ \Bigg] d_{kj}$$

$$= \ \mathbf{f}_l \ , \ l = 1, ..., N \ . \qquad (81)$$

Note that the index j spans the number of Polynomial Chaoses used, while the index k spans the number of basis vectors used in $\mathbf{C}^m$. Multiplying equation (81) by $\Psi_m[\{\xi_r\}]$, averaging throughout and noting that

$$<\Psi_j[\{\xi_r\}] \ \Psi_m[\{\xi_r\}]> \ = \ \delta_{jm} <\Psi_m^2[\{\xi_r\}]> , \qquad (82)$$

one can derive

$$\sum_{k=1}^{N} <\Psi_m^2[\{\xi_r\}]>\mathbf{L}_{kl} d_{km} + \sum_{j=0}^{P} \sum_{k=1}^{N} d_{kj} \sum_{i=1}^{M} <\xi_i(\theta)\Psi_j[\{\xi_r\}]\Psi_m[\{\xi_r\}]>\mathbf{R}_{ikl}$$

$$= <\mathbf{f}_l \ \Psi_m[\{\xi_r\}]>, \quad l = 1,...,N \ , \ m = 1,...,P \ . \qquad (83)$$

Introducing

$$c_{ijm} \equiv <\xi_i \ \Psi_j[\{\xi_r\}] \ \Psi_m[\{\xi_r\}]> , \qquad (84)$$

and assuming, without loss of generality, that the Polynomial Chaoses have been normalized, equation (83) becomes

$$\sum_{k=1}^{N} \mathbf{L}_{kl} \ d_{km} \ + \ \sum_{j=0}^{P} \sum_{k=1}^{N} d_{kj} \sum_{i=1}^{M} \mathbf{R}_{ikl} \ c_{ijm} \ = \ <f_l \ \Psi_m[\{\xi_r\}]> ,$$

$$l \ = \ 1, \ldots, N \ , \ m \ = \ 1, \ldots, P . \qquad (85)$$

For a large number of index combinations the coefficients c_{ijm} are identically zero. Equation (84) was implemented using the symbolic manipulation program MACSYMA (1986). Forming equation (83) for all P values of m, produces a set of $N \times P$ algebraic equations of the form

$$[\ \mathbf{G} \ + \ \mathbf{R} \ ] \ \mathbf{d} \ = \ \mathbf{h} , \qquad (86)$$

where $\mathbf{G}$ and $\mathbf{R}$ are block matrices of dimension $N \times P$. Their mj^{th} blocks are N-dimensional square matrices given by the equations

$$\mathbf{G}_{mj} \ = \ \delta_{mj} \ \mathbf{L} , \qquad (87)$$

and

$$\mathbf{R}_{mj} \ = \ \sum_{i=1}^{M} c_{ijm} \ \mathbf{R}_i . \qquad (88)$$

In equations (87) and (88), $\mathbf{L}$ and $\mathbf{R}_i$ denote N-dimensional square matrices whose kl^{th} element is given by equations (78) and (79), respectively. In equation (86), $\mathbf{h}$ signifies the $N \times M$ vector whose m^{th} block is given by the equation

$$\mathbf{h}_m \ = \ <f \ \Psi_m[\{\xi_r\}]> . \qquad (89)$$

The N-dimensional vectors $\mathbf{d}_m$ can be obtained as the subvectors of the solution to the deterministic algebraic problem given by equation (86). Once these coefficients are obtained, back substituting into equation (69) yields an expression of the response process in terms of the Polynomial Chaoses of the form

$$\mathbf{u} \ = \ \sum_{j=0}^{P} \mathbf{d}_j \ \Psi_j[\{\xi_r\}] . \qquad (90)$$

Based on equation (90), realizations of the random response vector can be computed from realizations of the random variables $\{\xi_r\}$. Also, statistical moments of the random response vector can be evaluated using the inner product defined in equation (2).

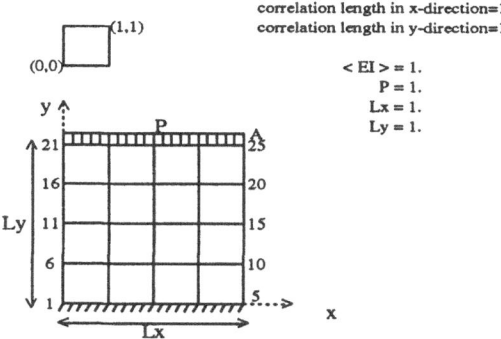

Figure 1. Plate with Random Rigidity; Exponential Covariance Model.

5. Numerical Examples

The preceding development of the stochastic finite element method was applied to a number of problems from engineering mechanics. The first step in the solution of any of these problems was the solution of the eigenvalue problem associated with the Karhunen-Loeve expansion. Following that, the coefficients in the Polynomial Chaos expansion for the solution process were computed. Finally, various statistics, as well as the probability distribution of the solution process were numerically evaluated. Figure (1) shows a thin plate whose modulus of elasticity is assumed to be a two-dimensional random process. The plate is analyzed using the stochastic finite element formulation described above. Figure (2) compares some of the coefficients in equation (90) for various levels of approximation; note the excellent convergence. Figure (3) shows the variation of the standard deviation of the response against the standard deviation of the material property again for various levels of approximation. Finally, figure (4) shows the probability distribution of the response variable at the free corner of the plate.

6. Conclusions

A method for the solution of differential equations with random processes as coefficients was discussed. The method relies on viewing the random aspect of the problem as an added dimension, and on treating random variables and processes as functions defined over that dimension. In this

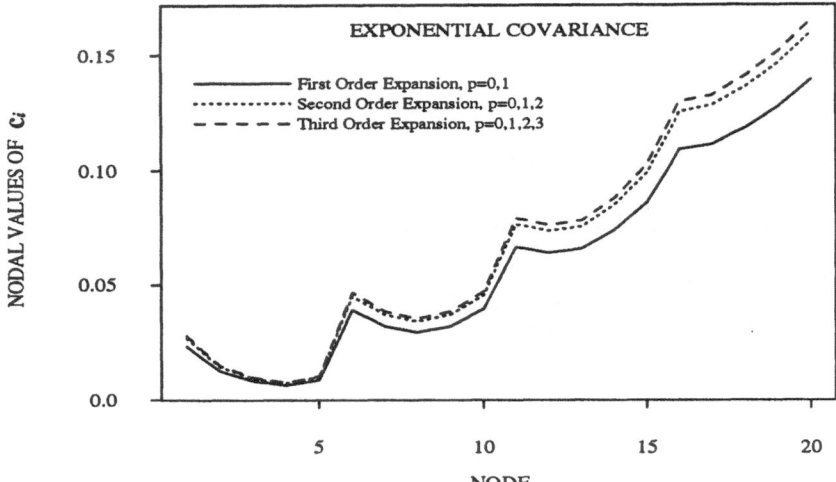

Figure 2.   Linear Interpolation of the Nodal Values of the Vector c_i of Equation (5.73) for the Rectangular Plate Stretching Problem, $i = 2$; Longitudinal Displacement Representation; 2 Terms in K-L Expansion, $M = 2$; $P = 3, 6, 10$.

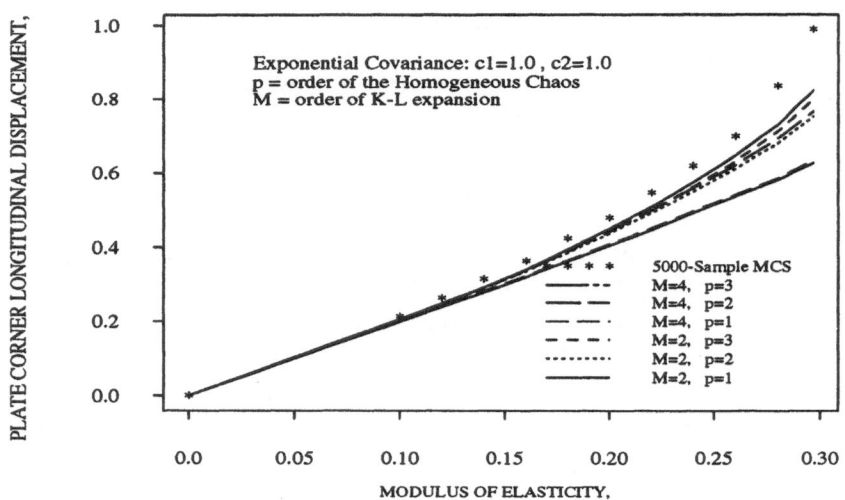

Figure 3.   Normalized Standard Deviation of Longitudinal Displacement at Corner A of the Rectangular Plate, versus Standard Deviation of the Modulus of Elasticity; $\sigma_{max} = 0.433$; Exponential Covariance; Polynomial Chaos Solution

manner, a formulation for the stochastic finite element method was derived which could be construed as a natural extension of the deterministic

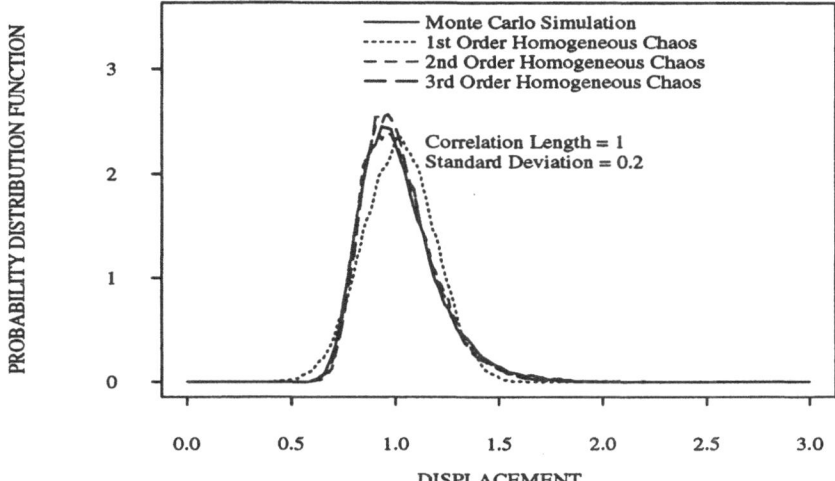

Figure 4. Longitudinal Displacement at the Free End of the Rectangular Plate; Probability Density Function Using 30,000-Sample MSC, and Using Third Order Homogeneous Chaos; Four Terms in the K-L Expansion; Exponential Covariance.

finite element method. Finite element representation along the random dimension was achieved via two spectral expansions. One of them was used to represent the coefficients of the differential equation which model the random material properties, the other was used to represent the random solution process. The new concepts were implemented using a number of computational models for simple engineering systems. The convergence of the discussed approximations was demonstrated numerically. Probability distribution functions of the response variables were obtained.

The present formulation can be viewed as a definite step towards a unification of various finite element techniques. Indeed it consists of generalizing the concepts of finite element approximation to abstract spaces, of which the usual euclidian space is a special case. The deterministic case can then be regarded as a digression of this formalism to the particular instance when the space of elementary events consists of a single element, and where the probability density function induced on the associated σ-algebra is the uniform distribution.

References

Cameron, R.H. and Martin, W.T. (1947) "The orthogonal development of nonlinear functionals in series of Fourier-Hermite functionals", *Ann. Math*, Vol. 48, pp. 385-392.

Chorin, A. J. (1971) "Hermite expansions in Monte-Carlo computation", *Journal of Com-*

putational Physics, Vol. 8, pp. 472-482.

Courant and Hilbert (1953) *Methods of Mathematical Physics,*, Interscience, New York, 1953.

Doob, J.L. *Stochastic Processes*, Wiley, New York.

Ghanem, R. and Spanos, P. (1991) *Stochastic Finite Elements: A Spectral Approach*, Springer-Verlag, New York.

Juncosa, M. (1945) "An integral equation related to the Bessel functions", *Duke Mathematical Journal*, Vol. 12, pp. 465-468.

Kac, M. and Siegert, A.J.F. (1947) "An explicit representation of a stationary gaussian process", *Ann. Math. Stat.*, Vol. 18, pp.438-442.

Kakutani, S. (1961) "Spectral analysis of stationary gaussian processes", *Proceedings of the Fourth Berkeley Symposium on Mathematical Statistics and Probability*, Neyman J. Editor, University of California, Vol. II, pp. 239-247.

Karhunen, K. (1960) "Uber lineare methoden in der wahrscheinlichkeitsrechnung", *Amer. Acad. Sci., Fennicade, Ser. A, I*, Vol. 37, pp. 3-79, 1947; (Translation: RAND Corporation, Santa Monica, California, Rep. T-131, Aug.).

Kolmogorov, A.N. (1950) *Foundations of the Theory of Probability*, Springer, 1933, (English Translation, Chelsea, New York.)

Kree, P. and Soize, C. (1986) *Mathematics of Random Phenomena*, MIA, Reidel Publishing, Boston, Massachussets.

Loeve, M. (1948) "Fonctions aleatoires du second ordre", supplement to P. Levy, *Processus Stochastic et Mouvement Brownien*, Paris, Gauthier Villars.

Loeve, M. (1977) *Probability Theory, 4th edition*, Springer-Verlag, Berlin.

MACSYMA, Reference Manual, Version 12 (1986) Symbolics Inc..

Maltz, F.H. and Hitzl, D.L. (1979) "Variance reduction in Monte-Carlo computations using multi-dimensional Hermite polynomials", *Journal of Computational Physics*, Vol. 32, pp. 345-376.

Oden J. T. (1979) *Applied Functional Analysis*, Prentice-Hall, Englewood Cliffs, New Jersey.

Slepian, D. and Pollak, H.O. (1961) "Prolate spheroidal wave functions; Fourier analysis and uncertainty - I", *Bell System Technical Journal*, pp. 43-63.

Van Trees, H.L. (1968) *Detection, Estimation and Modulation Theory, Part 1*, Wiley, New York.

Volterra, V. (1913) *Lecons sur les Equations Integrales et Integrodifferentielles*, Paris: Gauthier Villars.

Wiener, N. (1938) "The homogeneous chaos", *Amer. J. Math*, Vol. 60, pp. 897-936.

STOCHASTIC FINITE ELEMENTS VIA RESPONSE SURFACE: FATIGUE CRACK GROWTH PROBLEMS

P. COLOMBI

Department of Structural Engineering, Polytechnic of Milan
Piazza L. Da Vinci 32, I20133, Milan (Italy)

AND

L. FARAVELLI

Department of Structural Mechanics, University of Pavia
Via Abbiategrasso 211, I27100, Pavia (Italy)

Abstract. A stochastic finite element method based on an extended response surface technique is coupled with fracture mechanics concepts to evaluate the lifetime of a cracked structural component subjected to cyclic loading. For a complex structure the relationship between the fracture mechanics parameters and the crack depth can only be obtained by numerical approaches based on the discretization of the continuum into finite elements. As a consequence, any random field describing the stochastic nature of the input parameters has to be discretized into stochastic finite elements. An extended response surface approach is used to characterize in closed form the numerical input-output stochastic relationship. This information is used to define a fatigue crack growth model for the evaluation of the lifetime probability distribution function.

1. Introduction

Fracture mechanics [3] [8] [9] [28] [1] [4] [37] is a central topic in modern engineering and technology. Due to the randomness of microdefects, material properties and external loads, deterministic approaches are not effective in modelling the problem. Probabilistic fracture mechanics [36] [5] [6] [30] [31] [44] [29], which combines fracture mechanics and stochastic models [17] [32], provides a useful tool of analysis [21]. In the specific scheme adopted in this study, the randomness of the geometric input quantities is propagated by a

C. Guedes Soares (ed.), Probabilistic Methods for Structural Design, 313–338.
© *1997 Kluwer Academic Publishers.*

response surface technique, in order to give a stochastic characterization of the relation between the crack length and some fracture mechanics parameters. This relation is the basis for any subsequent evaluation of the fatigue lifetime. A significant scatter is in fact observed due to the sensitiveness of the fatigue crack growth to the values of parameters which cannot be accurately determined. Moreover crack trajectories for multiple experiments under identical loading conditions are not identical. This is due to the existence of random microdefects, such as voids, inclusions and microcracks in a heterogenous materials. The change of mechanical and geometrical parameters due to microdefect leads to a substantial difference in both the crack path and the fatigue life. In this chapter the stochastic relation between the crack length and the classical fracture mechanics parameters is first evaluated. This relation is then used in a fatigue crack growth model to evaluate the lifetime probability distribution function. A brief outline of the response surface approach to stochastic finite element problems is also provided.

2. Response Surface as Probabilistic Finite Element Method

The stochastic finite element approach adopted in this chapter is based on the evaluation of a response surface. This is done by regression analysis on the output of numerical experiments, appropriately planned [17] [23] [24] [26] [7]. The surface describes the output of a mechanical system as a function of the input variables modelled as random variables, stochastic processes or random fields. The reader is referred to [17] for details on the procedure which is just summarized in this section. The number of experiments to be performed becomes rather high when several input random variables are considered. The input random variables are therefore grouped into two classes as the result of a preliminary sensitivity analysis: primary random variables and secondary random variables. The primary random variables are the random variables whose randomness strongly influences the randomness of the structural response. The secondary random variables are the remaining ones. Let $\underline{x}$ be the vector of the principal random variables and y the structural response of interest. Suitable transformations Y of y and X_j of x_j are introduced in order to make the model more flexible. The relation between Y and X_j is described using a second order polynomial [17]. In matrix notation one has:

$$Y = \alpha_0 + \mathbf{X}^T \alpha_1 + \mathbf{X}^T \alpha_2 \mathbf{X} + \epsilon \tag{1}$$

The coefficients α_0, α_1 and α_2 are evaluated by a regression analysis over the results of the numerical experiments [17]. The term ϵ on the hand-side takes into account the model error and the effect of the secondary random variables as well as that of random vectors and stochastic processes. Let y_1 and y_2 be two response variables, and Y_1, Y_2 their appropriate transformations. When this stochastic finite element analysis is solved, the marginal probability distribution functions of both Y_1 and Y_2 are known. The variables Y_1 and Y_2 are generally correlated and, hence the model given in Eq. (1) for Y_1 can be updated in the form:

$$Y_1 = \alpha'_0 + \mathbf{X}^T \alpha'_1 + \mathbf{X}^T \alpha'_2 \mathbf{X} + c_1 Y_2 + \epsilon'_1 \tag{2}$$

where α'_0, α'_1, α'_2 and c_1 are the coefficients of the new regression problem. Classical statistical theory [25] provides the way to evaluate the correlation coefficient ρ_{Y_1,Y_2} between Y_1 and Y_2. If more information is necessary, one also estimates the joint probability distribution of Y_1 and Y_2. In terms of probability distribution function, it is given by :

$$P_{Y_1,Y_2}(\zeta_1, \zeta_2) = P_{Y_1|Y_2}(\zeta_1|\zeta_2) P_{Y_2}(\zeta_2) \tag{3}$$

The cumulative distribution function $P_{Y_2}(\zeta_2)$ is obtained applying level-2 reliability methods to Eq. (1) [17]. Moreover the conditional cumulative distribution function $P_{Y_1|Y_2}(\zeta_1|\zeta_2)$ is found by starting from the model given in Eq. (2) with $Y_2 = \zeta_2$.

3. Probabilistic Definition of the Fracture Mechanics Parameters

When a crack is present in a structural component, its analysis requires the knowledge of the stress intensity factor at the tip of that crack. The fracture mechanics parameters (the stress intensity factor K_I and the J integral) depend on the stress distribution in the region of the crack, on the geometry of the crack and on the stiffness of the structure. For simple geometries and loading systems, solutions are available in standard handbooks, but they do not cover complicated structural situations.

3.1. METHODS FOR ESTIMATING THE FRACTURE MECHANICS PARAMETERS

Different methods are available in the literature for the estimation of the fracture mechanics parameters [1]:

1. direct methods;
2. energy based methods;
3. line-spring methods.

In the direct method [1] [8] [28], a straightforward finite element analysis is performed with mesh refinement confined to an area at and around the crack-tip. The stress intensity factor is determined by manipulating the calculated displacements and stresses at the crack-tip. Special techniques are often required to reproduce the $1/\sqrt{r}$ crack tip singularity of the stress and strain field. If the stress σ_{yy} is known along a radial line ahead of the crack tip, the mode I stress intensity factor K_I is then evaluated as follows:

$$K_I = \lim_{x \to 0} \sigma_{yy} \cdot \sqrt{2\pi x} \tag{4}$$

where x is the distance from the crack tip. The value of K_I can be obtained from a plot as x approaches zero. The same procedure can be repeated using displacements instead of stresses.

The evaluation of the fracture mechanics parameters is obtained by energy methods [1] [8] directly from the definition of the energy release rate G:

$$G = \frac{dU}{da} = \lim_{\Delta a \to 0} \frac{U(a + \Delta a) - U(a)}{\Delta a} \tag{5}$$

where U is the strain energy per unit volume due to the loading. Once the value of the energy release rate is known, one evaluates the stress intensity factor K_I as:

$$
\begin{aligned}
K_I &= \sqrt{E \cdot G} &\text{(plane stress)} \\
K_I &= \sqrt{\frac{E}{1 - \nu^2} \cdot G} &\text{(plane strain)}
\end{aligned}
\tag{6}
$$

The J integral was introduced [34] as an alternative parameter for fracture mechanics problems. It is defined as:

$$J = \oint_C [U \, dy - \boldsymbol{\Gamma} \cdot \frac{\partial \mathbf{u}}{dx} ds] \tag{7}$$

where C is the path of the integral which encloses the crack, $\boldsymbol{\Gamma}$ is the outward traction acting on the contour around the crack, $\mathbf{u}$ is the displacement vector, ds is the increment along the contour path. For linear elastic problems $J = G$, and the stress intensity factor can be evaluated by Eq. (6).

For a thin shell with a surface crack, a simplifed analysis can be conducted by the line-spring method [35] [33]. Let t the local thickness of the shell, a the local depth of the crack and $2l$ the length of the crack. The part-through surface crack is idealized as a through-wall crack of length $2l$ with a series of one-dimensional springs across the crack faces. Because of the uncracked ligament of size $t - a$ in the surface crack problem there are a non zero membrane force N and bending moment M transmitted across the crack face. Let δ and θ represent respectively the relative displacement and rotation of the plate mid-surface across the crack faces. The line-spring method relates the local N and M to δ and θ at each point in the following manner:

$$\left\{ \begin{array}{c} \delta \\ \theta \end{array} \right\} = [\mathbf{C}_{ls}] \cdot \left\{ \begin{array}{c} N \\ M \end{array} \right\} \tag{8}$$

The evaluation of the local compliance matrix $\mathbf{C}_{ls}$ is the central point of the method. Its entries are obtained by modelling the springs as plane strain single-edged crack plate specimen of width t and crack depth a, subjected to axial force N and bending moment M. The local compliance of the single-edged crack plate specimen is set equal to $\mathbf{C}_{ls}$ (see Eq. (8)). The finite element analysis of the structural component provides the relative displacements of the two sides of the crack and hence, using the matrix $[\mathbf{C}_{ls}]^{-1}$, the forces in the truss elements. As these forces are evaluated, the stress intensity factors K_I is obtained as shown in [33] [34] [35]. The J integral can be eventually evaluated by Eq. (6).

3.2. THE NOZZLE-SAFE-END EXAMPLE

The structural system of Fig. 1, a nozzle to safe-end connection is studied as a reference example [16] [21].

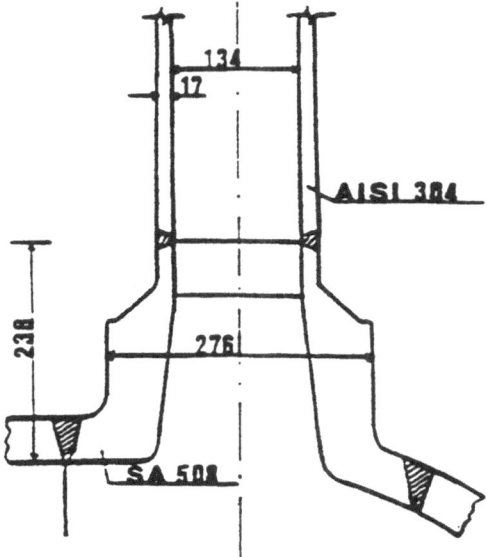

Figure 1.   Nozzle to safe-end connection of the 1:5 scale PWR vessel (A circumferential section; B longitudinal section).

The safe end is modelled with 76 isoparametric, 8-nodes shell elements, and the nozzle with 160 isoparametric, 20-nodes brick elements. The mesh plot (841 nodes) is given in Fig. 2. Fig. 3 shows a detail of the mesh around the crack tip.

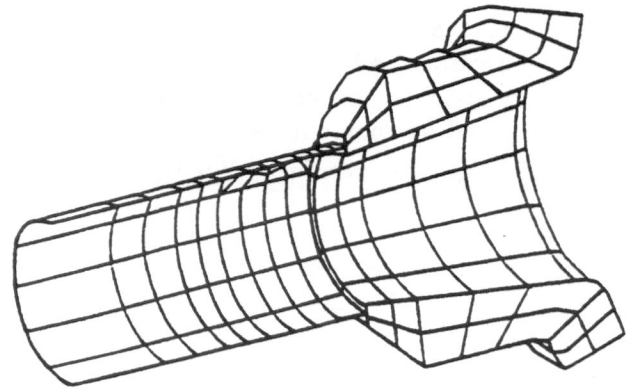

Figure 2.   The finite element mesh used for the structural analysis.

A semielliptical surface crack is supposed to be present in the welded connection between nozzle and safe-end. Line-spring elements are used to discretize the crack. The external loads consist of an internal pressure of 3 MPa and a distributed vertical load for a total of 78 KN at the end of the safe-end. This latter excitation represents the effect of an external cyclic excitation. Horizontal loads are also present at the end of the safe-end in order to reproduce the continuity conditions. The values of the fracture mechanics parameters were assessed by the finite element code ABAQUS [27] which includes the option of an elasto-plastic behaviour for the line-spring elements. The crack is modelled using 15 line-spring elements of different length. In this way the crack front is described with good accuracy for all the situations its probabilistic definition makes possible. In order to change the length of the crack, a mesh refinement in the region around the crack front is required. The vessel and the safe-end are built of two different materials: SA 508 (the nozzle and the weld between the nozzle and the safe-end) and AISI 304 (the safe-end). An elastoplastic material behaviour with isotropic hardening has been assumed for the finite-element analyses and a bilateral approximation is introduced for the stress-strain relationship. The value of the after yielding strain hardening coefficient is 3000 MPa for the SA 508, and 2800 MPa for the AISI 347. The other materials parameters

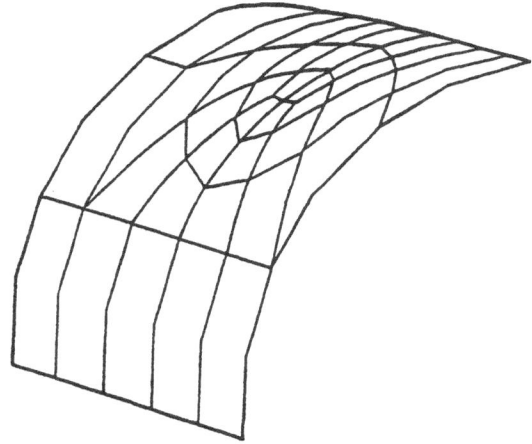

Figure 3. Detail of the finite element mesh around the crack tip.

(the Young's modulus E, Poisson's ratio ν, and the yielding stress σ_y) are listed in Table 1.

TABLE 1. Mechanical properties of the materials (SA 508 and AISI 304) used in the structural analysis of the nozzle to safe-end connection.

Material	E (GPa)	ν	σ_y (MPa)
SA 508	209	0.3	345
AISI	195	0.3	206

Since the objective of the analysis is the estimation of the probabilistic characteristics of the J integral for different values of the crack size a, the variables whose randomness influences the result are : E, σ_y, t and the major semi-axis b of the semielliptical crack. Their probabilistic definition is summarized in Table 2.

Only the major semi-axis b of the crack and the thickness t of the safe-end are considered as primary random variables. The other two random variables listed in Table 2, the Young's modulus E and the yielding stress σ_y of the safe-end, present in fact a very low randomness, and their influence

TABLE 2. Probabilistic definition of the input quantities assumed as random variables in the analysis of the nozzle to safe-end connection.

Physical quantity	Distribution	Mean Value	Standard deviation
Thickness t	Gaussian	17 mm	0.7 mm
Major semi-axis b	Gaussian	21.125 mm	3.25 mm
Young modulus E	Gaussian	195 GPa	3.90 GPa
Yielding stress σ_y	Gaussian	207 MPa	7 MPa

on the response surface is taken into account in the error term as described in [17]: they are regarded as secondary random variables. According to Fig. 2, the crack length b cannot be regarded as a continuous variable. Six levels of line-spring elements were considered in the analysis, the size of each class being 1.625 mm. Therefore the following discrete values of b can be retained: 16.25 mm, 17.875 mm, 19.5 mm, 21.125 mm, 22.75 mm, 24.375 mm and 26 mm. Different transformed variables J' are considered by the response surface code [17], and the one which optimizes the model accuracy is identified. The response surface model of J' can be written for a given crack depth:

$$J' = \theta_0 + \theta_1 t^* + \theta_2 b^* + \theta_3 (t^*)^2 + \theta_4 (b^*)^2 + \theta_5 b^* t^* + \epsilon \qquad (9)$$

where the star characterizes the standardized value of the variables. The coefficients of the model are evaluated by regression analysis over the results of numerical experiments conducted by changing the data file of the finite element code which performs the analysis of the structural system in Fig. 1. The modifications are introduced following a central composite design of experiments [17]. A polynomial interpolation of the J integral obtained using the nominal values of the input random variables listed in Table 2 is given in Fig. 4. The probabilistic equivalent of Fig. 4 is eventually given in Fig. 5.

In particular the probability distributions of the J integral for $a= 3$ and 10 mm, respectively, were computed by coupling Eq. (9) with a special form of the so-called level-2 reliability method [17].

3.3. THE CRACKED-PIPE EXAMPLE

As a second numerical example, consider an infinite long pipe with a surface crack in the axial direction [15] [18] [19]. The radius R and the thickness t of the pipe are 268 mm and 10 mm, respectively. The crack is assumed to be semielliptic: the nominal value of the major semiaxis b is 28.93 mm.

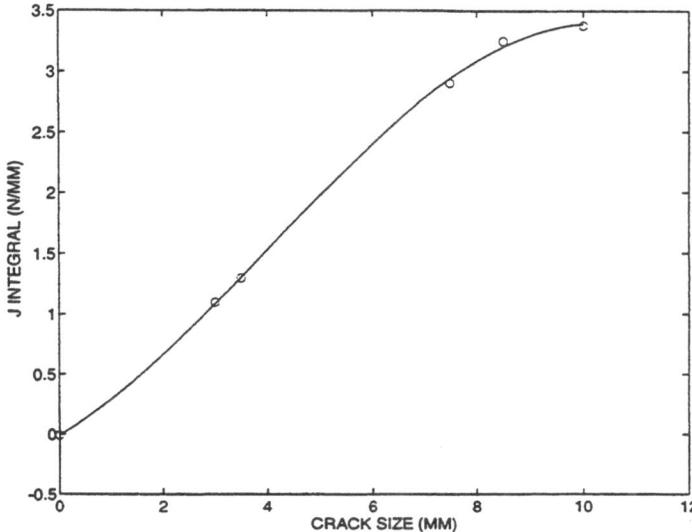

Figure 4. Interpolation curve for the J integral calculated from the nominal value of the input variables listed in Table 2.

Three different values of the depth a of the crack are considered. They correspond to the following values of the ratio a/t: 0.2 (shallow crack deep), 0.46 (moderate deep crack) and 0.7 (deep crack).

Due to the symmetry of the problem, only one half of the pipe was discretized into finite elements by the finite element code ABAQUS [27]. The mesh is given in Fig. 6, and was realized for a length equal to 15 times the semiaxis b of the crack. It was assumed that this length is sufficient to realize the condition of an infinite long pipeline. The special value of the ratio radius/thickness allows a discretization by shell elements. This makes possible to use line-spring elements for the computation of the J integral: 12 line-spring elements were used for the crack discretization. The line-spring elements have different lengths, as shown in Fig. 7.

The material behaviour is assumed to be elasto-plastic with isotropic hardening. The material characteristics are given in Table 3.

The external loads consist of an internal pressure of 12 MPa. Concentrated loads are present at the end of the part of the pipe considered in the analysis in order to reproduce the continuity condition. All these external actions are considered as deterministic quantities.

The input variables assumed as random in the analysis are : the elastic modulus E, the yield stress σ_y, the radius R, the thickness t and the semiaxis b of the crack. Their probabilistic definition is given in Table 4. Only

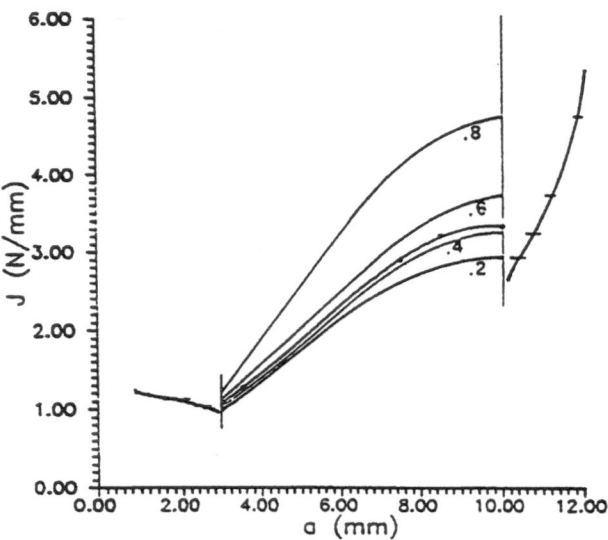

Figure 5. Probability distribution function of the J integral for the crack depth a equal to 3 and 10 mm, respectively.

TABLE 3. Mechanical properties of the material used in the analysis of the pipe.

E (GPa)	ν	σ_y (MPa)
206.8	0.3	482.5

the semiaxis b, the thickness t and the radius R of the pipe are regarded as

TABLE 4. Probabilistic definition of the input quantities assumed as random variables in the analysis of the pipe.

Physical quantity	Distribution	Mean Value	Standard deviation
Thickness t	Gaussian	10 mm	0.67 mm
Radius R	Lognormal	268 mm	2.68 mm
Major semi-axis b	Lognormal	28.93 mm	2.97 mm
Young modulus E	Gaussian	206.8 GPa	6.67 GPa
Yielding stress σ_y	Gaussian	482.5 MPa	14.2 MPa

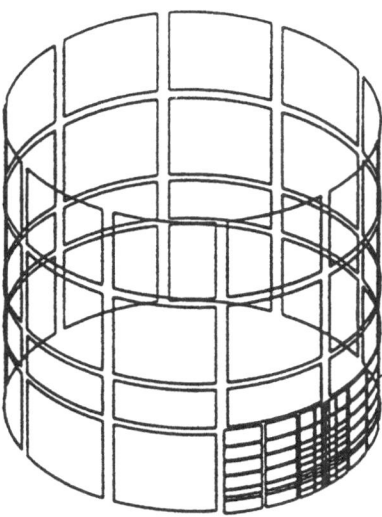

Figure 6. The finite element mesh used for the structural analysis of the pipe.

primary random variables. The mechanical properties of the material (the Young modulus E and the yielding stress σ_y) are regarded as secondary random variables, since it is assumed that the material is well controlled so that the randomness on the material parameters is small. Again the major semiaxis b of the crack can assume only discrete values. Eight classes were used in the analysis to describe the variability of this parameter.

First the cumulative distribution function of the J integral is evaluated for three different values of the crack depth (a/t equal to 0.2, 0.46 and 0.7, respectively) following the procedure illustrated in the previous example. The number of required numerical experiments is 20 for each of the considered crack depths. The results are shown in Fig. 8.

In order to characterize the stochastic relationship between the J integral and the crack depth, the J integral was computed using the following values for the crack depth a: 2 mm - 3 mm - 4 mm - 4.3 mm - 4.6 mm - 4.8 mm - 5 mm - 6 mm - 6.5 mm and 7 mm. In the structural analysis, the random input parameters listed in Table 4 were considered equal to their mean values. The numerical results (Fig. 9) were interpolated by a fourth order polynomial using the least square approach:

$$J(a) = -25.843 + 31.549a - 11.979a^2 + 2.135a^3 - 0.128a^4 \qquad (10)$$

Due to the randomness of the input parameters (see Table 4), the coef-

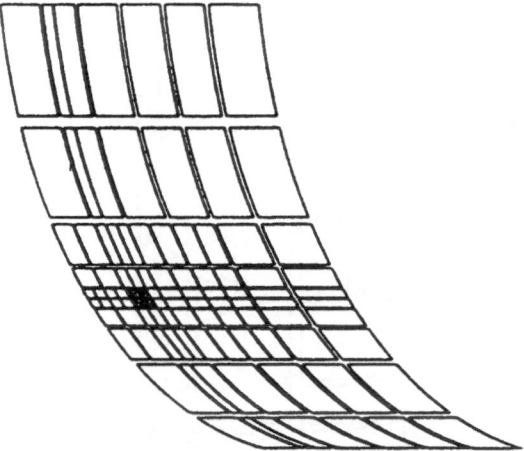

Figure 7. Detail of the finite element mesh around the crack tip for the analysis of the pipe.

ficients of the polynomial approximation of the J integral given in Eq. (10) are random [19]. Eq. (10) is then rewritten as:

$$J(a) = \underline{c}_0 + \underline{c}_1 a + \underline{c}_2 a^2 + \underline{c}_3 a^3 + \underline{c}_4 a^4 \tag{11}$$

where the coefficients $\underline{c}_0, \underline{c}_1, \underline{c}_2, \underline{c}_3$ e $\underline{c}_4$ are random, but correlated variables. Each of the J integral curves given by Eq. (11) is regarded as isoprobability curve.

The points with the same probability in Fig. 9 are interpolated by a fourth order polynomial (the first derivative of the polynomials at the extreme value of the crack depth range is kept constant).

4. Fatigue Crack Growth

The response surface stochastic finite element method can now be used to evaluate the lifetime distribution of a structural component subject to cyclic loads. The idea is to couple the probabilistic finite element method with a fatigue crack growth model [20] [12] [11] [21].

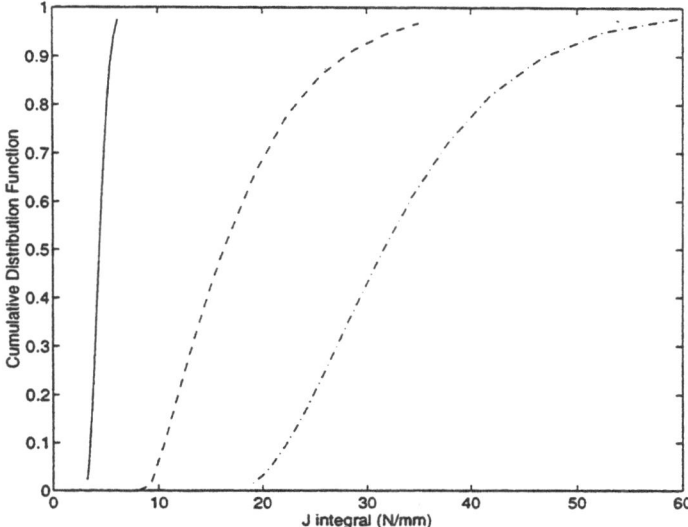

Figure 8. Cumulative distribution function of the J integral for three different values of the ratio a/t: $a/t=0.2$ (solid line), $a/t=0.46$ (dashed line) and $a/t=0.7$ (dashdot line).

4.1. FATIGUE LIFETIME DISTRIBUTION

The stochastic differential equation governing the crack growth is [14] [13]:

$$\begin{aligned}
d\underline{a}(N) &= C \cdot (\Delta K_{eq})^m [\mu_{\underline{x}} + \rho \cdot \cos\underline{\phi}(N)]dN \\
d\underline{\phi}(N) &= \sigma \cdot d\underline{w}(N)
\end{aligned} \tag{12}$$

with $\underline{\phi}(0) = 0$ and $\underline{a}(0) = a_0$. In Eq. (12) $\underline{w}(t)$ is a Wiener process, ρ and σ are model parameters and $\mu_{\underline{x}}$ is the mean value of the process $\underline{x}(N)$. The lifetime distribution for a given N is evaluated [14] [13] [12] by a technique based on Hermite moments [38] [39]. The equivalent stress intensity factor range $\Delta K_{eq}(a)$ is calculated for several cracks sizes, and the results fitted by a suitable polynomial interpolation:

$$\Delta K_{eq}(a) = c_0 + c_1 a + c_2 a^2 \tag{13}$$

Note that in this case a second order polynomial is sufficient to model the relation between $\Delta K_{eq}(a)$ and a. In the cracked-pipe example of the previous section, a fourth order polynomials was necessary. Of course the actual nature of the coefficients c_0, c_1 and c_2 is random due to the randomness of the input quantities of the finite element model. An accurate analysis requires the estimation of the randomness of these coefficients. This could

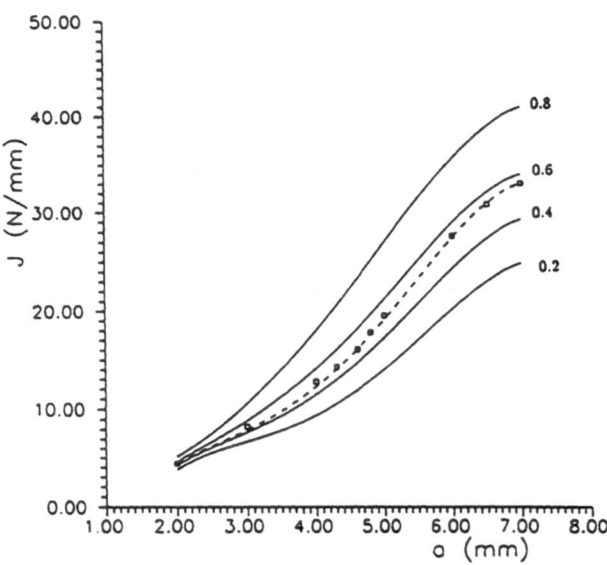

Figure 9. Polynomial approximation of the J integral for several probability levels. The dotted line is obtained using the mean values of the input random quantities listed in Table 4.

be done by a response surface scheme. In summary the proposed method for fatigue crack growth analysis consists of the following steps [11]:

1. the stress intensity factor is estimated for some values of the crack size by a suitable finite element code. The random geometrical variables which form the input for that code are introduced by their mean values;

2. the results are used to fit a suitable polynomial interpolation to the relationship between the root-mean-square of the stress intensity factor range and the crack size;

3. the coefficients c_0, c_1 and c_2 of the polynomial interpolation and of the initial crack length a_0 are regarded as random variables;

4. a set of experiments is planned, according to experiment design theory, in the space of the independent random variables into which the polynomial coefficients c_0, c_1 and c_2 and the initial crack length a_0 have been mapped;

5. the lifetime probability is evaluated, for given crack length a_f and given number of duty cycles, by the method based on Hermite moments;

6. the resulting lifetime probability is modelled by an appropriate response surface;

7. the cumulative distribution function of the lifetime probability is estimated by a level-2 reliability method.

4.2. NUMERICAL EXAMPLE

Consider again the welded connection between the nozzle and the safe-end of the PWR vessel introduced in the previous section (see Fig. 1). A crucial point is the identification of the position and the initial size of the crack. Two types of cracks can be considered:

1. fabrication cracks;
2. cracks nucleated during service.

The present work deals with the first category. The crack under investigation is planar, and its shape is semielliptical. The crack grows initially along the thickness. Once it has reached the leak point, it propagates along the surface until the critical dimension is achieved and catastrophic failure takes place [12]. Attention is focussed on the first stage, i.e. crack propagation occurs along the thickness. The dimension b of the larger radius is 34 mm, and is kept constant in this analysis. The following values for the dimension a of the other radius are considered for the calculation of the fracture mechanics parameters: 1 mm - 3 mm - 7 mm - 10 mm. Indeed the initial value of the crack size a_0 is assumed to be 3 mm.

The vertical force at the end of the safe-end is modelled as a narrow band Gaussian process with zero mean and standard deviation σ_S of 13.42 KN. The equivalent stress intensity factor range ΔK_{eq} (see Eq. (12)) is obtained using the equivalent loading range ΔS_{eq} in the structural analysis. For narrow band Gaussian process it is given by:

$$\Delta S_{eq} = (2\sqrt{2} \cdot \sigma_{\underline{S}} \cdot \Gamma(\frac{m}{2} + 1))^{1/m} \qquad (14)$$

With the value of σ_S listed above, an equivalent loading range equal to 42 KN is obtained. The crack front was modelled in the finite element mesh using 6 line-spring elements in order to describe accurately the shape of the crack front. The equivalent stress intensity factor range ΔK_{eq} was calculated for the crack depth listed above. The result is interpolated by a second order polynomial (see Fig. 10):

$$\Delta K_{eq}(a) = -0.15 \cdot a^2 + 3.86 \cdot a - 1.308 \qquad (15)$$

The fatigue crack growth for SA 508 can be described by the stochastic version of the Paris law given in Eq. (12).

The fatigue crack growth parameters C and m, and the filter parameters $\mu_{\underline{x}}$, ρ and σ, are determined from experimental results on compact tension specimens. Experimental results are reported in Fig. 11: $\log(\frac{da}{dN})$, the logarithm of the crack growth rate, versus $\log(\Delta K)$, the logarithm of the stress intensity factor range [22].

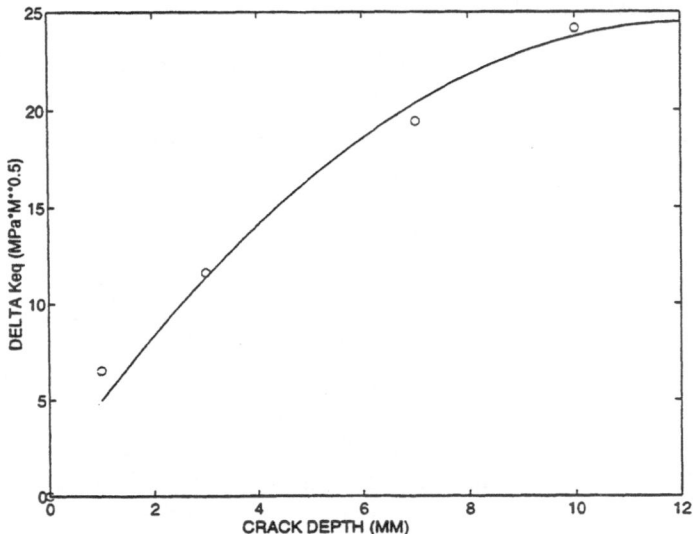

Figure 10.   Interpolation function for the crack depth.

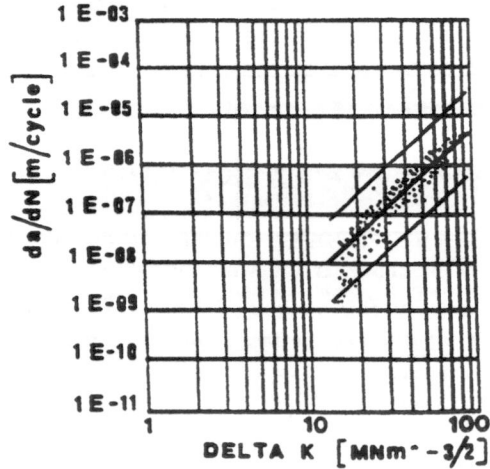

Figure 11.   Regression analysis and 95 % tolerance bands for the pressure vessel steels.

Experimental results are interpolated by a straight line, given in Fig. 11, by the least mean square approach. The slope of the best fit straight line represents the exponent m of the Paris law, while the intercept with the x-axis is the C coefficient. The relevant numerical values are:

$$\begin{aligned} C &= 3.67 \cdot 10^{-12} \quad \text{(MPa units)} \\ m &= 3.071 \end{aligned} \tag{16}$$

A regression analysis gave a residual standard deviation of 0.27. Classical normal-to-lognormal transformation formulae yielded the following model parameters:

$$\begin{aligned} m_{\underline{x}} &= 1.21 \\ \rho &= 1.17 \\ \sigma &= 4.5 \end{aligned} \tag{17}$$

It is assumed that the interpolation coefficient $\underline{c}_0$ is a Gaussian variable with mean value equal to -1.308 and coefficient of variation -0.3, while the second coefficient $\underline{c}_1$ is modelled as a lognormal random variable with median 3.86 and coefficient of variation 0.2. Let $\underline{a}_v$ be the value of the crack length at which the first derivative of the function $s_{\underline{K}}(a)$ is equal to zero. One has:

$$\underline{a}_v = \frac{-\underline{c}_1}{2 \cdot \underline{c}_2} \tag{18}$$

It is assumed that the random variable $\underline{a}_v$ has a mean value equal to 12.86 and a dispersion around the mean value given by a Gaussian random factor $\underline{\Omega}$ with mean value one and standard deviation .1. Then the third coefficient $\underline{c}_2$ is given by:

$$\underline{c}_2 = \frac{-\underline{c}_1}{2 \cdot 12.86 \cdot \underline{\Omega}} \tag{19}$$

Realization of the interpolation coefficient and of the initial crack length was selected according to experimental design theory. The initial crack length is idealized to be a Gaussian random variable with mean value equal to 3 mm and standard deviation 1 mm.

Experimental design theory was used to obtain 20 experimental points in the space of the random variables $\underline{c}_0$, $\underline{c}_1$, $\underline{a}_0$; $\underline{\Omega}$ was regarded as a secondary variable, due to its narrow variability. For each of the planned experiments, Eq. (13) is written. The relevant plots are given in Fig. 12. In this figure the solid line is the plot of Eq. (15).

The lifetime probability was computed for the planned experiments and for a value of the limit crack length a_f equal to 8 mm. The probability distribution of the lifetime probability was evaluated by a level-2 reliability

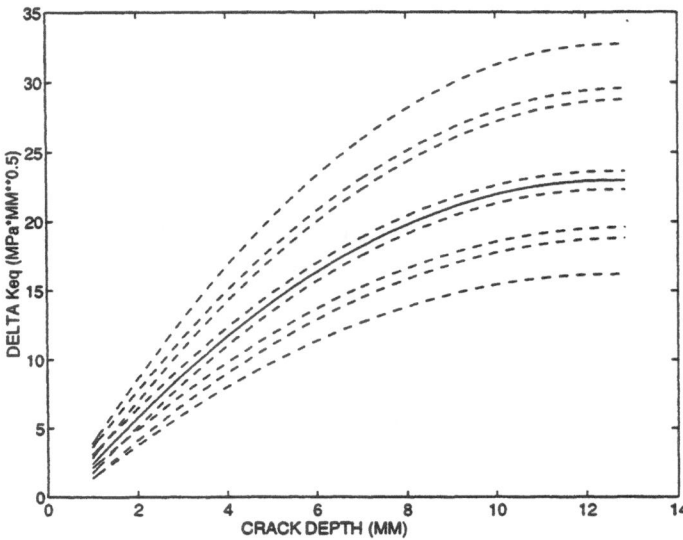

Figure 12. Equivalent stress intensity factor range vs. crack size for the values of the interpolation coefficients obtained from experimental design theory (the solid line shows the polynomial form obtained for the mean values of the interpolation coefficients).

method. The corresponding probability density is then derived. Plots of the probability density of the lifetime probability after 360000, 380000, 400000, 420000 and 440000 duty cycles are given in Fig. 13.

5. Response Surface vs. AMVFO

5.1. ADVANCED MEAN VALUE FIRST ORDER (AMVFO)

The advanced mean value first order concept (AMVFO) [40] [41] [42] [43] has been used in the literature to estimated the cumulative distribution function (CDF) of the number of cycles to failure N_f for a structural component under fatigue. Let $\mathbf{X}$ be the vector of random variables in the expression for N_f. With the advanced mean value concept, the CDF of $N_f(\mathbf{X})$ (N_f being an implicit function of $\mathbf{X}$), is evaluated by introducing a first-order Taylor series expansion of N_f about the mean value $\mu_{\mathbf{X}}$ of $\mathbf{X}$. This requires solutions at $\mu_{\mathbf{X}}$ and at small perturbations about $\mu_{\mathbf{X}}$ in order to evaluate the parameters of the linear function. The function $\bar{N}_f(\mathbf{X})$ is then replaced by a linear function $N'_f(\mathbf{X})$:

$$N'_f(\mathbf{X}) = c_0 + \sum_{i=1}^{k} c_i(X_i - \mu_{X_i}) \tag{20}$$

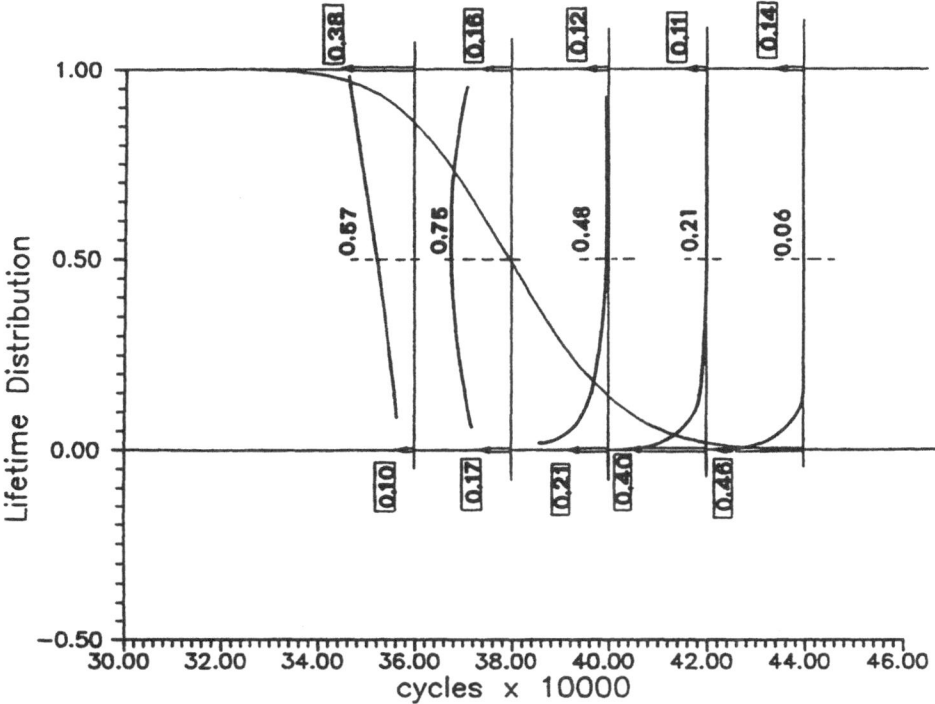

Figure 13. Probability density function of the lifetime probability for $a_f = 8$ mm and some values of N, namely N equal to 360000, 380000, 400000, 420000 and 440000 duty cycles.

where:

$$c_i = [\frac{\partial N_f}{\partial X_i}]_{\mu_{\underline{X}}} \tag{21}$$

The partial derivatives are computed using the perturbed data. The CDF is approximated at discrete points N_{f_i}. The number of points is arbitrary, but it was shown that three or four points on both side of the mean are adequate to define the CDF, when the probability of failure is expected to be a smooth function. A first estimate of the CDF is made from the linear form. The limit state function at each N_{f_i} is:

$$g_i(\mathbf{X}) = N'_f(\mathbf{X}) - N_{f_i} \tag{22}$$

The probability of failure is then computed by:

$$P_f(N_{f_i}) = \Phi(-\beta_i) \tag{23}$$

where β_i is the reliability index [2] associated with each g_i; $P_f(N_{f_i})$ is the probability of failure after N_{f_i} duty cycles, and $\Phi(\cdot)$ the standard normal CDF. This is the mean value first order (MVFO) method: it leads one to estimate the design point $\mathbf{X}^*_i$ for the linear approximation of Eq. (20). To improve the estimate of the CDF, the actual lifetime $N_f(\mathbf{X}^*_i)$ is evaluated at each design point of the sample space identified. These values are associated with the probability of failure $P_f(N_{f_i})$ obtained from Eq. (23): the corresponding plot is regarded as an improved form of the sought CDF. This improvement gives rise to the advanced mean value first order (AMVFO) method. The total number of function evaluations is given by:

$$J = n_v + n_p + 1 \tag{24}$$

where n_v is the total number of random variables and n_p is the number of points used to define the CDF. The estimate of $P_f(N_f)$ could be improved by introducing a new linear approximation for N_f at each design point $\mathbf{X}^*_i$. The reliability analysis is then performed for each of the linear functions; as a result the probability estimate at each value of N_{f_i} is more accurate. Still another improvement could be made by evaluating the real function at each design point. This process can be repeated, but experience suggests that the CDF obtained by the first function evaluation (AMVFO) is generally very close to the exact solution.

The following basic differences from the previously discussed response surface stochastic finite element method can be listed [21] [10]:

1. Eq. (20) is linear while Eq. (1) is quadratic;
2. Eq. (20) is written in the original space of the variables $\underline{\mathbf{X}}$ while Eq. (1) is written in a transformed space (generally the standardized one);
3. A deterministic error term η was not written explicitly in Eq. (20); however for a given vector $\mathbf{X}^*_i$, the probability $P_f(N_{f_i})$ is associated with $N_f(\mathbf{X}^*_i)$ instead of $N'_f(\mathbf{X}^*_i)$ where:

$$N_f(\mathbf{X}^*_i) = N'_f(\mathbf{X}^*_i) + \eta \tag{25}$$

Therefore, Eq. (20) assumes that η can be computed, while the error term ϵ in Eq. (1) is a random variable, the variance of which is estimated through statistical analysis. This makes it possible to incorporate in it the effect of all the uncertainties which affect N_f.

5.2. NUMERICAL EXAMPLE

A numerical example [10] is studied in order to emphasize the difference between the response surface finite element method and the AMVFO method. Upon integration of the Paris law, the number of cycles to failure is given by:

$$N_f = \frac{1}{C} \int_{a_0}^{a_f} \frac{da}{(S_0(a))^m \cdot (Y(a))^m \cdot (\pi a)^{m/2}} \tag{26}$$

where $S_0(a)$ is the equivalent stress range. It is a function of the crack depth because a threshold level ΔK_{th} is introduced for the stress intensity factor range ΔK. Below this level the crack will not propagate. The following espression is retained for the geometry factor $Y(a)$ in Eq. (26):

$$Y(a) = (\psi - 1) \cdot e^{-\gamma \cdot a^\nu} + 1 \tag{27}$$

where ψ is the theoretical stress concentration factor and γ and ν are some constants. Moreover assume that the long term stress range is Weibull distributed with shape parameter ξ and scale parameter δ given by:

$$\delta = \bar{S}_0 \cdot (\ln(N_t))^{-1/\xi} \tag{28}$$

where N_t is the total number of cycles and $\bar{S}_0$ is the design stress range defined by $P(S > \bar{S}_0) = \frac{1}{N_t}$. The numerical data of the terms in Eq. (26) are given in Table 5.

TABLE 5. Numerical data for the deterministic and random parameters in the fatigue crack growth equation.

Variable	Distribution	Mean Value (MPa units)	Standard deviation (MPa units)
$\bar{S}_0$	Type I extreme value	172.05	13.764
ΔK_{th}	Normal	121.394	12.1394
C	Weibull	1.2723E-13	1.2723E-14
a_0	Lognormal	25.4E-02	1.27E-02
ξ	Constant	1.	–
N_t	Constant	1.0E+8	–
m	Constant	3.	–
a_f	Normal	25.4	2.54E-01
ψ	Constant	3.3	–
γ	Constant	4.42	–
ν	Constant	0.44	–

Fig. 14 shows the CDF of the number of cycles to failure computed by the AMVFO method. The Monte Carlo solution is superimposed in order to check the accuracy of the computed CDF.

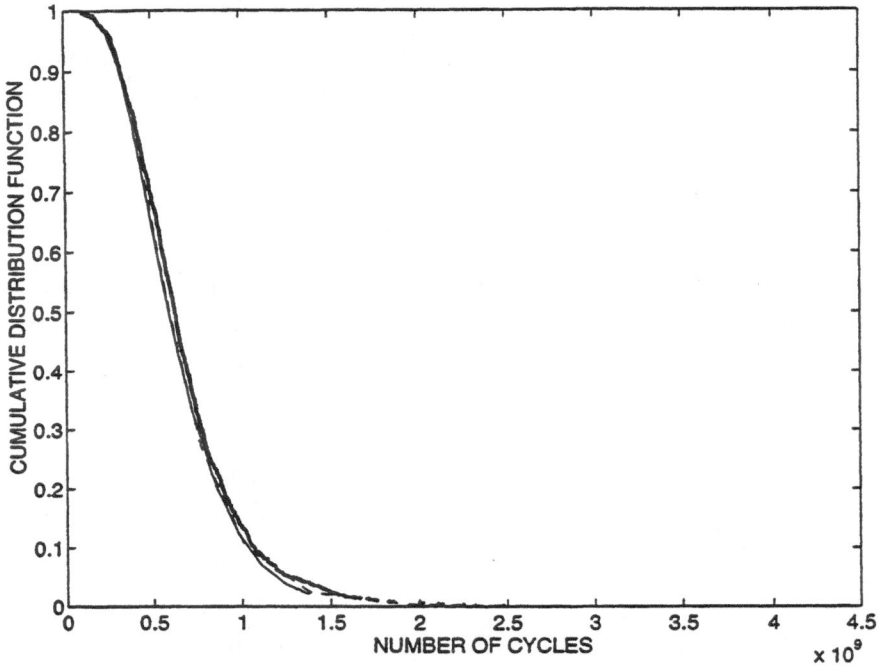

Figure 14. Comparison between the response surface solution (solid line), the AMVFO solution (dashed line) and the Monte Carlo solution.

Fig. 14 shows also the CDF of the number of cycles to failure computed by the response surface stochastic finite element method. In this case also, the comparison is excellent. Since for reliability purposes it is the low tail of the CDF which is of practical interest, the response surface can be shifted in order to give a good approximation of the low value of the number of cycles to failure. The results are given in Fig. 15, where the CDF is compared with the original response surface solution and the Monte Carlo plot. The match of this second solution with the simulation result is very good.

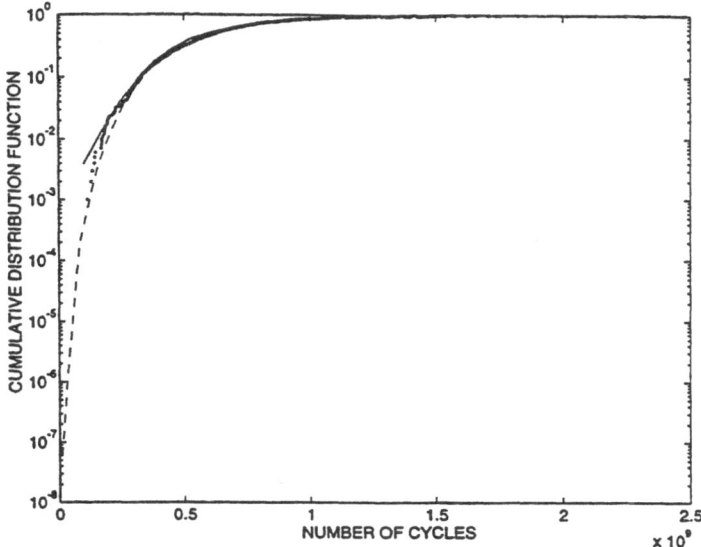

Figure 15. Comparison in logarithmic scale between the shifted response surface solution (dashed line), the original response surface solution (solid line) and the Monte Carlo solution.

6. Conclusions

In this contribution, the stochastic response surface method is applied to fracture mechanics problems. In the first part of the chapter, the method is used to characterize the stochastic relation between the fracture mechanics parameters (the stress intensity factor and the J integral) and the crack depth. For complex structures this relationship is modelled by a suitable polynomial with random coefficients. In this case a numerical scheme (finite element or boundary element method) must be used to evaluate the fracture mechanics parameters. The finite element code ABAQUS was used for this purpose. This relation defines the fatigue crack growth rate in any of the available fatigue crack propagation equations. In the second part of the chapter, this relation is used to evaluate the distribution of the number of cycles to failure. The randomness of the fatigue strength capacity is modelled by a Markov process, and a filter technique is adopted. The lifetime probability is evaluated by a method which makes use of Hermite moments and Itô calculus. The response surface approach is finally compared with the AMVFO method. The AMVFO method is shown to be a special form of the response surface scheme in which the error term is regarded as a deterministic function and, hence, estimable on the basis of specific numerical experiments.

Aknowledgement

This research has been supported by funds from both the Italian Ministry of University and Research (MURST) and the National Research Council (CNR).

References

1. Aliabadi, M.H. and Rooke, D.P. (1991) *Numerical Fracture Mechanics*, Kluwer Academic Publishers.
2. Augusti, G., Baratta, A. and Casciati, F. (1984) *Probabilistic Methods in Structural Engineering*, Chapman & Hall.
3. Bannantine, J.A., Comer, J.J. and Handrock J.L. (1989) *Fundamental of Metal Fatigue Analysis*, Prentice Hall.
4. Barsom J.M. and Rolfe S.T. (1977) *Fracture and Fatigue Control in Structures: Applications of Fracture Mechanics*, Prentice Hall.
5. Besterfield, G.H., Liu, W.K., Lawrence, M.A. and Belytschko, T.B. (1991) Fatigue Crack Growth Reliability by Probabilistic Finite Elements, *Comp. Meth. Appl. Mech. and Engng.*, **86**, 297-320.
6. Besterfield, G.H., Liu, W.K., Lawrence, M.A. and Belytschko, T.B. (1990) Brittle Fracture Reliability by Probabilistic Finite Elements, *J. Engng. Mech. (ASCE)*, **116**(3), 642-659.
7. Breitung, K. and Faravelli, L. (1994) Log-Likelihood Maximization and Response Surface in Reliability Assessment, *Nonlinear Dynamics*, 5, 273-285.
8. Broek, D. (1982) *Elementary Engineering Fracture Mechanics*, Martinus Nijhoff Publishers.
9. Broek, D. (1988) *The Practical Use of Fracture Mechanics*, Kluwer Academic Publishers.
10. Casciati, F. and Colombi, P. (1994) Fatigue Lifetime Prediction for Uncertain Systems, in P.D. Spanos and Y.T. Wu (eds.), *Probability Structural Mechanics : Advances in Structural Reliability Methods*, Springer-Verlag, pp. 87-99.
11. Casciati, F., Colombi, P. and Faravelli, L. (1991) Filter Technique for Stochastic Crack Growth, in P.D. Spanos and C.A. Brebbia (eds.), *Computational Stochastic Mechanics*, Computational Mechanical Pubblications, pp. 485-496.
12. Casciati, F., Colombi, P. and Faravelli, L. (1991) Stochastic Crack Growth by Filter Technique, in L. Esteva and S.R. Ruiz (eds.), *Sixth International Conference on Application of Statistics and Probability in Civil Engineering*, 1, pp. 71-81.
13. Casciati, F., Colombi, P. and Faravelli, L. (1991) Stochastic Crack Growth and Reliability Analysis, in C.G. Soares, C. Ostergaard, M.J. Baker, A. Pittaluga, M. Huter and P. Thof-Christensen (eds.), *10th International Conferences on Offshore Mechanics and Artic Engineering*, ASME, pp. 107-111.
14. Casciati, F., Colombi, P. and Faravelli, L. (1992) Fatigue Crack Size Probability Distribution via Filter Technique, *Fatigue Fract. Engng. Mater. Struct.*, 15(5), 463-475.
15. Casciati, F., Colombi, P. and Faravelli, L. (1993) Lifetime Prediction of Fatigue Sensitive Structural Elements, *Structural Safety*, 12, 105-111.
16. Casciati, F., Colombi, P. and Faravelli, L. (1992) Fatigue Lifetime Evaluation via Response Surface Methodology, in K.E. Petersen and B. Rasmussen (eds.), *Safety and Reliability '92*, Elsevier, pp. 157-166.
17. Casciati, F. and Faravelli, L. (1991) *Fragility Analysis of Complex Structural Systems*, Research Studies Press, Taunton.
18. Colombi, P. (1992) Propagazione delle Incertezze sui Parametri della Meccanica della Frattura (in Italian), *VIII Congresso Nazionale del Gruppo Italiano Frattura*, pp. 157-166.

19. Colombi, P. (1992) Problemi di Fatica per Strutture Complesse (in Italian), *XI Congresso Nazionale AIMETA*, pp. 550-561.
20. Colombi, P. and Faravelli L. (1991), Metodologie per la Valutazione della Vita Utile a Fatica (in Italian), *Problemi di Meccanica dei Materiali e delle Strutture*, pp. 117-126.
21. Colombi, P. (1995) Vita Residua di Componenti Strutturali Metallici Soggetti ad Eccitazione Stocastica (in Italian), Ph.D Dissertation, Polytechnic of Milan-University of Pavia.
22. Evans, W.J. and Bache, M.R. (1991) Crack Formation Lives and Crack Propagation Behaviour of Pressure Vessel Steel, *Private Communication*.
23. Faravelli, L. (1992) Structural Reliability via Response Surface, in N. Bellomo and F. Casciati (eds.), *Proc. Iutam Symposium on Nonlinear Stochastic Mechanics*, Springer-Verlag, pp. 213-223.
24. Faravelli, L. (1989) Response Surface Approach for Reliability Analysis, *J. of Eng. Mechanics (ASCE)*, **115**(12), 2763-2781.
25. Faravelli, L. (1989) Finite Element Analysis of Stochastic Nonlinear Continua, in W.K. Liu and T. Belytschko (eds.), *Computational Mechanics of Probabilistic and Reliability Analysis*, Elmepress, pp. 264-280.
26. Faravelli, L. (1994) Blocking problems in the analysis of random fields, *Probabilities and Materials*, in D. Breysse (ed.), Kluwer Academic Pubblishers, pp. 177-195.
27. Hibbit, Karlsson & Sorensen Inc. (1992) ABAQUS Manual: Vol. 1, User Manual; Vol. 2, Theory Manual; Vol. 3, Example Problems Manual.
28. Kanninen, M.F. and Popelar, C.H. (1985) *Advanced Fracture Mechanics*, Oxford University Press.
29. Lawrence, M.A., Liu, W.K., Besyerfield, G.H. and Belytschko, T.B. (1990) Fatigue Crack Growth Reliability, *J. Engng. Mech. (ASCE)*, **116**(3), 698-708.
30. Liu, W.K., Lua, Y.J., Chen, Y. and Belytcshko, T.B. (1994) Study of Three Reliability Methods for Fatigue Crack Growth, in P.D. Spanos and Y.T. Wu (eds.), *Probability Structural Mechanics: Advances in Structural Reliability Methods*, Springer-Verlag, pp. 319-334.
31. Lua, Y.J., Liu, W.K. and Belytschko, T.B. (1993) Curvilinear Fatigue Crack Reliability Analysis by Stochastic Boundary Element Method, *Int. J. of Num. Meth. in Engng.*, **36**, 3841-3858.
32. Madsen, H.O., Krenk, S. and Lind, N.C. (1986) *Methods of Structural Safety*, Prentice Hall.
33. Parks, D.M. (1981) The Inelastic Line-Spring: Estimates of Elastic-Plastic Fracture Mechanics Parameters for Surface Cracked Plates and Shells, *Jour. Pressure Vessel Tech.*, **103**, 246-254.
34. Rice, J.R. (1968) A Path Independent Integral and the Approximate Analysis of Strain Concentration by Notches and Cracks, *J. of Appl. Mech.*, 379-386.
35. Rice, J.R. and Levy, N. (1972) The Part-Trough Surface Crack in an Elastic Plate, *J. Applied Mech.*, 185-194.
36. Sobczyk, K. and Spencer Jr., B.F. (1992) *Random Fatigue: From Data to Theory*, Academic Press Inc.
37. Suresh, S. (1991) *Fatigue of Materials*, Cambridge University Press.
38. Winterstein, S.R. and Ness, O.B. (1989) Hermite Moment Analysis of Nonlinear Random Vibration, in W.K. Liu and T.B. Belytschko (eds.) *Computational Mechanics of Probability and Reliability Analysis*, Elmepress, pp. 452-478.
39. Winterstein, S.R. (1988) Non Linear Vibration Models for Extremes and Fatigue, *J. of Engng. Mech. (ASCE)*, **114**(10), 1772-1790.
40. Wirshing, P.H., Ortiz, K. and Chen, Y.N. (1987) Fracture Mechanics Fatigue Model in a Reliability Format, *6th International Conference on Offshore Mechanics and Artic Engineering*, ASME, pp. 331-337.
41. Wirshing, P.H., Torng, T.Y. and Martin, W.S. (1991) Advanced Fatigue Reliability Analysis, *Int. Jour. Fatigue*, **13**(5), 389-394.

42. Wu, Y.T., Burnside. O.H. and Dominquez, J. (1987) Efficient Probabilistic Fracture Mechanics Analysis, *IV International Conference on Numerical Methods in Fracture Mechanics*, Pineridge Press, pp. 85-100.

43. Wu, Y.T., Millwater, H.R. and Cruse, T.A. (1989) An Advanced Probability Structural Analysis Method for Implicit Performance Functions, *30th Structures, Structural Dynamics and Materials Conference*.

44. Zhang, Y. and Der Kiureghian, A. (1994) Reliability Against Fracture with Uncertain Crack Geometry, in P.D. Spanos and Y.T. Wu (eds.) *Probability Structural Mechanics: Advances in Structural Reliability Methods*, Springer-Verlag, pp. 582-594.

PROBABILITY BASED STRUCTURAL CODES: PAST AND FUTURE

J. FERRY BORGES
Laboratório Nacional de Engenharia Civil
Lisbon

1. Introduction

The evolution of structural codes during the present century is outlined. The description is centered on the design of buildings and public works of structural concrete and mainly reflects experience in Western Europe. Particular attention is paid to the introduction and generalised use of probabilistic concepts. Although most of the specific considerations refer to structural concrete, the fundamental concepts apply to other materials used in civil engineering : steel, masonry, timber, etc.

During the last decades international associations in civil engineering have made significant contributions to the practical implementation of research results. The guidance documents they have produced: recommendations, state of the art reports, manuals, codes of practice, etc., have been directly applied in practice, and have much contributed to the improvement of national and international standards.

However the most important step towards the harmonization and improvement of structural codes in the civil engineering field arose from the initiative of the Commission of the European Communities of preparing the set of Eurocodes.

In this paper, after a brief historical review of the probabilistic formulation of structural safety, the evolution of probability based guidance documents is discussed. The comments are mainly centered in the evolution of the CEB - FIP Model Codes. Finally the need to carry out reliability studies based on probabilistic data and to perform the logic analysis of code provisions is emphasized.

2. Probalilistic formulation of structural safety

In 1926 Max Mayer published the thesis "The Safety of Structures and their Design According to Ultimate Forces Instead of Allowable Stresses" (Mayer 1926). This thesis presents two main proposals aiming at rationalizing design: to consider pertinent limit states, particularly failure, and to idealize the variability of the different quantities such

C. Guedes Soares (ed.), Probabilistic Methods for Structural Design, 339–350.
© 1997 *Kluwer Academic Publishers.*

as mechanical properties, loads and dimensions according to probabilistic concepts.These proposals were far ahead of their time. Only in the forties did these problems start to be thoroughly discussed. However, even then probabilistic concepts were introduced mainly in qualitative terms.

A brief historical review of the evolution of probabilistic structural safety up to 1980 is presented in the third edition of the book "Structural Safety" Ferry Borges and Castanheta. Table 1 lists the papers published in this period, in chronological order.

In 1969 the International Association for Bridge and Structural Engineering, IABSE, organized a Symposium in London on Concepts of Safety of Structures and Methods of Design. The ICOSSAR conferences took place in Washington, 1969; Munich, 1977; Trondheim, 1981; Tokyo, 1985; San Francisco, 1989 and Innsbruck 1993. More than 450 papers were presented at the last two conferences.

Conferences on reliability which concentrate in soil mechanics have been held in Hong Kong, 1971; Aachen, 1975; Sidney, 1979; Florence, 1983 Vancouver, 1987 Mexico 1991 and Paris 1995. These conferences are organized by a permanent international commitee under the title "International Conferences on Applications of Statistics and Probability in Soil and Structural Engineering".

State of the art reviews on structural safety were published by the American Society of Civil Engineers (1972), Mathieu, (1980) and Bosshard, (1979). The International journal "Structural Safety" starts in 1982 (Vanmarcke, 1982).

Text books on structural reliability have been published by Leporati, (1979), Thoft-Christensen and Baker (1982), Thoft-Christensen and Murotsu (1986), Augusti. Baratta and Casciatti (1984).

In 1971 the Liaison Commitee, which coordinates the activity of seven international associations in civil engineering: CEB, CECM, CIB; FIP, IABSE, IASS and RILEM, decided to create the "Joint Committe on Structural Safety", JCSS, with the aims of improving the general knowledge in structural safety and providing sound bases for formulation of structural design recommendations. The JCSS has prepared several documents of general character such as:

- Common Unified Rules for Different Types of Construction and Materials, which constitutes Volume 1 of the International System of Unified Standard of Practice for Structures,
- General Principles on Reliability for Structural Design (JCSS, 1981a)
- General Principles on Quality Assurance for Structures (JCSS, 1981b)
- General Principles on Reliability for Structures. A Commentary on IS 2394 (JCSS, 1988).

During the last 20 years a large number of papers on structural reliability have been discussed at JCSS meetings, published in scientific journals and presented to national and international conferences. It is difficult to make a selection for a brief review.

PROBABILITY BASED CODES

TABLE 1. Chronological order of publication of papers on structural safety from 1926 to 1980

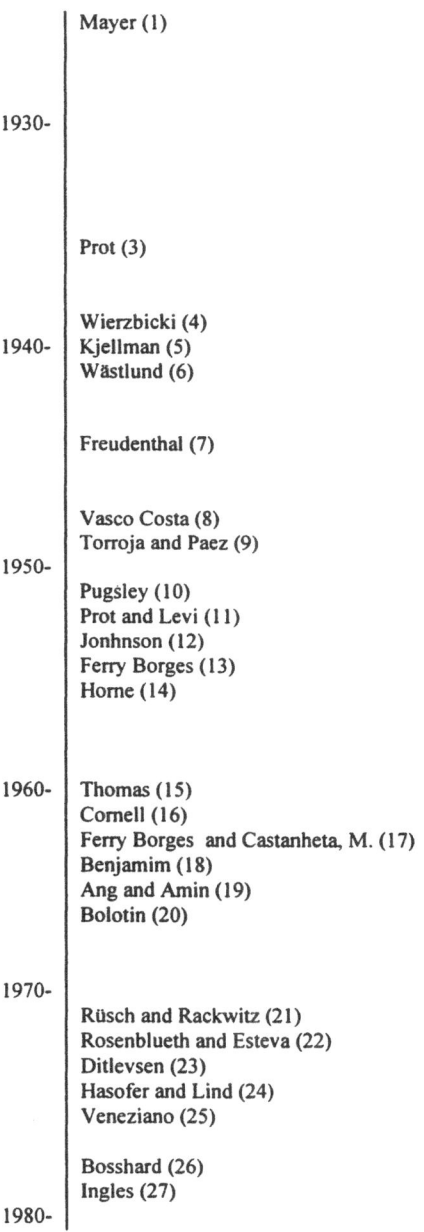

Mayer (1)

1930-

Prot (3)

Wierzbicki (4)
1940-   Kjellman (5)
Wästlund (6)

Freudenthal (7)

Vasco Costa (8)
Torroja and Paez (9)
1950-
Pugsley (10)
Prot and Levi (11)
Jonhnson (12)
Ferry Borges (13)
Horne (14)

1960-   Thomas (15)
Cornell (16)
Ferry Borges and Castanheta, M. (17)
Benjamim (18)
Ang and Amin (19)
Bolotin (20)

1970-

Rüsch and Rackwitz (21)
Rosenblueth and Esteva (22)
Ditlevsen (23)
Hasofer and Lind (24)
Veneziano (25)

Bosshard (26)
Ingles (27)
1980-

Recently, under the auspices of the Commission of the European Communities, there was created the European Safety and Reliability Association (38). This Association is intended to cover all branches of engineering with a very wide spectrum of activities: assistance to industry and consultants, organisation of courses and seminars, establishment of an information network between industrial, academic and professional organizations, advice on industrial, environmental and transport risks, etc.

All over the world, active research is going on in the field of structural safety reliability, and the domains covered are becoming larger and larger. This is well expressed by the list of themes dealt with in the last ICOSSAR conference: wind engineering, fuzzy logic for damage and safety assessment, seismic hazard estimation, fatigue fracture and damage analysis, structural systems reliability analysis, probabilistic analysis, bridges and buildings, error analysis and expert systems, engineering modelling of seismic ground motion, reliability based design materials, seismic damage estimation, structural reliability using expert opinions, industrial facilities, damage method and analysis, design codes and related issues, stochastic computational methods and stochastic dynamics, reliability-based design-methods, applied probabilistic analysis, nuclear structures, seismic structural damage, life systems, water delivery systems, loads and load combination, system identification and control of structures, aerospace structures, structural risk, systems identification, reliability based optimisation, inspection, quality control and quality assurance, space station freedom, offshore and marine structures, methods for systems reliability and random vibration.

However in an overall judgement of the situation, it is recognized that only a very small part of the results obtained in recent research is included in current guidance documents.

3. Evolution of probability based guidance documents

3.1. TYPES OF DOCUMENTS AND THEIR CONTENT

In 1985, the Economic Commission for Europe (Geneva) published a second report on Building Regulations in ECE Countries (ECE, 1985). This report describes the systems of building regulations in 25 countries. It indicates the legislative framework of construction, means of control and procedures for approval of buildings and building products, research and development work, and international cooperation in the field. It shows that the systems of building regulations in various countries are quite different.

The following brief description of the evolution of guidance documents in civil engineering concentrates on the domain of reinforced and prestressed concrete; this is a paradigm in the field.

At the beginning of the century the knowledge on structural concrete was mainly to be foun in treatises, of which those by Mörsch (1902) and by Saliger (1905) are good examples. The first codes dealing with the design and execution of concrete structures were published in Germany 1904 (42) and in France 1906 (43) respectively. Within a few years almost every European country had published its code. These codes were based on similar fundamental concepts (elastic design and allowable stresses they differed mainly in the amount of information included and in conditions their use (whether compulsory or not). Some were published by the national standardisation bodies, as norms, other were enforced as laws.

In Europe the international cooperation aiming at producing recommendations to designers and builders started in the fifties with the creation of the Comité Européen du Béton. The Bulletins d'Information of this Comité include the studies produced within the CEB, since 1957.

The first edition of the "Recommendations Pratiques a L'Usage des Constructeurs" was published in 1963 and applied to reinforced concrete only.

The second edition of the Recommendations, prepared in cooperation with the Féderation Internationale de la Précontrainte, was published in 1970; it covered both reinforced and prestressed concrete. Although the Recommendations included extensive commentaries, it was recognised that there was a need to prepare manuals which justify the rules in the Recommendations and include aids for their practical application.

To improve of the technological solution of different types of problems and types of construction, FIP published a set of "Codes of Good Practice". The third edition of the CEB-FIP Recommendations was published in 1978 under the general title "International System of Unified Standard of Practice for Structures". As mentioned, Volume I of this System was prepared by the JCSS in order to harmonize different structural codes by the use of a common background.

CEB, in cooperation with FIP, are actively preparing the next edition of the recommendations to be published with the title "CEB-FIP Model Code".

The basic content of these recommendations is about the same as that of previous editions. However there has been a continuous increase of the number of words in sucessive editions, as indicated in Table 2. The total number of words in the 1978 edition is more than the double that of the 1963 edition.

TABLE 2. Number of words in CEB-FIP Model Code editions approximate

	Recommendations 1963	Recommendations 1970	Model Code 1978
Text	14500	31000	29500
Comments	10500	13000	23000
Total	25000	44000	52500

The drafting committee of the Model Code is confronted with a dilemma: include all pertinent information, and consequently publish a very extensive document; or select the contents and exclude the secundary issues, to obtain a shorter document.

It is often indicated that a structural building code should be concerned only with the protection of public health, safety and general welfare, and that the corresponding clauses should be mandatory.

For the situation in U.S.A. and particularly the ACI Code, MacGregor presents a set of strategies for future code improvements (44).

Structural codes produced and published by international associations are by their nature, non compulsory. The legal status of the documents produced at national level differs markedly from country to country even within the European Communities.

A problem which deserves particular attention is to find the most convenient way to distribute guidance information to the participants in the building process. This problem is directly connected to the way this information should be organized. There is a consensus that structural codes should include requirements, performance criteria and prescriptive rules. Furthermore, general statements and definitions are needed to clarify the text.

The first predraft of the CEB-FIP Model Code 1990 follows this organisation. The chapters on the verification of the ultimate limit states and of the serviceability limit states, open the indication that for the structure as a whole and for its component parts its hall be demonstrated that the probability of limit states being reached is acceptably small. The following chapters indicate the design criteria to be used in the verification of these two types of limit states.

The method of partial coefficients is adopted. Numerical values are given to the partial coefficients and it is indicated how they should be introduced in the calculations. There is no indication about the way the partial coefficients are derived.

The draft under consideration includes in "Part I: Design Input Data", not only the idealization of material properties, but also the description of 11 generalised behaviour models on which the structural analysis and dimensioning are based. These models cover: bond stress-slip relationships, tension stiffening effects, pullout, anchorage, design stress-strain diagram reinforced concrete subjected to compression and transverse tension, data for confined concrete, moment curvature relationships, rotation capacity, concrete to concrete friction, and dowel action. Each of these models would correspond to a set of γ values to be derived e.g by means of a level 2 method, from the mean values and standard deviations of the basic variables included in the model and the aimed reliability index. However no mention of this derivation is presented.

Design procedures are dealt with in Part II, which includes 6 chapters: structural analysis, verification of ultimate limit states, verification of serviceability limit states, durability aspects, detailing and minimun measures.

The chapter on verification of ultimate limit states with the sentence: "It shall be demonstrated that for the structure as a whole, and for its components parts, the probability of the ultimate state being reached is acceptably small". This sentence is followed by the indication of the conditions under which ultimate limit states may be reached, making use of the generalised behaviour models presented in Part I.

The design of buildings and civil engineering works in structural concrete is covered by Eurocode 2. This Eurocode is complemented by Eurocode 8, which concerns the design of structures in seismic regions. Part 1 of Eurocode 2 gives the general basis for design and detailed rules, which are mainly applicable to ordinary buildings. In this Eurocode distinction is made between principles and application rules. The advantage of this classification is not clear.

Eurocode 2 was partly inspired by the 1978 CEB - FIP Model Code.

The Commission of the European Communities has decided to commit the drafting of Eurocodes to the European Committee for Standardisation, CEN.

The most comprehensive set of documents on concrete technology, and particularly on structural concrete, is produced by the American Concrete Institute, ACI. Due to its large international membership, ACI should be considered as a international association.

Another important problem concerns the liability for errors or omissions by the bodies who publish the guidance documents. All CEB documents include the sentence: "Although the CEB has done its best to ensure that any information given is accurate, no liability or responsibility of any kind (including liability for negligence) is accepted in this respect by the Comité, its members or its agents". The legal value of this statement should be clarified.

3.2. EVOLUTION OF PROBABILITY BASED CODES

Table 3 summarizes the recent evolution of probability based structural codes. The table is divided in to three columns, decribing the situation in the past (before the introduction of probabilistic concepts), in the present and in the future. The statements concerning the future express the points of view of the Author about the desirable evolution of codes. However it is recognized that in several cases the indicated statements are idealistic and not fully attainable.

The different structural codes are not uniform in their treatment of the various types of construction and material. However due to the activity of the JCSS, of the ISO and of the different international associations that have prepared model codes and other pre-standardisation documents, there is similarity among the fundamental concepts and the basic design rules. The enormous amount of the work carried out in the framework of the European Communities to draft the Eurocodes has greatly contributed to the improvement of this type of document, and to their compatibility.

Even so, some statement included in Table 3, concerning the present, are too optimistic. Codes often indicate that they follow the probabilistic approach; in fact they simply use the partial factors method without any effort to relate the numerical values of the partial factors to theoretical values of the probability of failure.

Consequently it is recommended that careful calibration studies based in sound probabilistic principles be carried out to justify the defined γ values.

TABLE 3. Evolution of probability based structural codes

Past	Present (1989)	Future
The deterministic approach did not provide rational basis for the concept of safety	The probabilistic approach is extensively used in the formulation of structural safety problems.	The probabilistic approach will be combined with fuzzy theory and strategic idealisation, according to the problems.
The fundamental decision rule consisted in limiting the stress to allowable values. Allowable stresses were based on past experience.	The fundamental decision rule consists in limiting the probability of failure or the reliability index.	Decisions will be based on the socio-economic formulation of the problem, paying special attention to the comparison of generalised cost and benefit.
The mechanical properties of materials were defined by minimal values. The meaning of these values was not clear.	The mechanical properties are defined by characteristic values with a clear probabilistic meaning. The same values are used as rejection limits in quality control rules.	The probability distributi-ons of mechanical properties will be completely defined by combining all information available. The quality of production control.
Internal forces and the resisting capacity of the structure were computed by deterministic theories.	The characteristic values of the resisting capacity of the structure are derived from characteristic values of the mechanical properties of the materials by deterministic theories. Conversion factors are introduced.	The probability distribution of resisting capacity is derived taking into account the randomness of all basic variables and of the models used in it determination.
There was a dilemma between the choice of elastic or plastic design methods.	Computers can solve any structural problem, whether with linear, plastic or non-linear behaviour.	There will be a clear understanding of the field of application of different structural theories.Randomness of structural behaviour will be duly considered.
Values of the actions to be used in design were fixed by simple global judgement.	Actions are modelled by occurrence and descriptive schemes.Numerical values are fixed in a regional scale.	Strategic definition of some actions will be adopted, after users have been conveniently informed.
The rules for the combination of actions were empirical.	Numerical values of the combination factors are fixed according to a sound theoretical approach.	Algorithms for safety and serviceability checking will have a sound probabilistic base.

4. Logic analysis of code provisions

The quality of codes depends both on their scientific and technical contents and on the way these contents are expressed (45).

The correct expression of the codes is important for their implementation. The following conditions should be satisfied:

- The text should be expressed in natural language that is clear, complete and without ambiguities.
- The logic al analysis of the text should be carried out through decision tables and/or decision trees in order to judge its consistency.

Furthermore the logical analysis should he regarded as a preliminary step to the programming of automated design (CAD) and of expert systems.

The logical analysis of structural codes has often been performed in the U.S.A (see e.g. Analysis of Tentative Seismic Design Provisions for Building).

The logical analysis of shear provisions in the drafts of Eurocodes 2 and 8 was recently carried out by Mary Mun and Ferry Borges (46).

A state of the art discussion of the use of expert systems in Civil Engineering has been presented by Fenves (47). It is to be expected that expert systems will be largely used in the future.

5. Conclusions

Results from research activity carried out all over the World should be transformed into guidance to the participants in the building process. This guidance may take different forms. The designation "standard" is often used as comprehending documents such as: codes, guides of good practice, recommendations, norms, specifications, agreements, etc. On the other hand standard has an official meaning; the result of the activity of national, regional or international standardisation organisations.

What we wish to emphasise is that international scientific and technical associations are in a particularly favourable position to draft documents at pre-standardisation level. These documents may be directly used as information to the practice and may constitute the background to the drafting on national and international standards.

As in other branches of human activity, digital computers play an increasing role in the information process in the construction sector. Thus when drafting guidance documents it should be remembered that they may be included in data bases and expert systems. It is to be expected that knowledge-based expert systems will integrate a very wide spectrum of information in a form easily accessible to users.

In most cases, the probabilistic approach will be the convenient concept to support models and to interpret data. However, in some cases, strategic decision theories and fuzzy set concepts may be particularly convenient also.

In general, international associations claim no liability in relation to the documents they produce. The legal value of these claims should be clarified.

At least, it is imperative that the documents produced by the international associations and the standardisation organisations be of high quality, not only from the scientific and technical points of view, but also in what concerns the clarity of the writting. The redaction should be checked by logic analysis, using decision tables or decision trees.

Some documents state that they are based on probabilistic concepts, when in fact they are based only on partial factors methods. In a true probabilistic method, the randomness of the basic variables should be defined and combined in order to obtain the numerical values of the partial factors.

As indicated in Table 3, it is expected that in the future more and more probabilistic data will be collected, not only to improve the definition of the basic variables, but also to improve the quantification of the models that relate them.

6. Acknowledgement

The cooperation of Dr. Sílvio Pires in the preparation of this text is kindly acknowledged.

7. References

1. Mayer, Max: *Die Sicherheit der Bauwerke und ihre Brechnung nach Grenzkräften anstatt nach Zulässigen Spanungen*, Vorlag von Julius Springer, Berlin 1926 (English and Spanish translations published by Intemac, Madrid, 1975).
2. Borges, J. Ferry and Castanheta, M.: *Structural Safety*, Course 3rd Edition, Laboratório Nacional de Engenharia Civil, Lisbon, 1985.
3. Prot, M.: *Note sur la notion de coefficient de sécurité*, Annales des Ponts et Chaussées, n° 27, Paris, 1936.
4. Wierzbicki, M.W.: *La sécurité des constructions comme un problème de probabilité*, Annales de L'Academie Polonaise de Sciences Techniques, Tome VII, 1939-45. (Spanish Trans. on Informes de la Construction, n° 7, Madrid, Enero de 1949.
5. Kjellman. W.: *Säkerhetsproblement ur principielle och teoretisk synpunkt*, Ingeniörs Veteskaps Akademien, Handlinger nr 156, Stockholm, 1940.
6. Wästlund.G.: *Säkerhetsproblement ur praktisk-konstruktiv synpunkt*, Ingeniörs Vetenskaps Akademien, Handlinger nr 156, Stockholm, 1940.
7. Freudenthal, A.M.: *The safety of structures*, Proceedings of the American Society of Civil Engineers, Vol. 71, n° 8, October, 1945.
8. Vasco Costa, F.: *Notions de probabilité dans l'étude de la sécurité des constructions*, Rapport final du 3ème Congrés de l' Association Internationale des Ponts et Charpentes, Liège, Septembre, 1948.
9. Torroja, E et Paez, A.: *La determinación del coeficiente de securidad en las distintas obras*, Instituto Técnico de la Construcción e del Cemento, 1949.
10. Pugsley, A.G.: *Concepts of safety in structural engineering*, Proc. Inst. Civ. Engrs., 1951.
11. Prot, M. et Levi, R.: *Conceptions modernes relatives à la sécurité des constructions*, Revue Générale des Chemins de Fer, Paris , Juin, 1951.
12. Johnson, A.I.: *Strength, safety and economical dimensions of structures*, Kungl. Tekniska Högskola, Institutionen for Byggnadsstatik, Meddelanden Nr. 12, Stockholm, 1953.
13. Borges, J.Ferry: *O dimensionamento de estruturas*, Laboratório Nacional de Engenharia Civil, Publ. n° 54, Lisboa, 1954.
14. Horne, M.R.: *Some results of the theory of probability in the estimation of design of design loads*, Proceedings of a Symposium on the Strength of Concrete Structures, London , May, 1956.
15. Thomas, F.G.: *Basic parameters and terminology in the consideration of structural safety*, CIB Bulletin n° 3, 1964.

16. Cornell, C:A.: *Bounds on the reliability of structural sytems*, Proc. of the Amer. Soc. of Civil Engineers, new York, 93 (STI), Feb., 1967.

17. Borges, J.Ferry and Castanheta, M.: *Structural Safety*, Course 101, 1st Edition, Laboratório Nacional de Engenharia Civil, Lisbon, 1968.

18. Benjmin, J.R.: *Probabilistic models for seismic force design*, Proc. of the Amer.Soc. of Civil Engineers, New York, 94 (ST 5) May, 1968.

19. Ang.A.H. and Amin, M.: *Reliability of structures and structural systems*, Proc. of the Amer. Soc. of Civil Engineers, New York, (EM2), April, 1968

20. Bolotin, V.V.: *Statistical Methods in Structural Mechanics*, Holden-Day series in mathematical physics, Holden-Day Inc., San Francisco, 1969.

21. Rüsch, H. and Rackwitz, R.: *Die Grundlagen der Sicherheitstheorie*, VDI - Bericht n° 42, Berlin, 1971.

22. Rosenblueth, E. and Esteva, L.: *Reliability Basis for some Mexican Codes* - ACI, Publication SP - 31, 1972.

23. Ditlevsen, O.: *Structural Reliability and the Invariance Problem*, Res. Rep. n° 22, Solid Mechanics Division, University of Waterloo, Ontario, 1973.

24. Hasofer, A.M., and Lind, M.C.: *An Exact and Invariant First-Order Reliability Format*, Proceedings ASCE, Journal Mech. Eng. Div. Volume 100, 1974.

25. Veneziano, D.: *Contributions of Second Moment Reliability Theory*. Res. Rep. R 74-33, Department of Civil Engineering Massachusetts Institute of Technology, Cambridge, Mass. 1974.

26. Bosshard, W.: *Structural Safety. A. Matter of Decision and Control*, International Association for Bridge and Structural Engineering, Peiodica, n° 2, Zürich, 1979.

27. Ingles, O.G.: *Safety in Civil Engineering - Its Perception and Promotion*, INICIV Report n° R-188, The University of New South Wales, Kensington, July 1979.

28. American Society for Civil Engineering; *Structural Safety*, A Structural Review, 1972.

29. Mathieu, H.: *Manuel Sécurité de Structures*, Bulletin d' Information n° 127, Comité Euro-International du Béton, Paris, Décembre 1979, Janvier 1980.

30. *Structural Safety* - An International Journal on Integrated Risk Assessment for Constructed Facilities. Elsevier Scientific Publishing Company, Amsterdam.

31. Leporati, E.: *The Assessment of Structural Safety: A Comparative Study of Evaluation and Use of Level 3, Level 2 and Level 1*. Series in Cement and Concrete Research, Vol. I, Research Studies Press, 1979.

32. Thoft-Christensen, P.and Baker, M.J.: *Structural Reliability Theory and its Applications*, Springer-Verlag, Berlin, 1982.

33. Thoft-Christensen, P. and Murotsu, Y. (Editors): *Application of Structural Systems Reliability Theory*, 1st ed., Springerverlag, Berlin, 1986.

34. Augusti, G., Baratta, A. and Casciati, F.: *Probabilistic Methods in Structural Engineering*. Chapman and Ltd, 1984.

35. Joint - Committee on Structural Safety - *General Principles on Reliability for Structures Design*, Reports, Vol. 35. International Association for Bridge and Structural Engineering, April 1981.

36. Joint- Committee on Structural Safety - *General Principles on Association for Bridge and Structural Engineering*, April 1981.

37. Joint-Committe on Structural Safety - *General Principles on Reliability for Structures*. A Comment on ISO 2394. CEB Bulletin d' Information n° 191, Lausanne, 1988.

38. *The European Safety and Reliability Association*, ESRA Newsletter, Vol. 1, n° 1, September 1984.

39. Economic Commission for Europe - *Building Regulations in ECE Countries*, second report. United Nations, New York, 1985.

40. Mörsch, E.: *Der Eisenbetonbau*. Theorie und Anwendung, Erste Auflage, Wayss und Freitag, A.G., Stuttgart, 1902.

41. Saliger, R.: *Der Eisenbetonbau, seine Berechnung und Gestaltung*. A. Kröner Verlag, Leipzig 1905.
42. *Bestimmungen für die Ausführung von Konstruktionen aus Eisenbeton bei Hochbauten*. Preuss. Minister der Öffentlicher Arbeiten, 1904.
43. *Instructions Relatives à l' Emploi du Béton Armé*. Circulaire du Ministre des Travaux Publics du 20 Octobre 1906, Paris.
44. Macgregor, J.G.: *A Simple Code - Dream or Possibility?* American Concrete Institute, SP 72-9, Detroit.
45. Harris, J.R. and Wright, R.N.: *Organization of Building Standards - Systematic Techniques for Scope and Arrangement*. NBS Building Science Series 136, National Bueau of Standards, Washington, 1981.
46. Mun, M. and Ferry Borges, J.: *Analysis of Shear Porvisions in the Drafts of Eurocodes 2 and 8*. Laboratório Nacional de Engenharia Civil, Lisbon, April 1988.
47. Fenves, S.J.: *Expert Systems in Civil Engineering State-of-the Art*. Proceedings of the 4th International Symposium on Robotics and Artificial Intelligence in Building Construction, Haifa, 1987.

RELIABILITY BASED SEISMIC DESIGN

FABIO CASCIATI AND ALBERTO CALLERIO
Università di Pavia
via Abbiategrasso 211, 27100 Pavia

1. Summary

After a brief overview of the so-called stochastic linearization methods for approaching structural dynamics problems, attention is focused on a special procedure for seismic fragility analysis which makes use of response surface techniques and level-2 reliability methods. The local amplification is considered through a boundary element idealization. The numerical examples investigate the goodness of the experiment plans adopted and, especially, the accuracy of the level-2 reliability idealization.

2. Introduction

In several practical cases, the engineer does not have a certain knowledge of the system under investigation. Moreover the actions on the system cannot always be specified in a deterministic way. Mainly when they result from environmental situations, these actions, like the seismic ground motion, are correctly idealized only by stochastic processes. System parameters and actions (input) provide the coefficients and the given terms, respectively, of the equations which govern the analytical idealization of the system behaviour. Their solution represent the response variables (output) of interest.

Modern engineering often deals with complex systems whose analytical model leads to equations of such complexity that their solution can be pursued only numerically. This situation can be summarized by saying that one is dealing with a "numerical input-output stochastic relationship".

In structural analysis, the classical numerical approach is to adopt a finite-element discretization of the structure. This led several researchers to speak of "stochastic finite elements" (SFEM), when one discretizes the domain of the structure, or "stochastic boundary elements" (SBEM) when one discretizes the boundary.

C. Guedes Soares (ed.), Probabilistic Methods for Structural Design, 351–375.
© 1997 *Kluwer Academic Publishers.*

An early state of the art report (Vanmarcke *et al.*, 1986), emphasized that these terms are used in the literature to identify at least three different problems:

- an explicit treatment of uncertainty in any quantity entering the structural analysis procedure. For instance, for a cantilever of random height h, "stochastic finite element" techniques aim at the evaluation of the uncertainty in the eigenfrequencies;
- a classical analysis of uncertainty associated within a discrete-element algorithm. The characteristic here is that the input for the discrete algorithm is a vector of random variables. One can distinguish linear and non-linear systems and, for both, either deterministic or random systems.
- the discretization of the parameter space of a random field of material properties and/or loads. With reference to a simple cantilever of length l, for instance, using SFEM techniques one discretizes into beam elements of length Δl. The stochastic process which describes the flexural stiffness is a random vector whose i-th entry is a moving average process in the range $((i - 1)\Delta l, i\Delta l)$. The stochastic description of this random vector can be pursued by introducing "variance function" and "scale of fluctuation" (Vanmarcke, 1983).

This chapter studies the characterization of the probabilistic properties of the response, given that the input-output relationship is of a numerical type. The general solution in this case can be pursued by the so-called Monte Carlo methods (Augusti *et al.*, 1984) (Der Kiureghian, 1983). By appropriate algorithms, the analyst simulates a realization of the input vector and in this way defines in a deterministic structural problem. Its numerical solution provides the corresponding values of the response variables of interest. The whole procedure is repeated till a satisfactory sample is obtained. The probabilistic properties are eventually inferred from the joint sample. The main disadvantages of such an approach are:

1. the accuracy depends on the sample size; for complicated systems, each deterministic computer run can require a large computational effort;
2. a minor modification of the input properties requires a new simulation;
3. anomalous behaviours may remain undetected.

The previous remarks suggest that one should develop analytical approaches to the problem. Of course, they can either interact with the numerical algorithm or regard this numerical algorithm as a black box. The selection between these two ways depends mainly on the degree of complication of the problem. The next section illustrates the "interacting" techniques for the cases where they can be proposed, while Section 4 is devoted to a more general procedure. Some operative features are discussed in the numerical example in Section 5.

3. Input-Output Numerical Relationships

3.1. NON-LINEAR SYSTEMS WITH RANDOM ACTIONS (FEM APPROACH)

3.1.1. *Governing Relations: Deterministic Systems*
Non-linearity can be introduced either by the presence of large displacements and deformations, or by non-linearities in the material constitutive law. In the latter case, for a structure discretized into finite elements, one expresses the generalized displacement $\mathbf{u}(\mathbf{x})$ in the i-th element in terms of the nodal displacements $\mathbf{u}_i$ by a matrix of shape functions $\Lambda_i(\mathbf{x})$:

$$\mathbf{u}(\mathbf{x}) = \Lambda_i(\mathbf{x})\mathbf{u}_i \tag{1}$$

The compatibility relation then provides the deformation vector $\varepsilon(\mathbf{x})$ (with $\mathbf{B}$ the compatibility matrix)

$$\varepsilon_i(\mathbf{x}) = \mathbf{B}_i\mathbf{u}_i \tag{2}$$

and the constitutive law has the incremental form ($\mathbf{D}$ = elastic matrix)

$$\dot{\sigma}_i(\mathbf{x}) = \mathbf{D}_i(\dot{\varepsilon}_i - \dot{\varepsilon}_{ip}) \tag{3}$$

where $\dot{\varepsilon}_p$ denotes the vector of the inelastic deformations which act as distortions in the determination of the stress vector $\sigma(\mathbf{x})$. The general form of the equilibrium equation is then:

$$\sum_i \left(\int_{V_i} \mathbf{B}_i^T \mathbf{D}_i \mathbf{B}_i dV \right) (d\mathbf{u}_i + cd\dot{\mathbf{u}}_i) + \left(\int_{V_i} \Lambda_i \Lambda_i^T \rho dV \right) d\ddot{\mathbf{u}}_i +$$
$$- \left(\int_{V_i} \mathbf{B}_i^T \mathbf{D}_i d\varepsilon_{ip} dV \right) - \left(\int_{V_i} \Lambda_i^T d\mathbf{g} dV \right) - \left(\int_{V_i} \Lambda_i^T d\mathbf{f} dA \right) = 0 \tag{4}$$

where ρ is the density, $\mathbf{g}$ is the body force and $\mathbf{f}$ is the surface force. The damping coefficient of the velocity $d\dot{\mathbf{u}}_i$ is assumed to be proportional to the stiffness by a factor c in Eq. (4). In a more compact form, one writes.

$$\mathbf{K}_i d\mathbf{u}_i + c\mathbf{K}_i d\dot{\mathbf{u}}_i + \mathbf{m}_i d\ddot{\mathbf{u}}_i - d\mathbf{W}_i - d\mathbf{W}_i^p = 0 \tag{5}$$

The term introducing the non-linearity is $d\mathbf{W}_i^p$; it is linearly related to $d\varepsilon_{ip}$ at the Gauss integration points. At any time instant it is a function of the previous time history, since the material constitutive law is no longer reversible.

When $\mathbf{W}$ is a vector of stochastic processes, Eq. (5) is well tackled by stochastic equivalent linearization techniques (Casciati *et al.*, 1991). At the k-th Gauss point, an univariate endochronic constitutive law $\dot{z}_k = f(\dot{\varepsilon}_k, z_k)$, where z is an auxiliary variable, is linearized in the form

$$\dot{z}_k = C_1\dot{\varepsilon} + C_2 z \tag{6}$$

where C_1 and C_2 are linearization coefficients depending on the response statistics. They are constant in the stationary case and time dependent for nonstationary $\mathbf{W}$ or degrading systems. For the case of a multivariate constitutive law see (Casciati *et al.*, 1991).

The terms $\mathbf{K}_i du_i$ and $d\mathbf{W}_i^p$ can then be expressed as linear functions of $\varepsilon, \dot{\varepsilon}$ and z. Elimination of z allows one to write Eqs. (5) and (6) in finite form as a linear system of first order differential equations (Casciati *et al.*, 1991)

$$\dot{\mathbf{d}} + \mathbf{L}\mathbf{d} = \mathbf{w} \tag{7}$$

with $\mathbf{d}^T = \{\mathbf{u}, \dot{\mathbf{u}}, \varepsilon\}$. In Eq. (7), some elements of $\mathbf{L}$ depends on the *a priori* unknown linearization coefficients C_1 and C_2. The covariance matrix of $\mathbf{d}$ is the solution of the matrix equation

$$d\boldsymbol{\Sigma}/dt + \mathbf{L}\boldsymbol{\Sigma} + \boldsymbol{\Sigma}\mathbf{L}^T = \boldsymbol{\Omega} \tag{8}$$

where $\boldsymbol{\Omega} = E[\mathbf{w}\mathbf{d}] + E[\mathbf{d}\mathbf{w}]$, and can be easily computed when the system is excited by a single white noise (e.g. the ground acceleration due to a seismic excitation). For the problems considered in this subsection, the task of the analyst is two fold:

1. to express the structural matrices which form $\mathbf{L}$ by finite element discretization and subsequent algebra;
2. to solve Eq. (8) by numerical integration, updating the linearization coefficients C_1 and C_2 in $\mathbf{L}$ at each step.

3.1.2. *Stochastic System*

Randomness in system parameters still leads one to write Eq. (5) but its coefficients become random. In numerical simulation, realizations of the random variables make it deterministic, and the solution is obtained by conducting, in parallel with the equilibrium force-displacement relation (Newton-Raphson schemes), the integration of the constitutive law by either forward (Euler) or backward (Nakagiri, 1985) difference schemes. An interesting alternative approach was provided by Liu and his coworkers (Liu *et al.*, 1986).

A response surface technique can also be adopted to approach the problem. Let $\mathbf{X}$ be a vector of random design variables X_i of the mechanical system under investigation, and Y any response variable. The vector $\mathbf{X}$ can be decomposed as two random vectors, $\mathbf{X}_v$ and $\mathbf{X}_s$:

$$\{\mathbf{X}\}^T = \{\{\mathbf{X}_v\}^T, \{\mathbf{X}_s\}^T\} \tag{9}$$

where $\mathbf{X}_v$ only contains random variables, i.e. the variability which is independent of time and space, and $\mathbf{X}_s$ the spatial and temporal fluctuations. Any component of the vector $\mathbf{X}_s$, say X_{si}, expressing the spatial variability of the design variable X_i, can be written in one of the two alternative forms:

$$X_{si}(t, x, y, z) = X_{svi} + X_{ssi}(t, x, y, z) \tag{10}$$
$$X_{si}(t, x, y, z) = X_{svi} \cdot X_{ssi}(t, x, y, z) \tag{11}$$

where X_{svi} is any random central value of X_{si}, and X_{ssi} denotes the deviations of X_{si} from the central value X_{svi}. Discretization of the structural system leads one to evaluate X_{ssi} as a random vector rather than a random field.

Appropriate transformations $\mathcal{Y}$ of Y and $\mathcal{X}_i$ of X_i can be found for which a low order polynomial relationship $\mathcal{Y}(\mathcal{X}_i)$ holds. The transformation $\mathcal{Y}$ of Y is selected in order to achieve the greater accuracy in the response surface modeling. The polynomial model of the dependence of $\mathcal{Y}$ on $\{\mathcal{X}\}$ is written in matrix notation as:

$$g(\{\mathcal{X}\}|\mathbf{X}_{ss}) = F_R(\{\mathcal{X}_v\}, \{\mathcal{X}_{sv}\}, \{\boldsymbol{\theta}\}) + \epsilon(\mathbf{X}_{ss}) \tag{12}$$

where $F_R(\cdot)$ is a polynomial with coefficients $\boldsymbol{\theta}$, and the random term ϵ takes into account the error due to the lack of fit and the randomness of the variables $\mathbf{X}_{ss}$ which do not appear explicitly in equation (12).

The coefficients $\boldsymbol{\theta}$ are found by regression analysis of results obtained from numerical structural analyses whose input parameters are selected in accordance with experiment design theory. The error can be studied by one-way ANOVA as shown in (Faravelli, 1989). It can also be decomposed by multi-way ANOVA.

The variables $\mathbf{X}_v$ and $\mathbf{X}_{sv}$ in particular can be mapped in the standardized space $\{\mathcal{Z}\}$ where all the variables are uncorrelated, and have zero mean and unit variance, in order to achieve an uniformity in the experiment design. This can be done by any transformation of classical reliability theory. It follows that the response surface model in the space $\mathcal{Z}$ can be given through the parametrized form:

$$g(\{\mathcal{Z}\}|\mathbf{X}_{ss}) = F_R(\{\mathcal{Z}\}, \{\boldsymbol{\theta}\}) + \epsilon(\mathbf{X}_{ss}) \tag{13}$$

The calculations are greatly simplified when ϵ is assumed to be nearly constant with $\mathbf{X}_{ss}$. The design of the experiments necessary for developing the regression analysis and the one-way ANOVA is discussed in detail in (Faravelli, 1989), where the validation of the model is also discussed.

3.2. BEM THEORY FOR LINEAR MEDIA

3.2.1. *Governing Relations: Deterministic Systems*
For a finite domain Ω of boundary Γ, the local dynamic equilibrium can be represented by the equation:

$$\sigma_{kj,j} + \rho b_k - \rho \ddot{u}_k = 0 \tag{14}$$

where σ_{kj} is the kj-th element of the stress tensor, ρ the mass density, ρb_k the k-th component of the body force, and $\ddot{u}_k$ the k-th component of the acceleration vector. An auxiliary elastostatic state defined over Ω, with displacement function u^*, must be introduced, following the boundary element theory. As shown in (Brebbia *et al.*, 1989), with a weighted residuals formulation, it is possible to write:

$$\int_\Omega \sigma_{kj,j} u_k^* d\Omega + \int_\Omega \rho b_k u_k^* d\Omega - \int_\Omega \rho \ddot{u}_k u_k^* d\Omega = 0 \tag{15}$$

The relation $\sigma_{kj} \varepsilon_{kj}^* = \sigma_{kj}^* \varepsilon_{kj}$, and the equilibrium equation of the new elastostatic state

$$\sigma_{kj,j}^* + \rho b_k^* = 0$$

allow an integration by parts to be performed; the result is the following reciprocity equation with respect to the original elastodynamic state:

$$\int_\Gamma p_k u_k^* d\Gamma + \int_\Omega \rho(b_k - \ddot{u}_k) u_k^* d\Omega = \int_\Gamma p_k^* u_k d\Gamma + \int_\Omega \rho b_k^* u_k d\Omega \tag{16}$$

where p_k and p_k^* are respectively the tractions on Γ for the elastodynamic and the auxiliary elastostatic problem.

The auxiliary elastostatic state u^* can be seen as the solution of a unit impulse, on the boundary point ξ, on a infinite domain and along the l-direction:

$$\rho b_k^* = \delta(\mathbf{x} - \boldsymbol{\xi}) \delta_{lk} \tag{17}$$

Displacements and tractions can be defined as:

$$u_k^* = u_{lk}^*(\mathbf{x} - \boldsymbol{\xi}) n_l$$
$$p_k^* = p_{lk}^*(\mathbf{x} - \boldsymbol{\xi}) n_l$$

If the body forces b_k are zero, Eq. (16) can be rewritten in the form:

$$c_{lk}(\boldsymbol{\xi}) u_k(\boldsymbol{\xi}) + \int_\Gamma p_{lk}^* u_k d\Gamma = \int_\Gamma u_{lk}^* p_k d\Gamma - \int_\Omega u_{lk}^* \rho \ddot{u}_k d\Omega \tag{18}$$

where the last integral is the only domain integral and represents the inertial term. Note that $c_{lk}(\boldsymbol{\xi}) = \delta_{lk}$ if $\boldsymbol{\xi} \in \Omega$, $c_{lk}(\boldsymbol{\xi}) = (1/2)\delta_{lk}$ if $\boldsymbol{\xi} \in \Gamma$, $c_{lk}(\boldsymbol{\xi}) = 0$ if $\boldsymbol{\xi} \notin \Omega$ (smooth boundary).

To calculate the integral representing the inertial term, a set of unknown coefficients, α_k^m, and a set of functions, $f^m(\mathbf{x})$, are introduced, giving:

$$u_k(\mathbf{x}, t) = \sum_{m=1}^{N} \alpha_k^m(t) f^m(\mathbf{x}) \tag{19}$$

A third elastostatic state over the infinite domain caused by body forces $\rho b_k^m = -f^m \delta_{kn}$, applied in the n-th given direction, now needs to be introduced. Displacements and tractions relevant to this third stress state can be written as:

$$u_k^m = \psi_{kn}^m n_n \qquad p_k^m = \eta_{kn}^m n_n$$

where n_n is the unit vector along the direction of the load. Therefore Eq. (18) can be rewritten as a sum of boundary integrals only:

$$c_{lk}(\boldsymbol{\xi})u_k(\boldsymbol{\xi}) + \int_\Gamma p_{lk}^* u_k d\Gamma = \int_\Gamma u_{lk}^* p_k d\Gamma$$

$$+\rho \sum_{m=1}^{M} \ddot{\alpha}_n^m \{ c_{lk}(\boldsymbol{\xi})\psi_{kn}^m(\boldsymbol{\xi}) - \int_\Gamma \eta_{kn}^m u_{lk}^* d\Gamma + \int_\Gamma p_{lk}^* \psi_{kn}^m d\Gamma \} \tag{20}$$

Since the BEM is a discrete method, the boundary Γ has to be discretized into elements; then Eq. (20) is reduced to a system of linear differential equations. Displacements and tractions will be expressed as functions of nodal quantities, by an interpolation matrix, say ϕ, of dimension $2 \times 2Q$, Q being the number of nodes for each element (2 for linear and 3 for quadratic elements):

$$\mathbf{u} = \phi \mathbf{u}^j \qquad \mathbf{p} = \phi \mathbf{p}^j$$

The discretization procedure leads, from Eq. (20), to the system of differential equations (Domínguez et al., 1993) mentioned before:

$$\sum_{j=1}^{N} \mathbf{H}^{ij} \mathbf{u}^j = \sum_{j=1}^{N} \mathbf{G}^{ij} \mathbf{p}^j + \rho \sum_{m=1}^{M} \sum_{j=1}^{N} (\mathbf{H}^{ij} \psi^{jm} - \mathbf{G}^{ij} \eta^{jm}) \ddot{\alpha}^m \tag{21}$$

where $\mathbf{H}^{ij}$ and $\mathbf{G}^{ij}$ are 2×2 matrices which correlate the i-th node with the j-th boundary node. Eq. (19) can be rewritten as:

$$\ddot{\mathbf{u}} = \mathbf{F}\ddot{\alpha} \tag{22}$$

in which $\mathbf{F}$ includes the independent functions $f^m(\mathbf{x})$, with m chosen equal to N, i.e. to the number of the boundary nodes. Once Eq. (22) has been inverted and substituted in (21), one has:

$$\mathbf{Hu} = \mathbf{Gp} + \rho(\mathbf{H}\psi - \mathbf{G}\eta)\mathbf{F}^{-1}\ddot{\mathbf{u}} \qquad (23)$$

If one puts $\mathbf{M} = \rho(\mathbf{G}\eta - \mathbf{H}\psi)\mathbf{F}^{-1}$, one can rewrite Equation (23) in the form:

$$\mathbf{M}\ddot{\mathbf{u}} + \mathbf{Hu} = \mathbf{Gp} \qquad (24)$$

which represents the dynamic equilbrium relation of the entire discretized structure (Domínguez *et al.*, 1993) (Cen *et al.*, 1989). Eventually, Equation (24) becomes

$$\mathbf{M}'\ddot{\mathbf{u}}' + \mathbf{H}'\mathbf{u}' = 0 \qquad (25)$$

for a body with a free boundary described by the degrees of freedom in $\mathbf{u}'$.

3.2.2. *Stochastic System*

For a random field description of the physical and mechanical properties of the medium, one could couple boundary element theory with the response surface technique mentioned before. The two-dimensional or three-dimensional domain of the structure could be discretized in cells Ω_i in which the stochastic fields of the physical and mechanical properties are assumed to be constant. Then the problem will be reduced to the assemblage of different problems defined on the homogeneus domains Ω_i, with boundaries Γ_i discretized by boundary elements (Brebbia *et al.*, 1989) (Cen *et al.*, 1989). In this way the resulting structure not only has elements defined on the real boundary, but also elements on the contact surfaces between different cells. The drawback is that, with the introduction of these cells, the main advantage of boundary elements over finite elements techniques, i.e no domain discretization, is lost.

The method proposed in (Burczynski, 1993) is an alternative approach to the elastostatic problem. The new approach proposed here was inspired by (Burczynski, 1993), but it is characterized by the following differences and extensions:

1. it can solve elastodynamics as well as elastostatics problems;

2. a direct computation of the cell contributions avoid any sort of numerical iteration;

3. the coupling with the response surface scheme gives the probabilistic description of the response in terms of probability distribution function.

Let there be two random fields $X_i(\mathbf{x})$: the Young's modulus $C(\mathbf{x}) = C^\circ + \bar{C}(\mathbf{x})$ and the soil density $\rho(\mathbf{x}) = \rho^\circ + \bar{\rho}(\mathbf{x})$. For a constitutive relationship:

$$\sigma_{kj}(\mathbf{x}) = [C^\circ_{kjmn} + \bar{C}_{kjmn}(\mathbf{x})]\varepsilon_{mn} \qquad (26)$$

the internal stresses becomes a stochastic field over the domain Ω. Following the BEM theory explained before, one can introduce an auxiliary elastostatic problem. It is then possible to rewrite the boundary Eq. (20) for C° and ρ°, with two additional domain integrals for the terms $\bar{C}(\mathbf{x})$ and $\bar{\rho}(\mathbf{x})$:

$$c_{lk}(\boldsymbol{\xi})u_k(\boldsymbol{\xi}) + \int_\Gamma p^\star_{lk}u_k d\Gamma = \int_\Gamma u^\star_{lk}p_k d\Gamma$$

$$+\rho \sum_{m=1}^{M} \ddot{\alpha}^m_n \{ c_{lk}(\boldsymbol{\xi})\psi^m_{kn}(\boldsymbol{\xi}) - \int_\Gamma \eta^m_{kn}u^\star_{lk}d\Gamma + \int_\Gamma p^\star_{lk}\psi^m_{kn}d\Gamma \}$$

$$- \sum_{m=1}^{M} \ddot{\alpha}^m_k \int_\Omega u^\star_{lk}\bar{\rho}(\mathbf{x},\gamma)f^m d\Omega - \int_\Omega \varepsilon^\star(\boldsymbol{\xi},\mathbf{x})\bar{\sigma}(\mathbf{x})d\Omega(\mathbf{x}) \qquad (27)$$

The second of the two domain integrals was calculated in (Burczynski, 1993) to solve the problem in terms of stresses at internal points, but the method proposed there requires an iterative solution.

An alternative direct method of computing the two domain integrals is proposed here. It makes use of the expansion for displacements and accelerations, with the same coefficients α_m and the same shape functions $f^m(\mathbf{x})$. From this expression and from the knowledge of $\mathbf{u}^\star$, it is easy to compute $\varepsilon^\star$ and ε, and hence $\bar{\sigma}$, at internal points by the compatibility relation:

$$\varepsilon_{ij} = \frac{1}{2}(u_{i,j} + u_{j,i}) \qquad (28)$$

and the constitutive relation:

$$\bar{\sigma}_{kj}(\mathbf{x}) = \bar{C}_{kjmn}(\mathbf{x})\varepsilon_{mn} \qquad (29)$$

Note that the previous equation just requires derivatives of $f^m(\mathbf{x})$ for $\bar{\sigma}$, since the α^m are independent of the spatial coordinates.

To evaluate the integral numerically, one or more points inside the single cell can be introduced, and then a sum over the cells performed. The number of points within a single cell should be selected on the basis of the significance of the fluctuation with respect with the actual value of the variable. For modest fluctuations, a single point will provide sufficient accuracy, as shown in the example. Putting the domain integrals which form the coefficients of the vectors α and $\ddot{\alpha}$ into two matrices $\mathbf{I}_C$ and $\mathbf{I}_\rho$, one finds that the classical governing relation (24) becomes:

$$(\mathbf{M} + \rho\mathbf{I}_\rho)\mathbf{F}^{-1}\ddot{\mathbf{u}} + (\mathbf{H} + \mathbf{I}_C\mathbf{F}^{-1})\mathbf{u} = \mathbf{Gp} \qquad (30)$$

3.3. ASSESSMENT OF THE PROBABILITY DISTRIBUTION

Once the regression coefficients are known and the variance s_ϵ^2 of the zero-mean variable ϵ has been calculated, a level-2 reliability method can be applied to (13) in order to derive the cumulative distribution function (CDF) of Y and hence its mean and variance. In the space of the variables $\zeta \equiv (\{\mathcal{Z}\}, \epsilon)$, the points

$$\zeta : \mathrm{Y}(\zeta) = y^\circ \qquad (31)$$

define a surface, say Ω. In standardized coordinates the distance from the origin $\delta(\zeta \in \Omega)$ to any point of Ω (i.e. satisfying (31)) can be written:

$$\delta(\zeta \in \Omega) = \{(\zeta - \mu_\zeta)^T \Sigma_\zeta^{-1} (\zeta - \mu_\zeta)\}^{\frac{1}{2}} \qquad (32)$$

ζ is assumed to be distributed like a joint normal probability distribution with mean vector μ_{ζ_i} and covariance matrix Σ_ζ. The minimum of this distance

$$\beta(y^\circ) = \min \delta(\zeta \in \Omega) \qquad (33)$$

decreases from infinity as y° increases from minus infinity.

According to the theory of level-2 reliability methods the cumulative distribution function of Y can be estimated as $\Phi(-\beta)$, Φ being the standard Gaussian distribution.

For the numerical calculation of $\beta(y^\circ)$, one can use any nonlinear optimization algorithm (Schittkowsi, 1985). An approach for determining the joint distribution of several response parameters has been proposed as an extension of the present procedure (Faravelli, 1988).

4. SFEM vs SBEM Example

The evaluation of the behaviour of soils under seismic loads leads one to study the way seismic load is transferred from the soil to buildings. Due to different constitutive properties of the soil, or presence of topographic irregularities, the subsequent seismic wave amplification is often dominant for structural reliability assessment.

In particular the scatter of physical and mechanical properties of a soil can be very large within one site. The modelling of those properties strongly depends on the way the appropriate tests are planned and carried out (Corsanego *et al.*, 1990). In general, each quantity relating to each variable of the problem must be modelled as a random field over the soil domain. The aim of this example is to evaluate the dynamic behaviour of a linear soil (Fig. 1) in terms of its eigenvalue and eigenvector properties, using a SBEM technique, and to compare the results with those achieved by SFEM.

Attention is focused on the propagation of the uncertainty regarding the physical and mechanical properties to the probability distribution of the system modal frequencies. The two techniques give comparable results, but the SBEM procedure requires a lower modelling and a reduced computation effort when compared with SFEM.

In Fig. 2 both SBEM and SFEM meshes are shown; it is clear that the former does not require any element in the domain, leading to a clear reduction in the dimension of the problem.

The soil lens shown in Fig. 1 was studied earlier in a deterministic context (Casciati *et al.*, 1993); the dimensions specified are in meters. The medium is linked with the rigid bedrock by bilateral supports. Moreover, the soil on each side of the lens does not offer resistance to horizontal displacements. Geometric and boundary conditions are supposed to be deterministic.

The mass density, ρ, and Young's modulus, C, are defined as random fields on the medium domain Ω. The probabilistic model, specified in Table 1, shows a higher value for the variance of the mass density than for the Young modulus of the soil, in order to take in account the larger uncertainty in evaluating that quantity. Nine cells are introduced over the domain, as sketched in Fig. 2a; within every cell of the structure the properties are assumed to be constant.

Table 2 provides the results, in terms of eigenvalues, of the 13 numerical experiments required by a response surface technique, given by the SBEM procedure. For comparison Table 2 shows the results of BEM and FEM analyses for a domain in which the mass density and the Young's modulus assume their mean values. The probability distribution of the eigenfrequencies is given in figure 3.

Figure 4 shows the variability of the transfer function with the simulated random fields, while Fig. 5 shows the sensitivity of the same function to the central value of the density. Fig. 6 shows the variability along two factorial points of the experiments design.

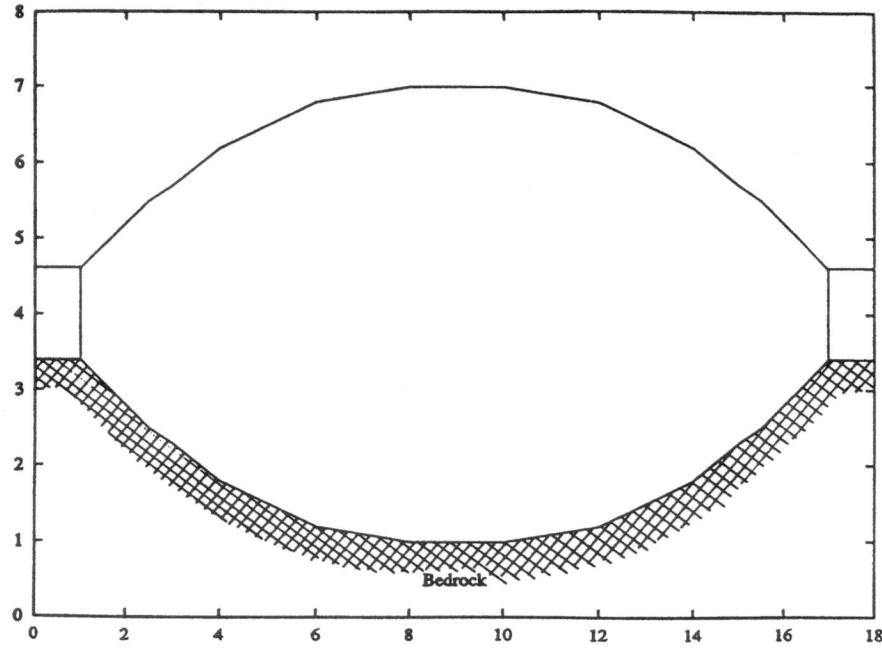

Figure 1. SBEM Example: soil lens.

TABLE 1. SBEM Example: probabilistic Model

Property	Distribution	Mean	Coeff. of Variation
Young Modulus C	Normal	25 MPa	0.05
Mass Density ρ	Normal	2000 Kg/m^3	0.10

5. A Fragility Analysis Example via SFEM

Seismic fragility has been studied in (Casciati *et al.*, 1985) and (Casciati *et al.*, 1991) with reference to a four storey three-span frame, shown in figure 7, which is also the object of the example presented here. All reinforced concrete columns and girders have the same cross section, 30.48x45.72 cm. by 30.48x50.80 cm., and are assumed to have equal bending stiffnesses in both loading directions. The geometric and boundary characteristics lead to a first natural period of 0.86 sec., with a damping ratio of 3.5%. For these values, the damping matrix $\mathbf{C}$ is $0.0097 \times \mathbf{K}_m$, where $\mathbf{K}_m$ is the global stiffness matrix. The reader is referred to (Casciati *et al.*, 1991) for a more

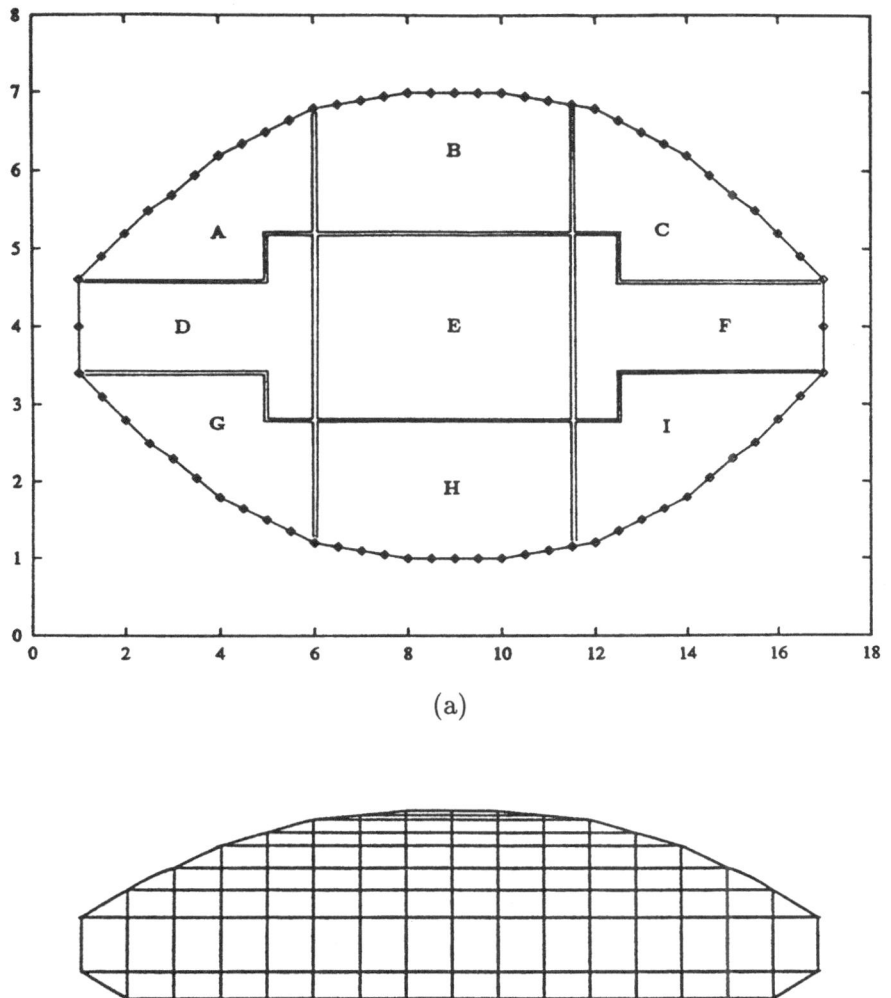

(a)

(b)

Figure 2.   SBEM Example: a) BEM mesh, with the discretization in cells of the domain required by the SBEM procedure here proposed b) FEM mesh.

TABLE 2. SBEM Example: results of the dynamic analyses in terms of modal frequencies.

Realiz.	Modal Frequencies (rad/sec)						
$n.$	ω_1	ω_2	ω_3	ω_4	ω_5	ω_6	ω_7
1	26.965	40.419	46.141	47.757	59.047	65.879	70.967
2	24.991	37.469	42.778	44.255	54.728	61.076	65.754
3	28.805	43.168	49.276	51.020	63.072	70.354	75.821
4	27.006	40.343	45.594	48.140	58.841	65.789	71.515
5	28.022	41.861	47.313	49.938	61.008	68.224	74.185
6	26.094	38.980	44.045	46.523	56.890	63.598	69.114
7	26.805	40.277	45.176	48.624	58.133	66.982	71.389
8	26.086	39.199	43.957	47.362	56.522	65.252	69.487
9	27.439	41.227	46.253	49.736	59.558	68.505	73.066
10	26.561	39.506	44.566	48.050	59.893	66.952	70.768
11	24.601	36.573	41.247	44.480	55.434	61.967	65.514
12	28.569	42.505	47.959	51.710	64.469	72.074	76.147
13	26.626	40.285	45.068	48.970	57.343	66.596	71.936
FEM om.	25.139	35.035	40.438	42.246	50.406	55.765	63.443
BEM om.	26.840	40.237	45.544	48.020	59.185	66.179	71.207

complete description of the geometrical and mechanical properties, such as the yield moment capacities and the dimensionless neutral axis position in the ultimate states.

Note that mass, stiffness, hardening and damping are vectors of strictly correlated quantities: each of them is thus described by a single random variable. By contrast, yielding moments and low-cycle fatigue resistance in the potential plastic hinges form two random vectors whose elements are assumed to be equicorrelated. They are the central values and the corresponding vector of deviations. Finally, the ground acceleration is a segment of white-noise appropriately filtered and modulated. The filtered white-noise gives rise to a Kanai-Tajimi type power spectral density function whose parameters are assumed to be random. The pseudo-stationary duration of the accelerograms is also assumed to be random.

The seismic intensity which causes the failure of the frame is the response variable of interest: its probability distribution is the so-called fra-

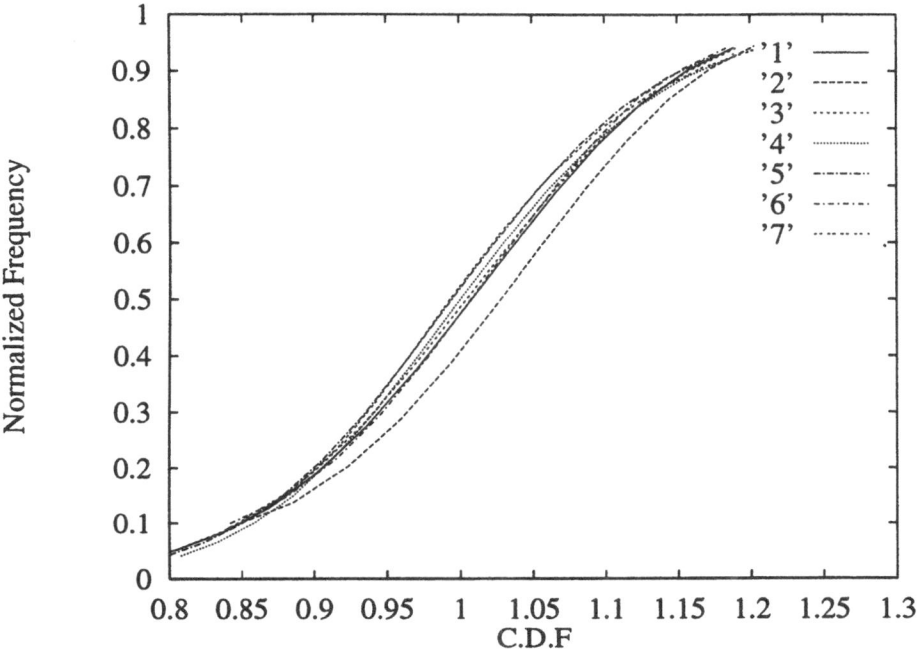

Figure 3. SBEM Example: CDF of the first seven modal frequencies, normalized to the omogeneus result

gility curve. However, accelerograms with the same intensity can have different features due to their stochastic nature and to the randomness which affects some of the model parameters.

In particular, the topographic irregularities can lead to a change in the frequency contents and to an amplification of the characteristics of the ground motion under the structure.

The seismic wave amplification, which occurs along the path from the bedrock to the foundation, is often dominant for a structural reliability assessment. Thus the scatter of the soil filtering properties should be investigated by the SBEM approach of the previous example.

In summary, six random variables (mass, stiffness, hardening, damping and the central values of yielding and resistance) and three deviation terms (yielding, resistance and accelerogram) are considered in the fragility analysis. The results of the computations in (Casciati *et al.*, 1991) are summarized in Fig. 8, which presents the fragility curves obtained in the two cases of deterministic and random coefficients by a standard experiment design. They are checked by simulation of a sample of size 20.

The polynomial form between the response variable and the logarithm of the six random variables is stated by regression analysis. It makes use of the

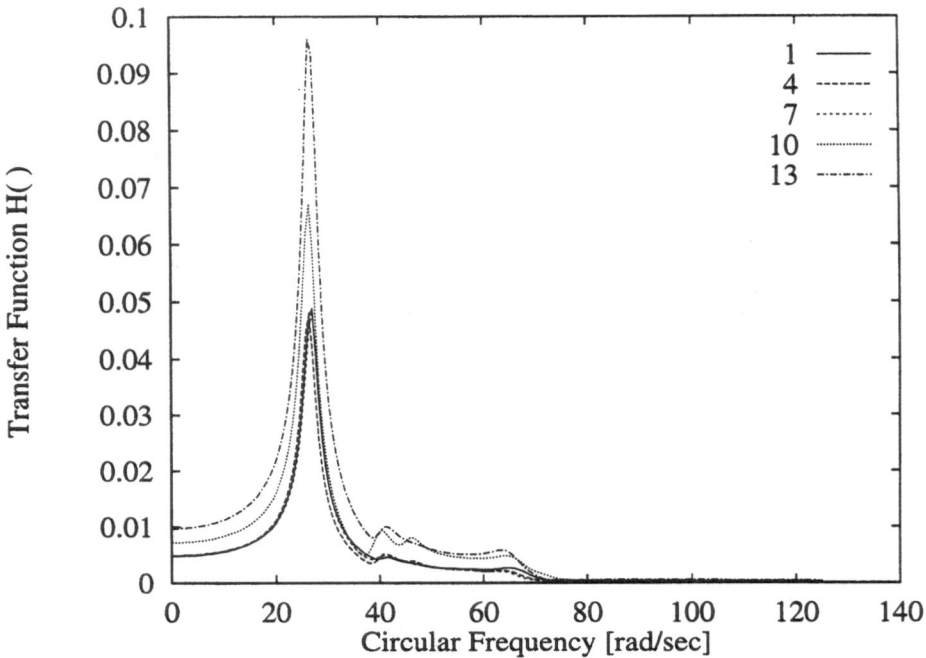

Figure 4. SBEM Example: variability of the transfer function for five simulated couples of random fields $\bar{p}$ and $\tilde{C}$

results of numerical experiments conducted according to two replications of a one-half factorial plane (32 by 2) in addition to three experiments (low, central and high level) along each axis in the variable space (6 by 3). A total of 82 experiments has therefore been carried out.

The curves of Figure 8 were obtained by the level-2 (β) reliability approach summarized in Eq. (33). This makes it possible to have estimates of the tails, i.e. to achieve results which would require an unbearable computational effort if pursued by simulation. Nevertheless, a large simulation (a sample of size 1000) was conducted in order to check the accuracy of the approximate approach, at least in the central part of the fragility curve. The results are presented in Fig. 9. It is worth noting the important role that the deviations play in this problem. When the error term is neglected (line a) the fragility curve is almost vertical in its central part, showing a low variability of the response. But, more significant uncertainty is indicated by the quite different behaviour of the result of the simulation including the error term (line b). The latter line is fitted well from the β-approach (line c) making use of a deterministic description of the coefficients of the polynomial form. By contrast, when these coefficients are regarded as random (line d) a greater content of uncertainty is found in the fragility curve.

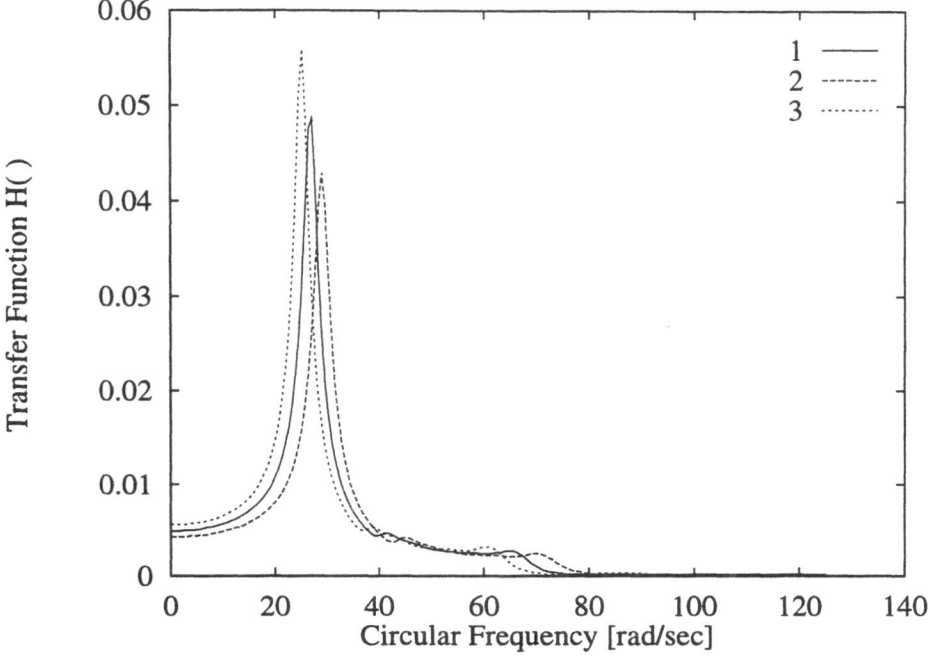

Figure 5.   SBEM Example: variability of the transfer function as the mean density value scan its definition range in the experiment design (1=central value; 2=lowest value; 3=higest value).

Note that the global slope of this curve was already overestimated in line c.

Another point of the procedure to be clarified is the dependence of the result on the selection of the realizations of the deviations in the numerical experiments. For each experiment of the composite plan introduced in view of performing regression analysis, in fact, one needs to specify the entire accelerogram and the whole of the random vectors of yielding moments and resistances. In the analysis, 16 different sets of deviations were simulated. Each of the two factorial designs was partitioned into 8 blocks and a set of deviations was associated with each block. Moreover, the first 6 sets were associated with the six axes in the random variable space for conducting the three experiments corresponding to each axis. The question is: does the way in which the 6 sets of deviations are selected affect the final result? The fragility analysis of the frame under investigation is sensitive especially to two random variables: the yielding moment and the resistance. The set of deviation number 4 was originally associated with the first variable. Figures 10a and b shows the perfect agreement with this original result (solid line) of the fragility curve obtained with the sets number 10 and 14, still associ-

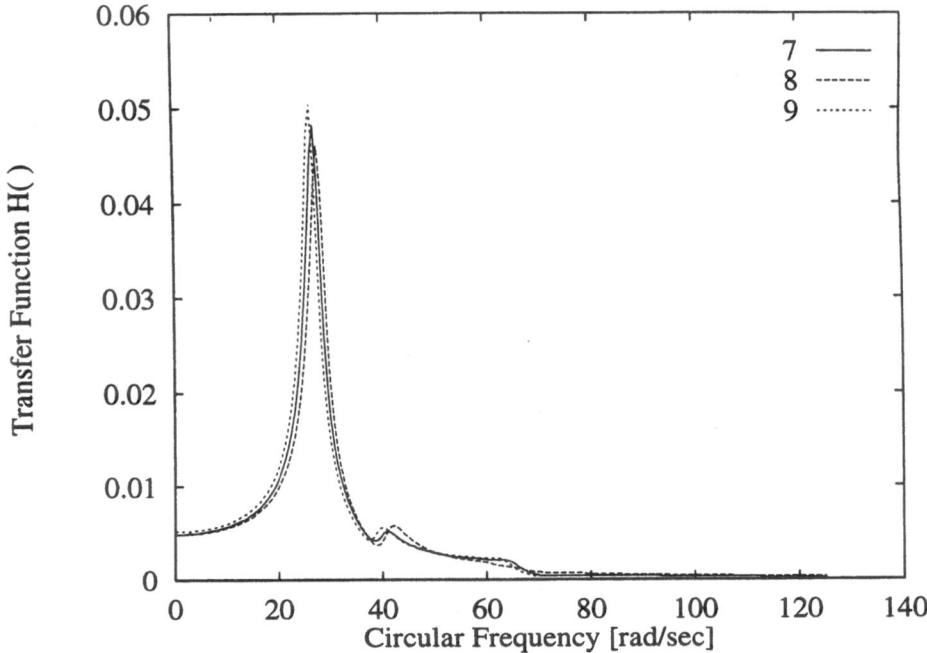

Figure 6. SBEM Example: Variability of the transfer function as the means of ρ° and C° span between two opposite fractional points (7=central value; 8=negative conbination; 9=positive combination)

ated with the yielding moment axis. By contrast, some minor discrepancy can be detected in Figs. 11a and b, where the resistence axis, originally associated with the set number 6 is associated with the sets numbered 13 and 16 respectively. These discrepancies, however, are quite unsignificant in the lower tail which is the area of practical interest. The sets 4, 10 and 14 contribute differently to the error term of the regression. However, they were selected in such a way to have a consistent average effect. Similarly, the sets 13 and 16 were associated with the set number 6. A further improvement could be based on the following idea. At the beginning, all the variables are supposed to have comparable significances, and experiment design theory provides plans of experiments which have optimal properties in this situation. But, when the analysis is accomplished, one knows the variables to which the result is more sensitive, in the present case, the yielding moment and the resistance. Therefore, accuracy could be increased by conducting a supplementary set of experiments, for instance, by doubling or tripling the analyses along the corresponding axes. This has been done here by associating the sets 10 and 14 with the yielding moment axis,

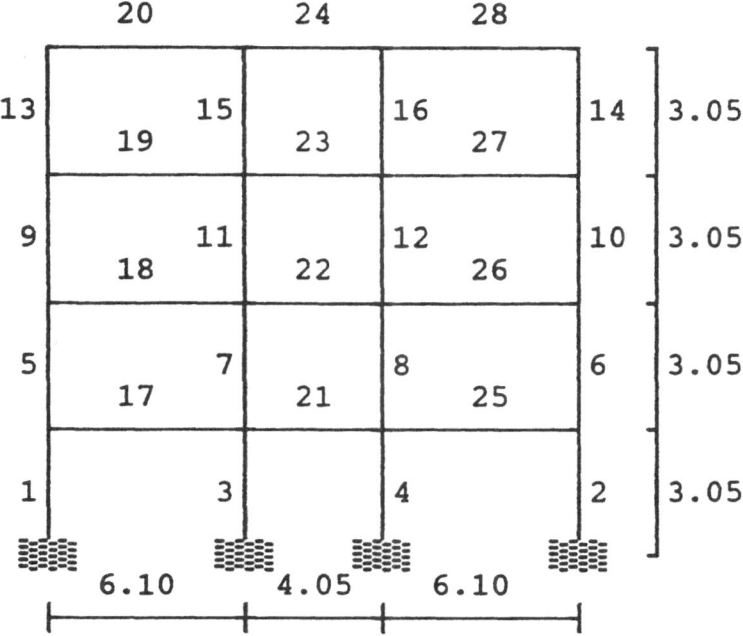

Figure 7. SFEM Example: geometric layout and boundary condition of the four-stories frame under investigation.

and the sets 13 and 16 with the resistance axis. The results are summarized in Fig. 12. No consequence is obtained by further investigation along the yielding moment axis. By contrast, a shift on the left is found by adding to the original 82 experiments two sets of three experiments along the resistance axis: the first for the set of deviation number 13 and the second for the number 16. The same result is found by including all the additional 12 experiments along yielding moment and resistance axes. Since the effect of the single additional analysis (see Fig. 11) is not so significant, the authors came to the conclusion that the policy of adding additional experiments to the original plan can only lead to distortion, because it indirectly amplifies the typical effect of the single variable.

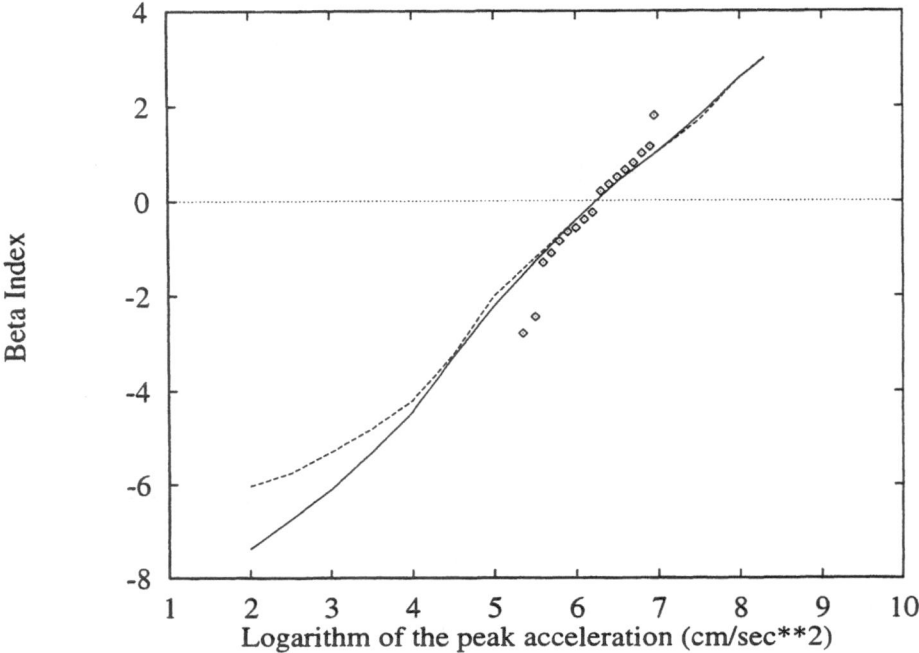

Figure 8. SFEM Example: result of the fragility analysis of a four storey three-span frame. Composite experiment plan with 82 experiments; β-approach for assessing the probability distribution. Deterministic (solid line) and random (dashed line) coefficients of the polynomial form. The points are obtained by simulation.

6. Conclusions

This chapter emphasizes the reasons that prevent one applying standard probabilistic methods of system theory to complex linear and non-linear structural systems. For non-linear structures, in particular, the input-output relationship is generally so complex that the solution can be pursued only in a numerical way, especially for dynamic analysis.

Response surface techniques can be usefully employed to provide an approximate simple input-output relationship for a probabilistic analysis of the response variables. For this purpose, level-2 reliability methods give good accuracy on the central values, and allow the analyst to estimate the tails of the probabilistic distributions. The latter task would require an unbearable computation effort if approached by a raw simulation.

The numerical examples produced results testing the accuracy of the level-2 approach, for both finite element and boundary element methods. Moreover, the direct applicability of experiment plans from the theory of experiment design has been checked by detailed analyses.

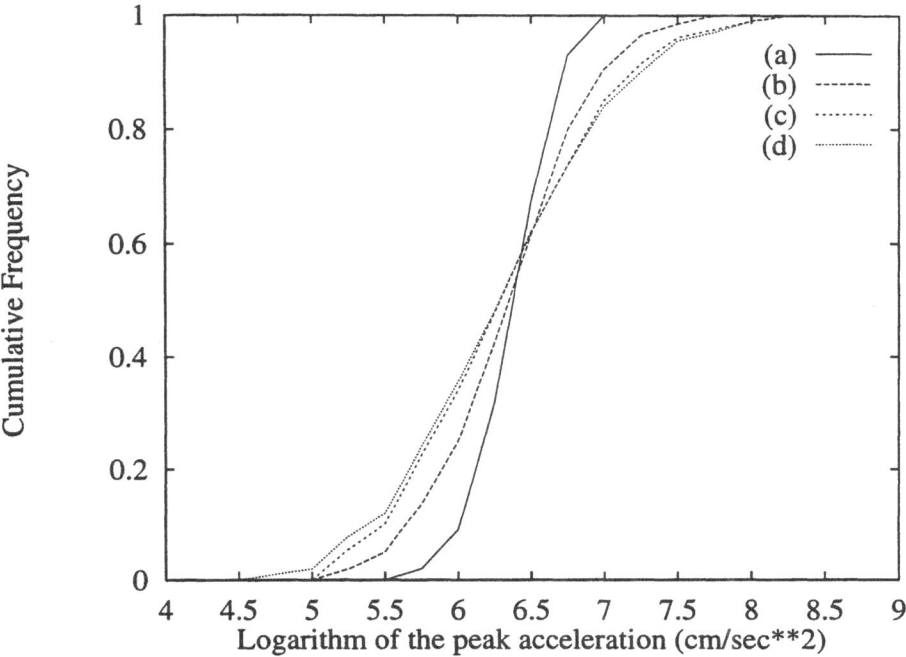

Figure 9. SFEM Example: check of the accuracy of the β-approach: a)simulation of 1000 values from the polinomial form obtained by regression analysis, in which the error term is neglected; b)as a) including the error term; c) solid line of figure 8; d) dashed line of figure 8.

Acknowledgement

The authors are grateful to the Italian Research Council (CNR) and to the Italian Ministry of University and Technological Research (MURST) for partial funding of this study.

References

Augusti G., Baratta A. and Casciati, F. (1984) *Probabilistic Methods in Structural Engineering*, Chapman & Hall, London

Brebbia, C.A. and Domínguez, J. (1989) *Boundary Elements – An Introductory Course*, Computational Mechanical Publications.

Burczynski, T. (1993) Stochastic Boundary Element Methods: Computational Methodology and Applications, in Spanos, P.D. and Wu, Y.-T. (eds.), *Probabilistic Structural Mechanics: Advances in Structural Reliability Methods*, Springer-Verlag, Berlin, pp. 42-55

Callerio, A., Casciati, F. and Faravelli, L. (1994) Dynamic Analysis by Stochastic Boundary Elements, Proc. Second International Conference on Computational Stochastic Mechanics, Athens, Greece, pp. 497–502

Callerio, A., Casciati, F. and Faravelli, L. (1995) Dynamic Analysis of Stochastic Media by Boundary Elements, Proc. IABEM '95 (International Association for Boundary

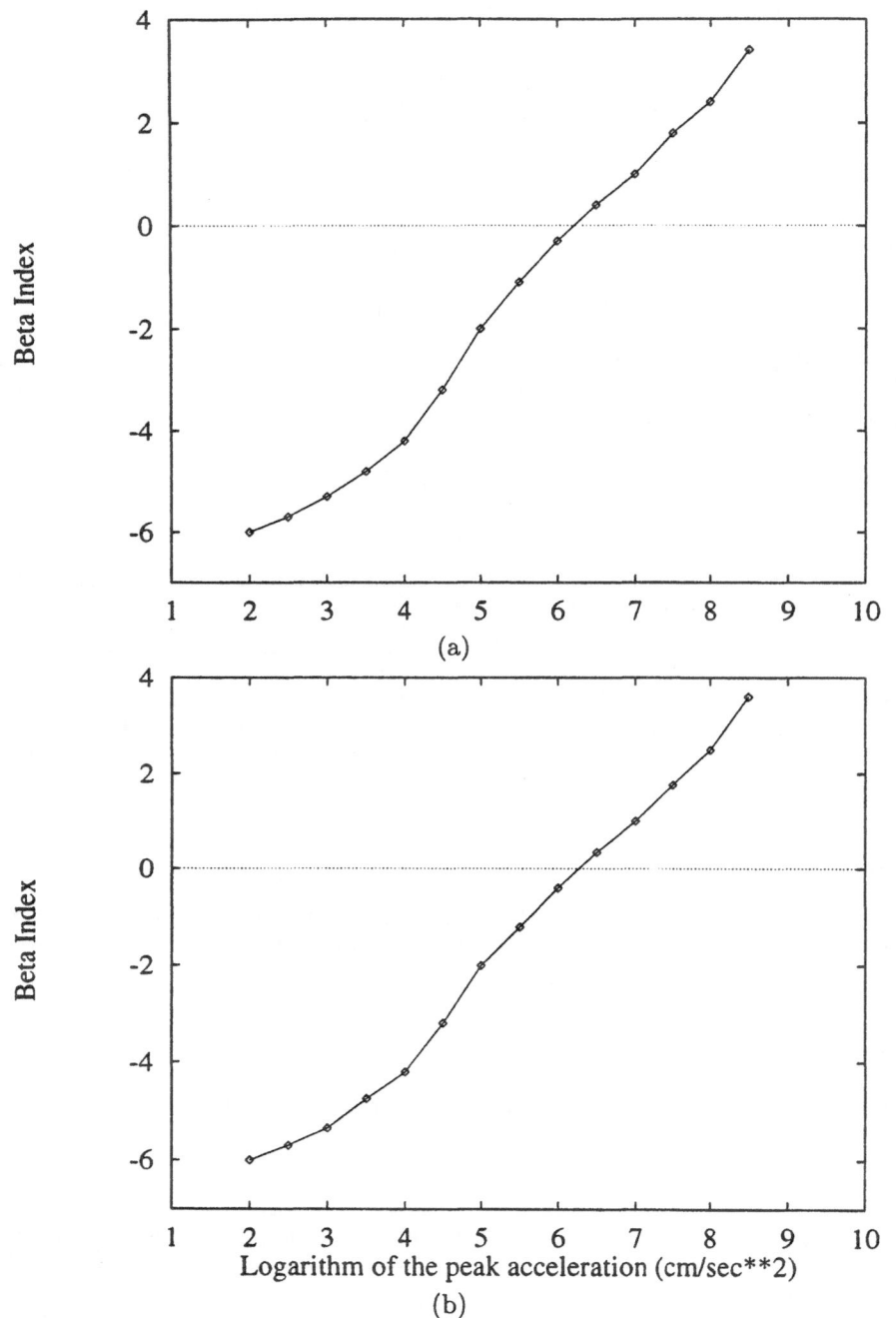

Figure 10. SFEM Example: comparison of the result of the β-approach in the basic case (dashed line) with the curve obtained considering: a) the set of deviations 10 instead of 4 and b) the set of deviations 14 instead of 4. Case with random regression coefficients.

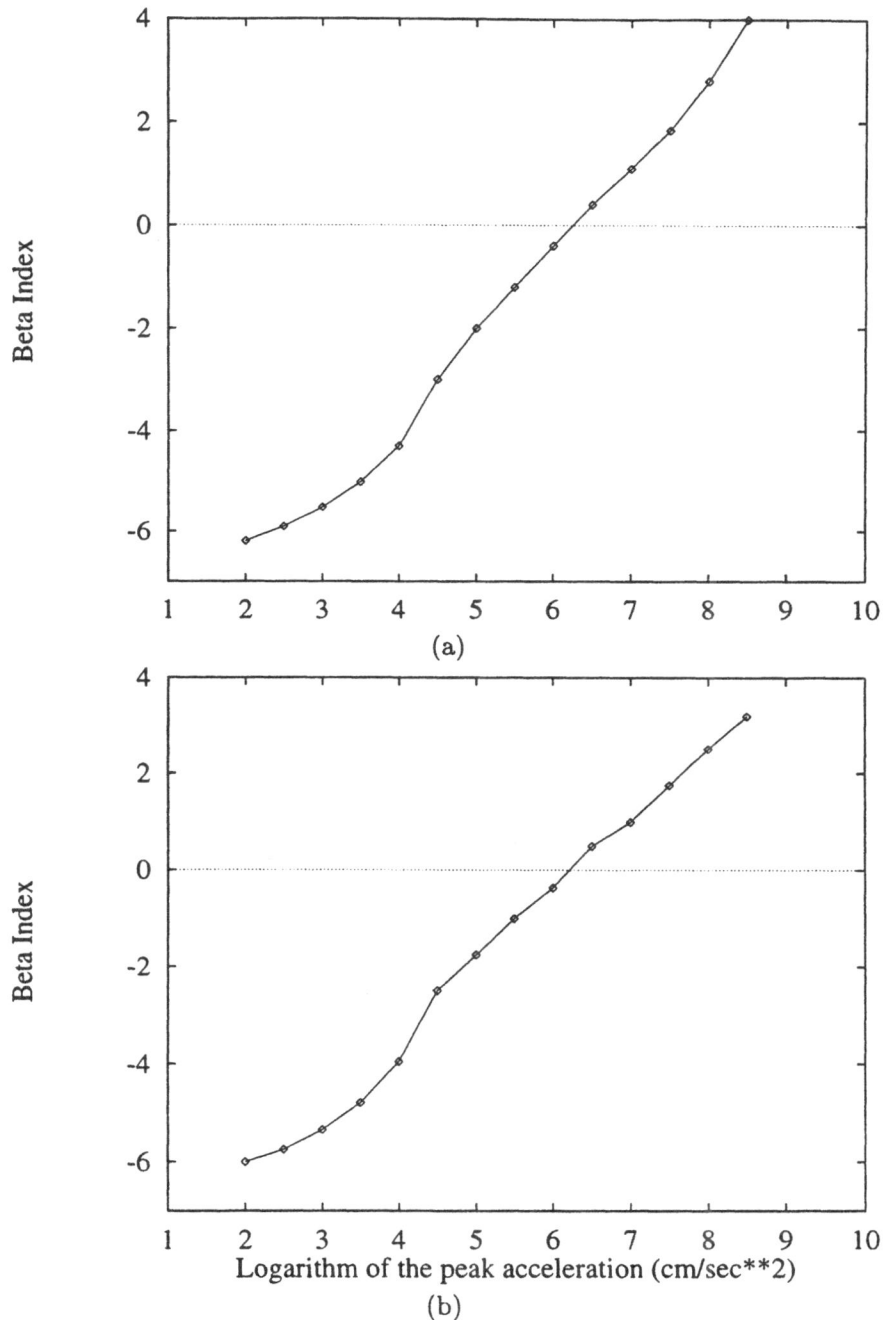

Figure 11. SFEM Example: comparison of the result of the β-approach in the basic case (dashed line) with the curve obtained considering: a) the set of deviations 13 instead of 6 and b) the set of deviations 16 instead of 6. Case with random regression coefficients.

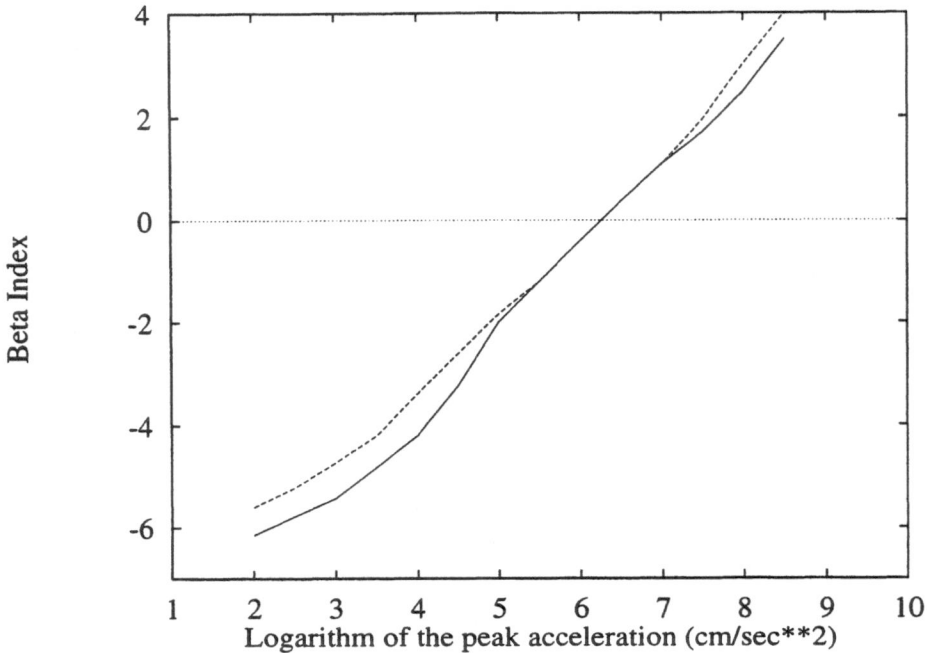

Figure 12. SFEM Example: result of the fragility analysis with improved experiment plans. The solid line is the dashed line of Fig. 8. It coincides with the fragility line obtained by conducting 88 experiments; 6 additional experiments along the yielding moment axis having been added using the sets of deviations 10 and 14. The dashed line was obtained in the same way, adding 6 experiments along the resistance axis, using the sets of deviations 13 and 16. This fragility curve coincides with the one reached after the analysis of the whole 94 experiments.

Element Methods Meeting), Mauna Laui, Haway

Casciati, F. and Faravelli, L. (1985) Methods of Nonlinear Stochastic Dynamics for the Assessment of Structural Fragility, *Nuclear Engineering and Design*, 90, pp. 341–356

Casciati, F., Faravelli, L., Fontana, M. and Marcellini, A. (1991) Local Amplification by Boundary Elements, Proc. 4th International Conference on Seismic Zonation, Stanford University, Stanford, California, USA

Casciati, F. and Faravelli, L. (1991) *Fragility Analysis of Complex Structural Systems*, Research Studies Press, Taunton

Casciati, F. and Faravelli, L. (1991) Randomness Effects in Crashworthiness Analysis, *International Journal of Non-Linear Mechanics*, pp. 827–834.

Casciati, F., Faravelli, L. and Callerio, A. (1993) Alcune Considerazioni sugli Elementi di Contorno Stocastici (in Italian), Atti del Convegno Nazionale del Gruppo AIMETA di Meccanica Stocastica, Taormina, Italy

Cen, Z., Wang, I-C., Sun, H. and Wang, F. (1989) Vibration Analysis by using Component Mode Synthesis and Boundary Element Method, Proc. 7th International Modal Analysis Conference, Las Vegas, Nevada, USA, pp. 59–62

Corsanego, A. and Lagomarsino, S. (1990) A Response Surface Technique for Fuzzy-Random Evaluations of Local Seismic Hazard, in Casciati, F., Elishakoff, I., Roberts, J.B. (eds.), *Nonlinear Structural Systems under Random Conditions*, Elsevier

Der Kiureghian, A. (1983) Numerical Methods in Structural Reliability, Proc 4th ICASP, Firenze, pp. 769–784

Domínguez, J. and Saez, A. (1993) Dual Reciprocity Approach for Transient Elastodynamics, in *Boundary Elements in Dynamics*, Computational Mechanics Publications

Faravelli, L. (1988) *Sicurezza Strutturale* (in Italian), Pitagora Editrice, Bologna

Faravelli, L. (1988) Response Correlation in Stochastic Finite Element Analysis, *Meccanica*

Faravelli, L. (1989) A Response Surface Approach to Reliability Assessment, *Journal of Engineering Mechanics*, ASCE

Faravelli, L. (1989) Finite Element Analysis of Stochastic Nonlinear Continua, in Liu W.K. and Belytschko T., *Computational Mechanics of Probabilistic and Reliability Analysis*, Elmpress, Lausanne, pp. 263–280

Liu, W.K., Belytschko, T. and Mani, A. (1986) Random Field Finite Element, *Int. J. for Numerical Methods in Engineering*, 23, pp. 1831–1845

Myers, R.M. (1971) *Response Surface Methodology*, Allyn and Bacon, Boston

Nakagiri, S. (1985) *Stochastic Finite Element Method: an Introduction* (In Japanese), Baifiukan

Petersen, R.G. (1985) Design and Analysis of Experiments, M. Decker Inc., New York

Schittkowski, K. (1985/6) *NLPQL: A Fortran Subroutine Solving Constrained Nonlinear Programming Problems Annals of Operations Research* 5, pp. 485–500.

Spanos, P.D. and Ghanem, R. (1991) Boundary Element Formulation for Random Vibration Problems, *Journal of Engineering Mechanics*, p. 409.

Vanmarcke, E. (1983) *Random Fields*, MIT Press

Vanmarcke, E., Shinozuka, M., Nakagiri, S., Schueller, G.I. and Grigoriu, M. (1986) Random Fields and Stochastic Finite Elements, *Structural Safety* 3, pp. 143–166

RISK BASED STRUCTURAL MAINTENANCE PLANNING

M.H. FABER
COWIconsult
Dept. of Assessment & Rehabilitation of Structures
Lyngby Denmark.

1. Introduction

Risk management has become increasingly important to industry and society during the last decades. This is largely due to an urge for increased efficiency and competitiveness by industry itself, and to increasing standards regarding personnel safety and environmental preservation, imposed by society.

Due to their economic significance and potential safety hazards, buildings and industrial structures should be designed, maintained and demolished according to life cycle risk evaluations. In this way, risk optimal decisions regarding the design, maintenance and demolition of structures may be based on an overall evaluation of the economic consequences, personnel safety hazards and the likelihood of environmental damages, evaluated over the design service life of the structure.

These remarks apply not only to unique and large structures such as large strait crossings, offshore production installations and nuclear power installations, which are associated with significant risk, but also with the more common types of structures such as highway bridges, dams and break waters, which are insignificant by themselves, but due their large numbers are associated with significant risk.

In recent years, a substantial effort has been devoted to economic risk based decision making in structural maintenance planning. There are theoretical developments contained in e.g. Vrijling [1], Madsen et al. [2], Fujita et al. [3], Sørensen et al. [4] and Sørensen & Thoft-Christensen [5]; and practicable decision support software tools as reported in e.g. Sørensen et al. [6] and Faber et al. [7]).

The following sections, which are based on the work of Faber et al. [8], address the basic theoretical framework for risk based maintenance planning for structures, and illustrate its application to real life engineering problems by an example.

2. General Problem Framework

The basic problem in risk management for structures is to make decisions regarding design and maintenance of the structures, such that the overall life cycle costs of the operation of

C. Guedes Soares (ed.), Probabilistic Methods for Structural Design, 377–402.
© 1997 *Kluwer Academic Publishers.*

the structures are minimized, and such that the personnel safety hazard is kept within the limits specified by legislation or society. As the available information regarding e.g. loading, material properties and deterioration processes in general is incomplete or uncertain, the decision problem is a decision problem subject to uncertain information.

Two risk management situations are normally distinguished for practical reasons: designing new structures and maintaining or demolishing existing structures.

In design of new structures, the design parameters, such as member dimensions, are chosen by evaluating their influence on the design costs and future maintenance costs. Optimal design parameters may hence be identified as the design parameters which minimize the overall costs, including design costs, expected costs of failure and expected costs of future maintenance.

For existing structures the optimal inspection, repair and reinforcing actions are based on evaluations of their influence on the immediate repair or reinforcing costs, the expected failure costs and the expected maintenance costs.

Thus there are no major differences between designing a new structure, and maintaining an existing structure, as the design parameters and the repair or reinforcing parameters may be treated in the same manner - simply as decision variables influencing the life cycle failure and maintenance costs. For this reason only maintenance of existing structures is addressed directly in the following.

Structural maintenance planning usually involves one or more (re-)assessment analyses and actions followed by decisions on requalification, rehabilitation or demolition of the structure. Due to the interrelation between the use of the structure, the present and the future state of the structure and the safety of the structure, decisions regarding requalification and rehabilitation cannot be made without at the same time specifying a corresponding strategy for the future maintenance of the structure.

Reassessment may be seen as an adaptive process of refining the state of knowledge about the present and the future state of the structure. Typically a structural reassessment may thus involve a review of project documentation, inspection of the structure, testing of materials, testing of structural performance, refined numerical analysis and planning of future inspections. The adaptivity in the refinement of the state of knowledge is introduced as the decision on whether or not to collect more information is at all times based on all the existing information (prior information) and the (pre-posterior) expected life cycle costs reductions (including expected maintenance and failure costs) achieved by the planned but not yet actually collected information. Depending on the actually achieved knowledge it may or may not turn out to be feasible to gather more information.

The maintenance plan i.e. the plan prescribing the future inspections and repair actions, is highly influenced by the costs associated with the engineering structure. Therefore it is important to identify the optimal inspection and repair plan, i.e. the plan minimizing the total expected life cycle costs associated with inspection, repair and failure of the structure with respect to inspection methods, inspection times and repair actions, and at the same time continue to fulfil all requirements on reliability.

Having identified the optimal inspection and maintenance plan one must investigate how this plan and its associated total expected costs are influenced if the assumed parameters for the inspection and maintenance planning are perturbed. Such investigations will indicate which parameters are the more significant to the inspection and maintenance plan, and may lead to rejection of a particular optimal inspection and maintenance plan as being too sensitive to certain parameters.

3. Formulation of the Maintenance Decision Problem

In practical decision problems such as reassessment and maintenance planning for structures the number of alternative actions can be extremely large, and it is expedient to have a framework for the systematic analysis of the corresponding consequences.

Bayesian life cycle risk (decision) analysis, as described by Raiffa & Schlaifer [9] and Benjamin & Cornell [10] forms such a fremework. They discuss structural systems subject to uncertain and subjective information about loading, material properties, damage conditions, damage accumulation laws and structural behaviour. Within this framework, maintenance strategies minimizing the overall life cycle costs of the operation of the structure may be identified.

Decision problems are conveniently represented by decision trees as illustrated in Figure 1.

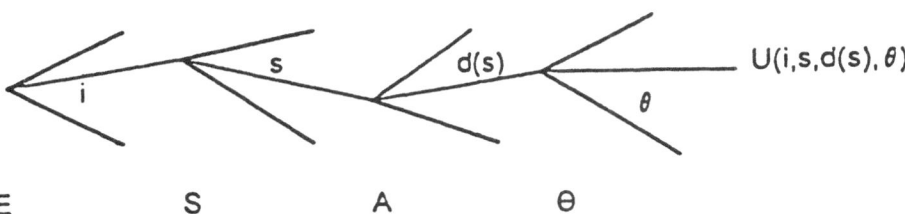

$$E \qquad S \qquad A \qquad \Theta$$

Figure 1 Decision tree used in Bayesian decision analysis

Generally speaking, the decision making problem is to choose an experiment (inspection method, inspection time etc.) e from the space of possible experiments E yielding a random outcome s (e.g. observed damage state) of possible experiment outcomes S which can be used by the decision maker to take an action a (e.g. repair, no repair) out of the possible available actions A. When the decision maker has taken an action this will lead to a random outcome of the event Q (e.g. failure, no failure) out of the possible states of the nature q. The performed experiment and the chosen action, together with the outcome of the experiment and the event determines the corresponding utility. The part of the decision tree starting with choosing an action *a* based on the experiment outcome *s* is also called a terminal analysis or posterior analysis because the statistics of the utility can be estimated by using known statistics about the nature, whereas the complete analysis is called a pre-posterior analysis, because the experiment outcomes are still unknown.

In order to perform a decision analysis, the following information and operations must be included: information concerning the alternative experiments *E* and actions *A*; assignment of a utility function u(*e,s,a,q*) on the space *E* x *S* x *A* x *Q*;
assignment of the probability $P_{Q,S}(q,s'e)$ on the space *Q* x *S*. This joint probability measure determines the four probabilities of importance :

a   The marginal probability measure $P'_Q(Q)$ on the state of the nature. This is normally referred to as a prior probability, in the sense that the decision maker assigns the probability measure to *Q* prior to knowing the outcome *s* of the experiment *e*.

b   The conditional probability measure $P_S(s'Q,e)$ on the outcome of the nature, referred to as the sample likelihood representing the new information obtained by the experiment.

c   New information can be combined with prior probabilities of the state of the nature by applying Bayes' rule

$$P''_\Theta(\theta|s) = \frac{P_S(s|\theta,e)\, P'_\Theta(\theta)}{\sum_\Theta\, P_S(s|\theta,e)\, P'_\Theta(\theta)} \tag{1}$$

The conditional probability measure on the state of the nature is called the posterior probability, posterior in the sense that the probability measure is assigned to q after (posterior) to knowing the outcome *s* of the experiment *e*.

d   The marginal probability measure $P_{\underline{S}}(s|e)$ on the outcome of the nature of a given experiment *e*.

The decision problem can then be stated as ; Given E,S,A, θ, u and $P_{q,S}(Q,s|e)$ how must one choose an experiment *e* yielding an outcome *s* based on which action *a* is taken, in such a way that the utility *u* is maximized.

There are two equivalent ways to formulate the analysis leading to the maximum utility, namely the so-called extensive and the normal form of the analysis.

Kroon [11] discusses the advantages of the two different formulations. The normal fom is the the more computational convenient for practical applications, and it is this that will now be described.

In the normal form, a decision rule *d* is specified which prescribes the action that must be taken for all possible outcomes of the experiment *e*. For every experiment *e*, the optimal decision rule *d* can be selected. By doing this for all possible experiments *e*, the optimal experiment can be selected. The decision rule for a specific experiment *e*, is a mapping carrying s in S into $d(s)$ in A. For a selected experiment e the expected utility is

$$u(e,d) = E_{\Theta, S|e}\left[u(e,s,d,(s),\Theta)\right]$$
$$= E'_{\Theta}\left[E_{S|e,\theta}\left[u(e,s,d(s),\Theta)\right]\right]$$

$$(2)$$

the optimal experiment e and the optimal decision rule can now be identified by solving

$$\max_{e}\max_{d} E'_{\Theta}\left[E_{S|e,\theta}\left[u(e,s,d)(s),\Theta\right]\right]$$

$$(3)$$

The complete analysis is called pre-posterior because a number of posterior analysis are performed, conditionally upon the experiment and the outcome of the experiment.

If the utility function is related to the life cycle total costs equation (3) can equivalently be formulated as an optimization problem on which the total expected life cycle costs is minimized:

$$\min_{i}\min_{d} E_{\theta, s:i}\left[C_{TOTAL}(i,s,d(s),\Theta)\right]$$

$$(4)$$

where $C_{TOTAL}(.,i,s,d(s),\Theta)$ is the total cost, equal to minus the utility function. The experiment e is now described by the inspection vector $i = N, \Delta t, q$ where N is the total number of inspections, $\Delta t = (\Delta t_1, \Delta t_2, \dots \Delta t_N)$ are the time intervals between inspections, and $q = (q_1, \dots q_N)$ are the inspection qualities.

If the total expected life cycle costs are divided into inspection, repair and failure costs, and a constraint related to a minimum of reliability is added, then the optimization problem is

$$\min_{i,d} C(i,d) = C_{IN}(i,d) + C_F(i,d)$$

$$(5)$$

$$s.t.\ \ \beta(T_L, i, d) \geq \beta_{min}$$

$$(6)$$

$C(z, i, d)$ is the total expected life cycle cost evaluated over the lifetime $(T_{N+1} = T_L)$ C_{IN} is the expected inspection cost, C_R is the expected cost of repair and C_F is the expected failure cost.
$\beta(T)$ is the generalized reliability index defined by:

$$\beta(T) = -\Phi^{-1}(P_T(T)) \tag{7}$$

where Φ is the standardized normal distribution function and $P_F(T)$ is the probability of failure in the time interval $[0,T]$

The constraint on the minimum reliability (6) is included to take account to reliability requirements from legislation and society, even though the consequences of structural failure may be included in the costs of failure.

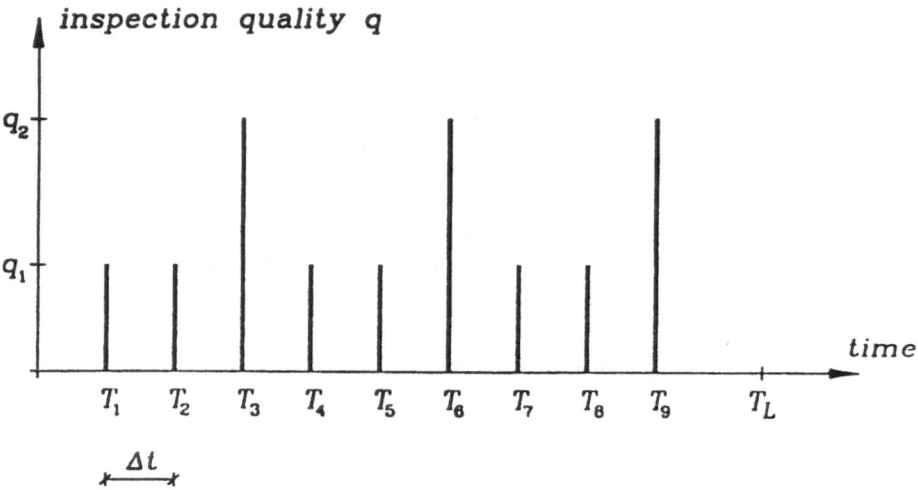

Figure 2 Typical inspection plan for bridge structures.

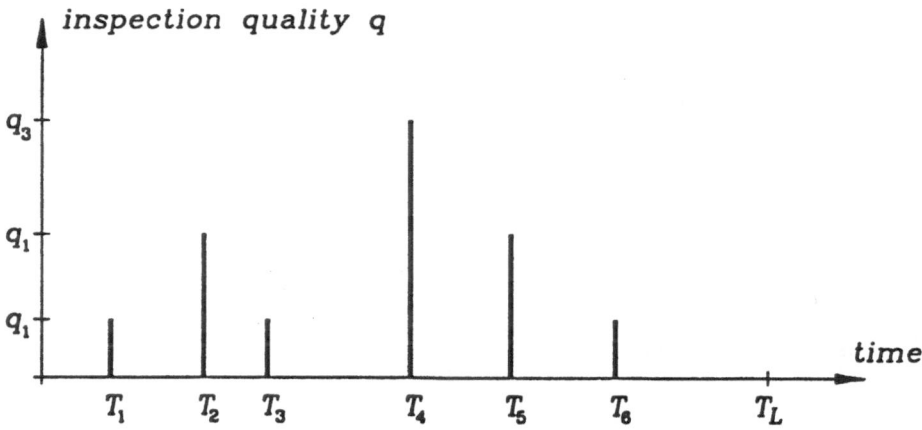

Figure 3 Typical inspection plan for offshore jacket structures.

Other constraints, e.g. on the maximum of the individual costs or direct bounds on the optimization variables, can be included in the problem, if necessary.

In inspection and maintenance planning for bridges, the inspection scheme will typically be as shown in figure 2. Two or three different types of inspection with qualities q_1 and q_2 are performed with fixed time intervals. For offshore steel jacket platforms, a typical inspection plan is shown in figure 3. The inspections are performed at non-uniform time intervals and a number of different inspection techniques may be used.

4. Modelling of Costs

In (4) the expected inspection, repair and failure costs must be modelled as functions of the decision variables. Modelling of the expected inspection, repair and failure costs depends on the structural modelling, and whether the maintenance plan is performed for the structural components individually or jointly. The total capitalized expected inspection costs are modelled by

$$C_{IN}(z,i,d) = \sum_{i=1}^{N} C_{IN_i}(q)(1 - P_F)(T_i) \; \frac{1}{(1+r)^{T_i}} \tag{8}$$

The i'th term represents the capitalized inspection costs at the i'th inspection when failure has not occurred earlier. Here it is assumed that if failure occurs, then the component cannot be repaired. $C_{IN_i}(q)$ is the inspection cost of the i'th inspection, $P_F(T_i)$ is the probability of failure in the time interval $[0, T_i]$ and r is the real rate of interest. $(1 - P_F(T_i))$ is usually close to one.

$$C_R(z,i,d) = \sum_{i=1}^{N} C_{R_i} P_{R_i} \frac{1}{(1+r)^{T_i}} \tag{9}$$

is the total capitalized expected repair costs. The i'th term represents the capitalized expected repair costs at the i'th inspection. C_{R_i} is the cost of a repair at the i'th inspection and P_{R_i} is the probability of performing a repair after the i'th inspection when failure has not occurred earlier. The total capitalized expected costs due to failure are determined from.

$$\begin{aligned} C_F(z,i,d) &= P_{S_0} \int_0^{T_L} C_F(\tau) f_T(\tau) \frac{1}{(1+r)^{\tau}} \, dr \\ &\approx \sum_{i=1}^{N+1} C_F(T_i) P_F(T_i) \frac{1}{(1+r)^{T_i}} - C_F(T_{i-1}) P_F(T_{i-1}) \frac{1}{(1+r)^{T_{i-1}}} \\ &\leq \sum_{i=1}^{N+1} C_F(T_i)(P_F(T_i) - P_F(T_{i-1})) \frac{1}{(1+r)^{T_i}} \end{aligned} \tag{10}$$

$C_F(T)$ is the cost of failure at the time T. $f_T(\tau)$ is the probability density function of the time to the first failure of the component, conditional on the event that the component is in a safe state at the time $T = 0$. The first line in (10) gives the integrated capitalized costs due to failure, which in the second line is approximated using the trapezoidal rule. If $C_F(\tau)$ is a non-increasing function of time then the upper limit in the last line is obtained. The approximation is assumed to be good if N is large.

5. Assessment of Failure and Repair Probabilities

From the expression of the total expected life cycle costs (equation (10)), it is seen that the expected costs may be evaluated as a summation of the products of the marginal costs and the marginal probabilities associated with the events of inspection, repair and failure.

Whereas the marginal costs are readily assessed, more consideration is necessary in order to estimate the marginal probabilities.

Normally it is adequate and sufficiently precise to represent the uncertainties in the decision problem by stochastic variables. The events may then be modelled by limit state functions and the marginal probabilities for the events of inspection repair and failure in equation (10) may be appropriately expressed in terms of intersections of the events of inspection, repair and failure. All branches of the repair event tree at the n inspection times of the chosen maintenance strategy must in principle be included when calculating the probabilities. In time-invariant reliability problems the probabilities can be estimated by FORM/SORM techniques as described in e.g. Madsen et al. [12].

If repair is assumed to be performed when a defect is detected and has a measured size a larger than a critical level a_r, then the total number of repair realization (branches) is 3^N, see figure 4. This is shown in figure 4, where 0 signifies that no defect has been detected (no repair); 1, that a defect has been detected but is too small to be repaired; and 2, that a defect has been repaired.

The probability of failure in the time interval $[0, T_i]$ is for $0 \leq T \leq T_1$:

$$P_F(T) = P(M_F(0) > 0 \cap M_F(T) \leq 0) \tag{11}$$

where $M_F(T)$ is the event margin modelling failure at the time T.
For $T_1 < T \leq T_2$:

$$P_F(T) = P_F(T_1) + P(M_F(0) > 0 \cap B^0 \cap M_F^0(T) \leq 0)$$
$$+ P(M_F(0) > 0 \cap B^1 \cap M_F^0(T) \leq 0) + P(M_F(0) > 0 \cap B^2 \cap M_F^2(T) \leq 0) \tag{12}$$

where B^0, B^1 and B^2 are the events corresponding respectively to no detection, detection and no repair, and detection and repair at the first inspection.

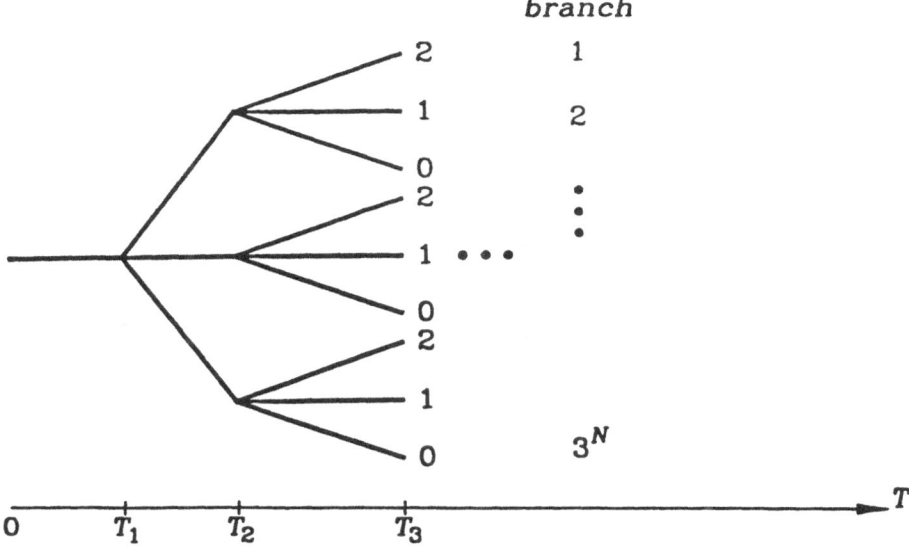

Figure 4 Repair realizations.

$$B^0 = \{M_F(T_1) > 0 \cap M_D(T_1) > 0\}$$
$$B^1 = \{M_F(T_1) > 0 \cap M_D(T_1) \le 0 \cap M_R(T_1) > 0\} \tag{13}$$
$$B^2 = \{M_F(T_1) > 0 \cap M_D(T_1) \le 0 \cap M_R(T_1) \le 0\}$$

$M_D(T_1)$ is the safety margin modelling detection of a defect, and $M_R(T_1)$ is the safety margin modelling repair. $M_F^0(T)$ and $M_F^2(T)$ are safety margins with respect to failure at the time $T > T_1$ corresponding to no repair, and repair at the first inspection. Similar expressions are obtained for $T > T_2$.

The probability of repair at the time T_i is determined in a similar manner, e.g.

$$P_{R1} = P\big(M_F(0)\big) > 0 \cap M_D\big(T_1\big) \le 0 \cap M_R\big(T_1 \le 0\big) \tag{14}$$

and

$$
\begin{aligned}
P_{R2} = \; & P(M_F(0) > 0 \cap B^0 \cap M_D^0(T_2) \le 0 \cap M_R^0(T_2) \le 0) \\
+ \; & P(M_F(0) > 0 \cap B^1 \cap M_D^0(T_2) \le 0 \cap M_R^0(T_2) \le 0) \\
+ \; & P(M_F(0) > 0 \cap B^2 \cap M_D^2(T_2) \le 0 \cap M_R^2(T_2) \le 0)
\end{aligned}
\tag{15}
$$

Similar expression are obtained for $T_i > T_2$.

Only inspection and repair of components are treated above, but similar considerations can be made for systems. In this case estimations of costs and probabilities are modified such that the events leading to systems failure are incorporated. The problem has been treated in Faber et al. [13] where a detailed description can be found.

Updating of Inspection Plans

When new information from measurements and inspections is obtained, the reliability estimates of the structure can be updated using Bayesian techniques see e.g. Lindley [14] and Madsen [15]. During the lifetime of the structure, two types of information are likely to be collected. The first type is information about the functional relationship between uncertain variables. For off-shore structures, for example, this functional relationship can be for crack depths. The second type of information is observations of one or more of the stochastic variables, e.g. for offshore structures, measurements of the significant wave height, the wave period, the thickness of marine growth etc. This type of information consists of actual samples of the uncertain basic variables.

Both types of information are taken into account by the use of Bayes' theorem see e.g. Lindley [14].

The information gathered during an inspection can be expressed in terms of event margins, and updating with respect to this information can be regarded as general event updating. The general information is assumed to be modelled by inequality and equality events. Updating of the probability of failure can be performed using Bayesian methods, as shown in Madsen [14] and Rackwitz & Schrupp [16].

The safety margin modelling failure of a single component is denoted by M. Let a single inequality event I be modelled by the event margin H, i.e. $I = \{H \leq 0\}$. The probability of failure of the component can then be updated by

$$P_f^U = P(M \leq 0 | H \leq 0) = \frac{P(M \leq 0 \cap H \leq 0)}{P(H \leq 0)} \tag{16}$$

If more than one inequality event are available, the updating can be performed in a similar way. Let $I_1 = \{H_1 \leq 0\}, ... , I_N = \{H_N \leq 0\}$ model N inequality events. The updated probability of failure is

$$P_f^U = P(M \leq 0 | H_1 \leq 0 \cap ... \cap H_N \leq 0)$$
$$= \frac{P(M \leq 0 \cap H_1 \leq 0 \cap ... \cap H_N \leq 0)}{P(H_1 \leq 0 \cap ... \cap H_n \leq 0)} \tag{17}$$

An equality event E is modelled by the event (safety) margin H, i.e. $E = \{H = 0\}$. The probability of failure of a single element can then be updated by

$$P_f^U = P(M \leq 0 | H = 0) = \frac{P(M \leq 0 \cap H = 0)}{P(H = 0)} \tag{18}$$

Schall et al. [17] show how (18) can be evaluated. If more than one equality event are available, the updating can be performed in a similar way.

Next, updating of stochastic variables is considered. An uncertain quantity modelled by a stochastic variable X is considered. A density function $f_X(x, P)$ for X is established on prior information, where P is a vector of parameters defining the distribution for X.

If one (or more) of the parameters P is treated as an uncertain parameter (Stochastic variable), then $f_X(x, P)$ is actually a conditional density function: $f_X(x|P)$. in the following P is for simplicity assumed to consist of only one parameter P. The prior density function of P is denoted $f_P(p)$.

Assume that an experiment or inspection is performed. n realizations of the stochastic variable X are obtained and are denoted by $\chi = (\chi_1, \chi_2, ..., \chi_n)$. The measurements are assumed to be independent. The posterior density function $f_P(p|x)$ of the uncertain parameter P, taking into account the realizations, is defined by

$$f_P''(p|\chi) = \frac{f_n(\chi|p) f_P'(p)}{\int f_n(\chi|p) f_P'(p) dp}$$

(19)

where $f_n(\chi|p) = \prod_{i=1}^{n} f_x(\chi_i|p)$.

The updated density function of the stochastic variable X, taking into account the realization c is denoted the predictive density function, and is obtained by

$$f_X(x|\chi) = \int f_X(x|p) f_P''(p|\chi) dp$$

(20)

If the prior distributions are chosen such that both the prior and the posterior distribution belong to the same family of distributions, then they are called conjugated. E.g. Raiffa & Schlaifer [9] derive conjugated prior, posterior and predictive distribution functions for a number of distributions.

Based on the updated probabilities and the predictive density functions, updated optimal inspection plans can be determined. If the inspection plan is updated after each inspection it is really only the next inspection time and quality which are important. In this way an adaptive inspection procedure is obtained.

Simplified Inspection Planning
The numerical effort in determining an optimal inspection plan can be significantly decreased in a number of ways:

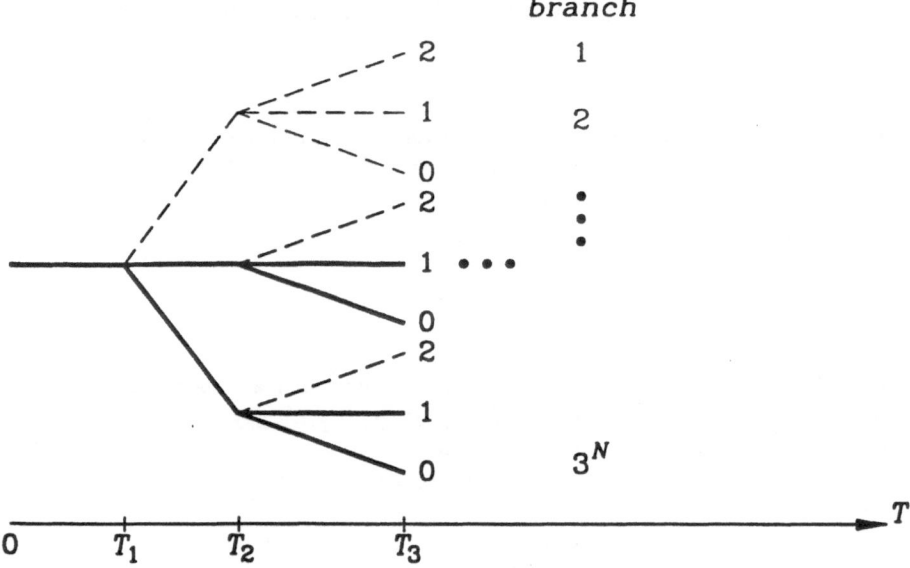

Figure 5 Repair realizations corresponding to no repair.

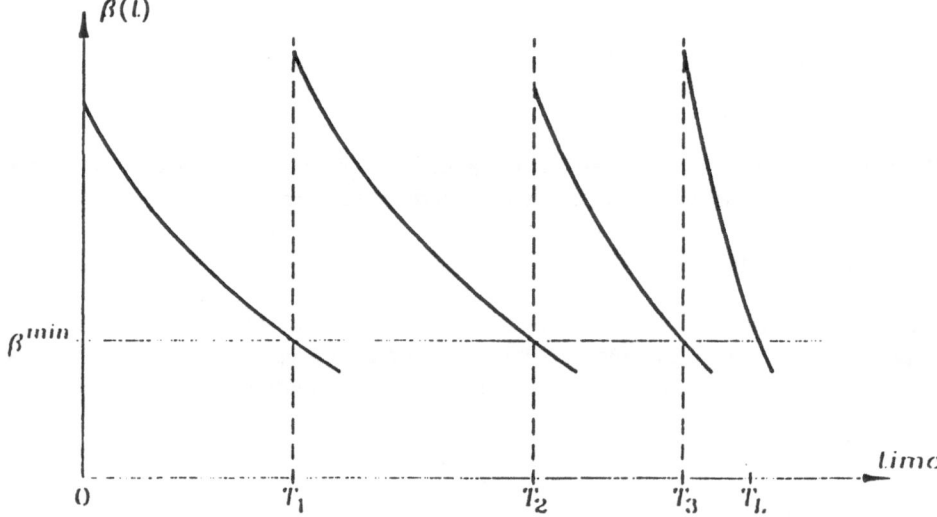

Figure 6 Inspection plan based on the reliability level alone.

· Only branches in the repair events corresponding to no repair are used to determine the probabilities, see figure 5. This is a reasonable approximation if the reliability level is high and only very few repairs are expected. This approximation must be checked for each practical application.

· Only the next time, quality of inspection and repair criteria for the next inspection are treated as optimization variables. The number of optimization variables and the complexity of the probabilities are significantly reduced. This approach is sometimes referred to as the adaptive approach, see e.g. Fujita et al. [3].

· The inspection planning can be based on reliability requirements only, i.e. cost considerations are neglected. The approximation is illustrated in figure 6. A minimum reliability level is chosen, for example based on code requirements. This level is represented by B_{min} in figure 6. The generalized reliability index $B(t)$ is determined as a function of time for the component. When $B(t)$ decreases to B_{min} an inspection has to be performed. The next inspection time is determined by assuming that no defects are found by the inspection. Based on this assumption, and taking into account the inspection uncertainty, the next inspection time is found as the time T_2 where the updated reliability index decreases to B_{min}. In this way the total inspection plan is determined. Inspection planning based on this approach has been performed for some platforms in the Danish part of the North Sea see e.g. Pedersen et al. [18]

· In (5)-(6) the optimization variables are assumed to be continuous and the optimization problem has to be solved by a mathematical optimization algorithm. If the possible inspection times and qualities and repair actions are discretized into a small number of possibilities, the numerical effort to solve (5)-(6) is reduced significantly, because the objective function and the constraint have to be calculated only a small number of times. This approach is used in the example in section 7.

6. Assessment of Sensitivities

The model of costs and uncertainties on which the optimal maintenance plan is based, is generally incomplete and influenced by subjective judgement. Furthermore the model may contain assumptions which depend on purely political considerations and therefore could change with time. Therefore it is important for the decision maker to be able to estimate how uncertain the estimated expenditures are, i.e. the probability that the budget for future maintenance will be exceeded by a certain percentage. It is also of considerable interest to be able to assess the sensitivity of the expenditures with respect to model assumptions which the decision maker does not control, and the uncertain parameters not modelled by the stochastic variables. Information of these influences may lead to a maintenance plan which does not yield the lowest expected costs, but is rather a more dependable plan.

Typical uncertainties not included in the models are: those associated with the costs of inspection, repair and failure: uncertainties associated with the parameters of the distributions in the applied stochastic models; or neglected physical and model uncertainties.

The uncertain quantities can be divided into the following groups:

a) quantities used to estimate the probabilities of failure, $P_F(T)$.

b) quantities used to estimate the costs, e.g. the real rate of interest r and the coefficients C_{IN}, C_R, C_F.

The uncertainty related to these quantities can be taken into account by

1) modelling the quantities by stochastic variables (or stochastic processes). This is typically done for the quantities in group a) and the uncertainty is then taken into account in the reliability calculations using e.g. FORM/SORM,

2) performing sensitivity analyses. This is typically done for quantities which are not known very well and which are not modelled by stochastic variables. Examples are the cost coefficients in group b) and the statistical parameters used to model the stochastic variables. The result of a sensitivity analysis with respect to a parameter p is conveniently measured by the elasticity

$$e = \frac{dC}{dp} \frac{p}{C} \qquad (21)$$

where C can be

-   the probability of failure within the time interval $[0, T]$

-   the total expected costs or e.g. the expected costs of inspections.

-   an optimization (decision) variable, e.g. the next inspection time.

In the following we assume that C is the total expected life cycle cost. With reference to section 3 the total expected cost considering an optimal inspection and maintenance strategy, can be written in the following form

$$\min_{x} C(x, p, q) = C_0(x, p) + \sum_{j} C_j(x, p) P_j(x, q)$$

$$\text{s.t. } P_f(x, q) \leq P_f^{max} \qquad (22)$$

where x are decision variables, p are quantities defining the cost and q are quantities defining the stochastic model. P_j denotes a probability (failure or repair), P_f denotes a probability of failure and P_f^{max} is the maximum accepted probability of failure, related to β_{min} in equation (6) and (7). The summation in equation (23) is made over the number of terms constituting equations (8)-(10). We are primarily interested in evaluating the derivatives of C with respect to the elements in p and q, but depending on the purpose we might also want to evaluate the derivatives of the decision variables x with respect to p and q. The evaluation of these is, however, more involved. The sensitivity of the total expected costs C with respect to the elements in p and q is obtained from Haftka et al. [19] and Enevoldsen [20]

$$\frac{dC}{dq_i} = \sum_j C_j \frac{\delta P_j}{\delta q_i} + \lambda \frac{\delta P_f}{\delta q_i} \tag{23}$$

$$\frac{dC}{dp_i} = \sum_j P_j \frac{\delta C_j}{\delta p_i} + \gamma \frac{\delta C_0}{\delta p_i} \tag{24}$$

where γ is the Lagrangian multiplier associated with the constraint in equation (22). The partial derivatives of the probabilities P_j and P_f can be determined semi-analytically, as shown in Madsen [21]. The partial derivatives of the cost coefficients C_0 and C_j can effectively be determined analytically or numerically.

The sensitivity of the decision vector x is measured by the partial derivatives of the elements in the decision vector x with respect to p_i and q_i. These can be calculated using the formulas given below which are obtained by use of the Kuhn Tucker conditions for the optimization problem defined in equation (22).

$$\frac{\delta x}{\delta p_i} \qquad \text{are obtained from}$$

$$\begin{bmatrix} A & B \\ B^T & 0 \end{bmatrix} \begin{bmatrix} \dfrac{\delta x}{\delta p_i} \\[2mm] \dfrac{\delta \gamma}{\delta p_i} \end{bmatrix} = \begin{bmatrix} C \\ 0 \end{bmatrix} \tag{25}$$

The elements in the matrix **A** and the vectors **B** and **C** are

$$A_{rs} = \frac{\delta^2 C_0}{\delta x_r \delta x_s} + \sum_j \left(P_j \frac{\delta^2 C_j}{\delta x_r \delta x_s} + 2 \frac{\delta P_j}{\delta x_r} \frac{\delta C_j}{\delta x_s} + C_j \frac{\delta^2 P_j}{\delta x_r \delta x_s} \right) + \lambda \frac{\delta^2 P_f}{\delta x_r \delta x_s} \tag{26}$$

$$B_r = \frac{\delta P_j}{\delta x_r} \tag{27}$$

$$C_r = \frac{\delta^2 C_0}{\delta x_r \delta p_i} - \sum_i \left(\frac{\delta^2 C_j}{\delta x_r \delta p_i} + \frac{\delta P_j}{\delta x_r} \frac{\delta C_j}{\delta p_i} \right) \tag{28}$$

$$\frac{\delta x}{\delta p_i} \qquad \text{are obtained from}$$

$$\begin{bmatrix} A & B \\ B^T & 0 \end{bmatrix} \begin{bmatrix} \dfrac{\delta x}{\delta q_i} \\ \dfrac{\delta \gamma}{\delta q_i} \end{bmatrix} = \begin{bmatrix} D \\ E \end{bmatrix} \tag{29}$$

where the elements in the vector **D** and in E are

$$D_r = \sum_j \left(-\frac{\delta C_j}{\delta x_r} \frac{\delta P_j}{\delta P_i} - C_j \frac{\delta^2 P_j}{\delta x_r \delta q_i} \right) - \lambda \frac{\delta^2 P_f}{\delta x_r \delta q_i} \tag{30}$$

$$E = \frac{\delta P_f}{\delta q_i} \tag{31}$$

It is seen that the sensitivity of the objective function (the total expected life cycle cost) with respect to some parameters can be determined on the basis of the first order sensitivity coefficients of the probabilities and of the cost functions, see (23)-(24). However, calculation of the sensitivities of the decision parameters with respect to some parameters is much more complicated because it involves estimation of both the second order sensitivity coefficients of the probabilities (which are the most complicated to estimate see e.g. Enevoldsen [20]), and the cost functions.

7. Example

To illustrate the practical aspects of the application of cost optimal inspection and maintenance planning and especially the use of sensitivity measures as decision tools an example from the offshore industry is presented in the following. The example is concerned with an offshore structure of the jacket type with tubular structural steel members connected by welded joints. As part of the inspection and maintenance planning for the entire structural system, the inspection and maintenance planning with respect to fatigue crack growth of a single joint is considered. It is assumed that a spectral stress analysis has been performed and that the results of this analysis are given in terms of a Weighted Average fatigue Stress Range (*WASR*) see e.g. FACTS [22], and a corresponding expected number of fatigue load cycles. In order to describe the events of failure, repair and inspection observations, it is necessary to model the crack growth. For this purpose the software module FACTS [22] is used. The probabilistic model of the variables used in the following maintenance planning is shown in table 1.

Three inspection techniques are considered. In table 1 the variable POD_i refers to the stochastic variable modelling the probability of detection for inspection technique no i, $i = 1, 2, 3$, and ε_i refers to the stochastic variable modelling the measuring uncertainty connected with the different inspection techniques. All stochastic variables in table 1 are assumed to be independent. The inspection techniques are defined in table 2.

Two repair strategies are considered. If the observed crack depth at an inspection is smaller than a certain fraction γ of the chord thickness, grinding will be used. Otherwise

Variable	Distribution	m	s
Initial chord thickness t_C	Weibull	0.04	0.005
Initial crack depth	Normal	$0.8 \cdot 10^{-3}$	$1.0 \cdot 10^{-4}$
Initial crack length	Normal	$8.0 \cdot 10^{-3}$	$8.0 \cdot 10^{-4}$
Chord thickness after grinding	Weibull	0.04	0.005
Crack depth after grinding	Normal	$0.8 \cdot 10^{-3}$	$1.0 \cdot 10^{-4}$
Crack length after grinding	Normal	$8.0 \cdot 10^{-3}$	$8.0 \cdot 10^{-4}$
Chord thickness after welding	Weibull	0.04	0.005
Crack depth after welding	Normal	$0.0 \cdot 10^{-3}$	$1.0 \cdot 10^{-4}$
Crack length after welding	Normal	$8.0 \cdot 10^{-3}$	$8.0 \cdot 10^{-4}$
Initial Paris C	Log-Normal	$0.45 \cdot 10^{-11}$	$0.31 \cdot 10^{-11}$
Initial Paris m	Deterministic	3.1	
Paris C after welding	Log-Normal	$0.45 \cdot 10^{-11}$	$0.31 \cdot 10^{-11}$
Paris m after welding	Deterministic	3.1	
Weigh. Aver. Str. Range ($WASR$)	Log-Normal	30.0	4.0
POD_1	Exponential	0.0013	0.0013
POD_2	Exponential	0.002	0.002
POD_3	Exponential	0.013	0.013
e_1	Normal	0.0	$0.5 \cdot 10^{-3}$
e_2	Normal	0.0	$1.5 \cdot 10^{-3}$
e_3	Normal	0.0	$2.5 \cdot 10^{-3}$
Design lifetime T_L	Deterministic	35 years	
Stress cycles per year	Deterministic	$6 \ 10^6$	
Real rate of interest r	Deterministic	0.02	

Table 1.   Statistical models for crack growth parameters (all dimensions in m and MPa). m: expected value and s: standard deviation.

the crack will be repaired by welding. The fractions used in this example are $\gamma = 1.25$ and 0.50.

The simplified approach for inspection and maintenance planning explained in section 5 is used: only the next inspection time, method and repair strategy are optimized. It is assumed that inspections can be performed in one weather window each year. The following discretized inspection times (in years from the inspection planning time) are considered: $T_1 = 2, 4, 6, 8, ..., 32, 34$.

The decision variables are thus the inspection time T_1, the inspection technique $I = 1, 2, 3$ and the repair criterion $R = 1, 2$. The possible inspection methods and repair strategies are summarized in table 2.

The following cost models are used, see (8)-(10):

$$C_{IN}(T_1, I, R) = C_{IN_0}(I)(1 - P_F(T_1))\frac{1}{(1+r)^{T_1}}$$

$$C_F(T_1, I, R) = C_{F_0}(P_F(T_1) - P_F(T_0))\frac{1}{(1+r)^{T_1}} +$$

$$C_{F_0}(P_F(T_L, I, R) - P_F(T_1))\frac{1}{(1+r)^{T_L}}$$

$$C_R(T_1, I, R) = C_{RGRIND} P_{RGRIND}(T_1)\frac{1}{(1+r)^{T_1}} + C_{RWELD} P_{RWELD}(T_1)\frac{1}{(1+r)^{T_1}}$$

where P_{RGRIND} and P_{RWELD} are the probabilities that repair is performed by grinding and welding, respectively, T_0 is the time where the inspection planning is performed. The cost coefficients are shown in table 3.

Strategy	Inspection Method	Repair Strategy
1	$I = 1$: method A	$R = 1$: weld if crack ³ 25% of t_C
2	$I = 2$: method B	$R = 1$: weld if crack ³ 25% of t_C
3	$I = 3$: method C	$R = 1$: weld if crack ³ 25% of t_C
4	$I = 1$: method A	$R = 2$: weld if crack ³ 50% of t_C
5	$I = 2$: method B	$R = 2$: weld if crack ³ 50% of t_C
6	$I = 3$: method C	$R = 2$: weld if crack ³ 50% of t_C

Table 2.   Inspection and repair options.

The optimal inspection and maintenance plan for the considered joint is selected as the inspection and repair option from table 2 and the time instant between 2 and 34 years, which leads to the smallest expected total costs. The expected total life cycle costs are evaluated by the software module PREDICT [23]. The expected costs corresponding to the individual inspection and maintenance options are plotted as functions of the inspection time in figures 7 - 12. The total expected costs and the optimal inspection times corresponding to the different maintenance strategies are shown in table 4 and figure 13, where the total expected costs from figures 7 - 12 are plotted.

		cost (ECU)
C_{IN_o}	I=1: method A	$0.5 \cdot 10^6$
C_{IN_o}	I=2: Method B	$0.25 \cdot 10^6$
C_{IN_o}	I=3: method C	$0.1 \cdot 10^6$
C_{R_1}	Grind repair	$0.25 \cdot 10^6$
C_{R_2}	Weld repair	$5.0 \cdot 10^6$
C_{F_0}	Failure	$100.0 \cdot 10^6$

Table 3.   Costs of inspection, repair and failure.

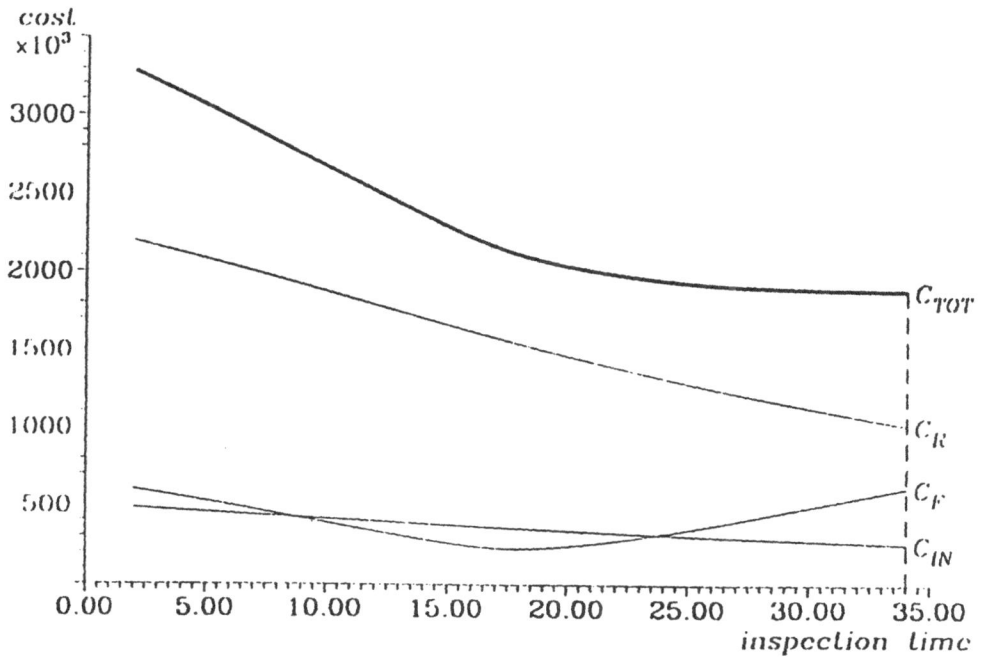

Figure 7 Expected costs corresponding to I=1 and R=1.

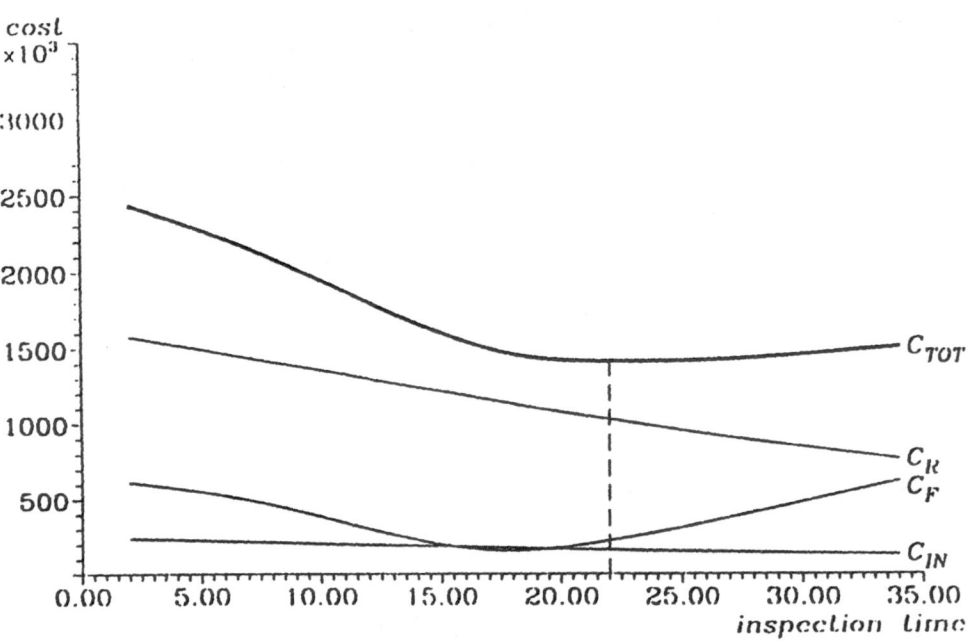

Figure 8 Expected costs corresponding to I=2 and R=1.

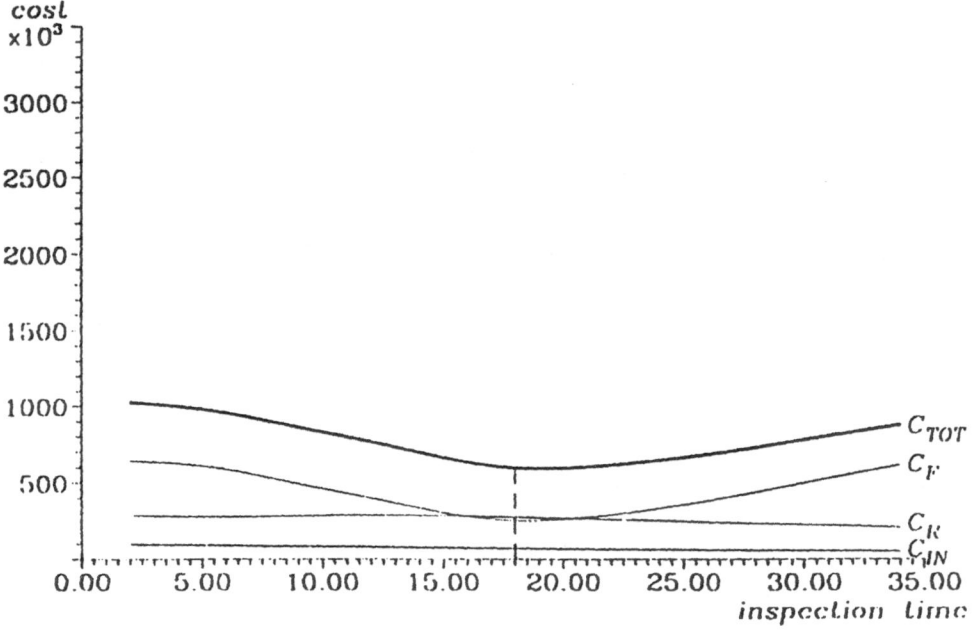

Figure 9 Expected costs corresponding to I=3 and R=1.

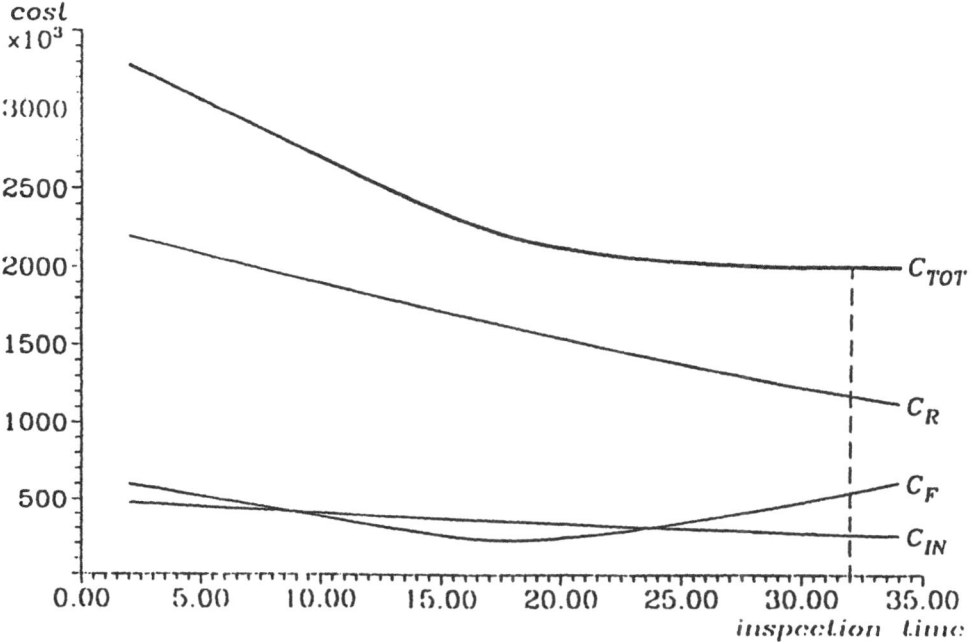

Figure 10 Expected costs corresponding to I=1 and R=2.

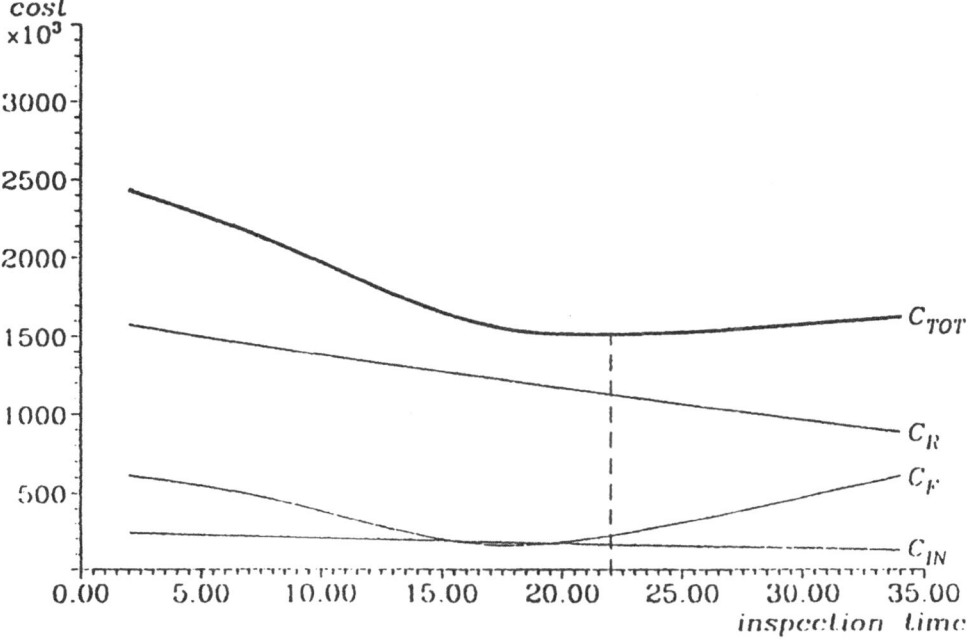

Figure 11 Expected costs corresponding to I=2 and R=2.

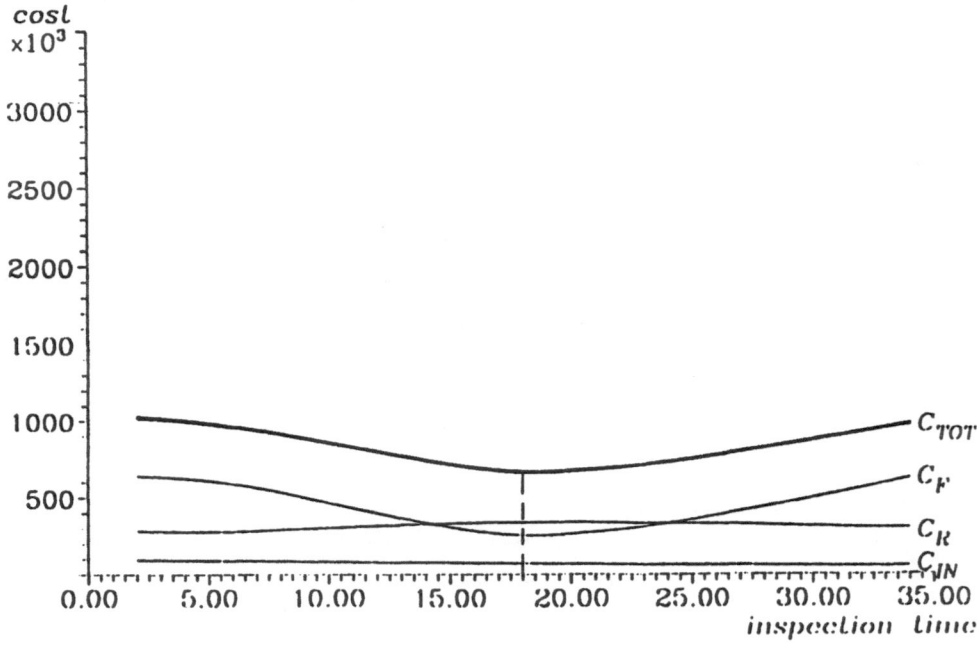

Figure 12 Expected costs corresponding to I=3 and R=2.

Strategy	T_1	Expected total costs (ECU)
$I=1\ R=1$	34	$1.885 \cdot 10^6$
$I=2\ R=1$	22	$1.404 \cdot 10^6$
$I=3\ R=1$	18	$0.591 \cdot 10^6$
$I=1\ R=2$	32	$2.002 \cdot 10^6$
$I=2\ R=2$	22	$1.511 \cdot 10^6$
$I=3\ R=2$	18	$0.651 \cdot 10^6$

Table 4.   Expected total costs of the individual maintenance strategies.

From figure 13 and table 4 it is seen that the strategy with $I = 3$ and $R = 1$ gives the lowest total expected costs.

In order to investigate the sensitivities of the total expected costs as described in section 3, a number of sensitivity studies has been performed with respect to the costs of inspection, repair and failure. The result of this study is given in table 5, where the elasticities e_p defined as

$$e_p = \frac{dE[C_T]}{dp} \frac{p}{E[C_T]}$$

are given corresponding to the individual inspection strategies as defined in table 3.

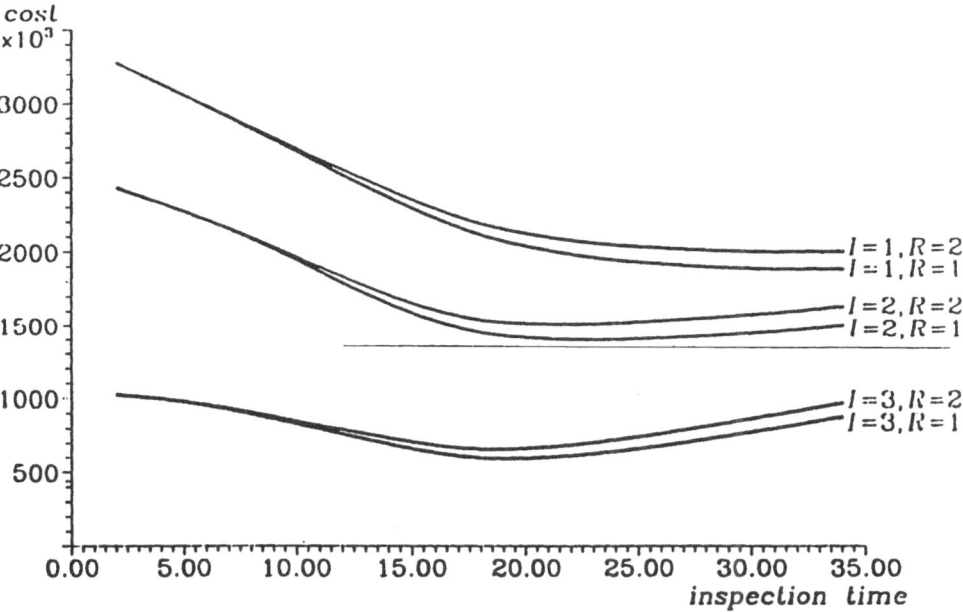

Figure 13 Total expected costs corresponding to the 6 strategies.

Strategy	T_1	$e_{C_{F_0}}$	$e_{C_{R\ GRIND}}$	$e_{C_{R\ WELD}}$	$e_{C_{IN_0}}$
$I=1\ R=1$	34	0.33	0.01	0.53	0.14
$I=2\ R=1$	22	0.16	0.01	0.72	0.12
$I=3\ R=1$	18	0.42	0.01	0.46	0.12
$I=1\ R=2$	32	0.28	0.002	0.59	0.13
$I=2\ R=2$	22	0.15	0.002	0.74	0.11
$I=3\ R=2$	18	0.38	0.002	0.51	0.11

Table 5.   Elasticities of costs of inspection, repair and failure.

For the optimal maintenance strategy ($I=3\ R=1$) a study of the elasticities with respect to some of the variables from table 2 has been performed. The results are given in table 6.

Variable p	$e_p = \dfrac{dE|C_T|}{dp} \dfrac{p}{E|C_T|}$
T_L: Design lifetime	1.33
m$WASR$: Weighted average stress range	4.55
r: Real rate of interest	-0.42
g: Grind / Weld criteria	4.37
Expected value of initial crack depth and length	0.27

Table 6.   Elasticities of expected total costs with respect to variables from table 2.

It should be noted that the results from the example have been produced on the assumptions associated with the simplified inspection and maintenance strategy as explained in section 5. Experience from other examples see e.g. Madsen and Sørensen [21] show that when several inspections are planned simultaneously, the first inspection time is earlier in comparison to the case where only one inspection (the case in the example) is planned. From figures 7-12 it is seen that the expected inspection costs are decreasing as functions of the time to the next inspection. This is due to the influence of the real rate of interest r. The same trend is seen for the repair costs whereas the failure costs are seen to have a significant minimum. The sensitivity study shows that for the optimal inspection strategy ($I = 3$ & $R = 1$), the elasticities with respect to failure and the weld repair costs are dominating. Regarding the elasticities of the remaining parameters it is seen that the mean value of the weighted average stress range (μ_{WASR}) and the criteria for weld repair (γ) are the most significant, but also the design lifetime (T_L) is important.

8.   Conclusions

Risk based maintenance planning has by now reached a state where it is not only a theoretical possibility, rather a practical tool. The methodology is based on a unification of modern reliability methods with the well developed Bayesian decision theory, which previously has found the most applications in economics.

Based on the methodology, it is possible to evaluate the economic consequences associated with a particular maintenance strategy, and even to identify the maintenance strategy which minimizes the total expected life cycle costs of the structure, including costs of inspections, costs of repairs and costs of failure. The maintenance strategy may be so selected that specific safety requirements, which are not included in the costs of failure, may be fulfilled.

As a valuable additional decision tool the optimal maintenance strategy may be evaluated by sensitivity analysis in order to assess how sensitive the optimal strategy is with respect to the uncertainty modelling and the estimates of marginal costs. Thereby decisions may be made regarding how the modelling is best refined such that the optimal strategy becomes more dependable or even to identify strategies which may not yield the lowest life cycle costs but which are less sensitive to the model assumptions.

9. References

1   Vrijling, J.K. (1989). *Some Considerations on the Acceptable Probability of Failure*, Proceedings ICOSSAR'89, pp. 1919-1926.

2   Madsen, H.O. and Sørensen, J.D.: *Probability-Based Optimization of Fatigue Design Inspection and Maintenance*. Presented at Int. Symp. on Offshore Structures, July 1990, University of Glasgow.

3   Fujita, M., Schall, G. and Rackwitz, R.: *Adaptive Reliability Based Inspection Strategies for Structures Subject to Fatigue*. Proceedings ICOSSAR 89, San Francisco 1989, pp. 1619-1626.

4   Sørensen, J.D., Faber, M.H., Thoft-Christensen, P. and Rackwitz, R.: *Modelling in Optimal Inspection and Repair*. Proceedings of OMAE 1991, Stavanger, Norway, pp. 281-288.

5   Sørensen, J.D. and Thoft-Christensen, P.: *Inspection Strategies for Concrete Bridges*. Proc. IFIP WG 7.5 1988, Vol. 48, pp. 325-335, Springer-Verlag.

6   Sørensen, J.D., Faber, M.H., Rackwitz, R. and Thoft-Christensen, P.: *Reliability Analysis of an Offshore Structure. A case Study - II*. Proceedings of OMAE 1992, Calgary, Canada.

7   Faber, M.H., Dharmavasan, S., and Dijkstra, O.D.: *Integrated Analysis Methodology for Reassessment and Maintenance of Offshore Structures*. In proc. 13th OMAE Conference on Offshore Mechanics and Arctic Engineering, Houston, Texas, February 27 - March 3, 1994.

8   Faber, M.H., Kroon, I.B. and Sørensen, J.D. (1993): *Sensitivities in Structural Maintenance Planning*. Accepted for publication in Reliability Engineering & System Safety.

9   Raiffa, H. and Schlaifer, R.: *Applied Statistical Decision Theory*. Harward University Press, Cambridge, Mass., 1961.

10 Benjamin, J.R. and Cornell, C.A.: *Probability, Statistics and Decision for Civil Engineers*. MacGraw-Hill, 1970.

11 Kroon, I.B., Ph.D.-Thesis June 1994: *Decision Theory Applied to Structural Egineering Problems*.

12 Madsen, H.O., Krenk, S. and Lind, N.C.: *Methods of Structural Safety*. Prentice-Hall, 1986.

13 Faber, M.H., Sørensen, J.D. and Kroon, I.B.: *Optimal Inspection Strategies for Offshore Structural Systems*. Proceedings of OMAE 1992, Calgary, Canada.

14 Lindley, D.V.: *Introduction to Probability and Statistics from a Bayesian Viewpoint, Vol 1+2*. Cambridge University Press, Cambridge 1976.

15 Madsen, H.O.: *Model Updating in Reliability Theory*. Proceedings of ICASP5, pp. 564-577, 1987.

16 Rackwitz, R. and Schrupp, K.: *Quality Control, Proof Testing and Structural Reliability*. Structural Safety, Vol. 2, 1985, pp. 239-244.

17 Schall, G., Gollwitzer, S. and Rackwitz, R.: *Integration of Multinormal Densities on Surfaces*. Proc. 2nd IFIP WG7.5 Conf., P. Thoft-Christensen (Ed.). Lecture Notes in Engineering, Springer Verlag, Vol. 48, pp. 235-248, 1988.

18 Pedersen, C., Nielsen, J.A., Riber, J.P., Madsen, H.O., Krenk, S.: *Reliability Based Inspection Planning for the Tyra Field*. Proceedings of OMAE 1992, Calgary, Canada.

19 Haftka, R.T. & M.P. Kamat: *Elements of Structural Optimization*. Martinus Nijhoff, The Hague, 1985.

20 Enevoldsen, I.: *Sensitivity Analysis of a Reliability-Based Optimal Solution*. Structural Reliability Theory, Paper No. 101, the University of Aalborg, 1992 - accepted for publication in ASCE, Journal of Engineering Mechanics.

21 Madsen, H.O.: *Sensitivity Factors for Parallel Systems*. Report DIAB, Danish Engineering Academy, Lyngby, Denmark, 1990.

22 Technical Manual for FACTS (fatigue crack growth analysis software for offshore structures). TSC (Technical Software Consultants), UK, 1990.

23 Technical Manual for PREDICT (inspection and maintenance scheduling software for offshore structures). PRISM (Platform Reliability Inspection Scheduling and Maintenance Ltd.), London, UK, 1993.

Mechanics

SOLID MECHANICS AND ITS APPLICATIONS

Series Editor: G.M.L. Gladwell

Aims and Scope of the Series

The fundamental questions arising in mechanics are: *Why?, How?*, and *How much?* The aim of this series is to provide lucid accounts written by authoritative researchers giving vision and insight in answering these questions on the subject of mechanics as it relates to solids. The scope of the series covers the entire spectrum of solid mechanics. Thus it includes the foundation of mechanics; variational formulations; computational mechanics; statics, kinematics and dynamics of rigid and elastic bodies; vibrations of solids and structures; dynamical systems and chaos; the theories of elasticity, plasticity and viscoelasticity; composite materials; rods, beams, shells and membranes; structural control and stability; soils, rocks and geomechanics; fracture; tribology; experimental mechanics; biomechanics and machine design.

Kluwer Academic Publishers – Dordrecht / Boston / London

Mechanics

SOLID MECHANICS AND ITS APPLICATIONS
Series Editor: G.M.L. Gladwell

Kluwer Academic Publishers – Dordrecht / Boston / London

Mechanics

SOLID MECHANICS AND ITS APPLICATIONS

Series Editor: G.M.L. Gladwell

Kluwer Academic Publishers – Dordrecht / Boston / London

Mechanics

FLUID MECHANICS AND ITS APPLICATIONS

Series Editor: R. Moreau

Aims and Scope of the Series

The purpose of this series is to focus on subjects in which fluid mechanics plays a fundamental role. As well as the more traditional applications of aeronautics, hydraulics, heat and mass transfer etc., books will be published dealing with topics which are currently in a state of rapid development, such as turbulence, suspensions and multiphase fluids, super and hypersonic flows and numerical modelling techniques. It is a widely held view that it is the interdisciplinary subjects that will receive intense scientific attention, bringing them to the forefront of technological advancement. Fluids have the ability to transport matter and its properties as well as transmit force, therefore fluid mechanics is a subject that is particularly open to cross fertilisation with other sciences and disciplines of engineering. The subject of fluid mechanics will be highly relevant in domains such as chemical, metallurgical, biological and ecological engineering. This series is particularly open to such new multidisciplinary domains.

Kluwer Academic Publishers – Dordrecht / Boston / London

Mechanics

FLUID MECHANICS AND ITS APPLICATIONS
Series Editor: R. Moreau

Kluwer Academic Publishers – Dordrecht / Boston / London

Mechanics

FLUID **MECHANICS AND ITS APPLICATIONS**

    Series Editor: R. Moreau

41. L. Fulachier, J.L. Lumley and F. Anselmet (eds.): *IUTAM Symposium on Variable Density Low-Speed Turbulent Flows.* Proceedings of the IUTAM Symposium held in Marseille, France. 1997   ISBN 0-7923-4602-5

Kluwer Academic Publishers – Dordrecht / Boston / London